PHARMACEUTICAL CALCULATIONS

FOURTH EDITION

PHARMACEUTICAL CALCULATIONS

JOEL L. ZATZ
Rutgers University

MARIA GLAUCIA TEIXEIRA
University of Wyoming

A JOHN WILEY & SONS, INC., PUBLICATION

Published by John Wiley & Sons, Inc., Hoboken, New Jersey.
Published simultaneously in Canada.

For general information on our other products and services please contact our Customer Care Department within the U.S. at 877-762-2974, outside the U.S. at 317-572-3993 or fax 317-572-4002.

Wiley also publishes its books in a variety of electronic formats. Some content that appears in print, however, may not be available in electronic format.

Library of Congress Cataloging-in-Publication Data:

Zatz, Joel L., 1935–
Pharmaceutical calculations / Joel L. Zatz, Maria Glaucia Teixeira. – 4th ed.
p. cm.
ISBN 978-0-471-43353-8
1. Pharmaceutical arithmetic. I. Teixeira, Maria Glaucia. II. Title.
RS57 .Z37 2005
615'.14'01513 – dc22

2003027189

11

To Arline
my partner in love and friendship
J.L.Z.

To my husband, my best friend
for his patience, encouragement, and permanent support
M.G.T.

CONTENTS

PREFACE

Our goal in preparing a new edition of *Pharmaceutical Calculations* was to keep pace with changes in pharmacy practice while retaining the philosophy, approach, and essential features of previous editions, which have been used successfully by pharmacy students and practitioners for over 30 years.

New clinically oriented chapters or chapter sections deal with calculations for injectables, parenteral nutrition, pediatric dosing, reconstitution of dry powders, body mass index, chemotherapy, and biologics for immunization. New material has been added on the prescription and mediation order, pH and buffers, intravenous admixtures, TPN, and isotonicity. In addition, many new examples have been added. Terms and abbreviations used by pharmacists have been updated and obsolete designations discarded. The appendix has been expanded with more short topics and supplementary information.

As in the preceding edition, each new chapter begins with a list of learning objectives. While dimensional analysis and direct approaches in problem solving are emphasized, alternate methodology is also described. Accuracy requirements, which depend on the particular application, are indicated and estimation as a means of preventing gross errors is encouraged.

This text can be used for individual self-instruction or it can serve as a primary or supplementary text in a formal course. Its programmed format permits self-paced learning. Solutions to all problems are given, providing students with immediate feedback regarding their progress. A set of review questions at the end of each chapter allows students to see how well they have mastered the subject matter.

Joel L. Zatz
Rutgers University, Piscataway, NJ

M. Glaucia Teixeira
University of Wyoming, Laramie, WY

TO THE STUDENT

This is probably not your first encounter with a programmed text. In that case, you will recognize that each chapter contains a series of sections, called frames, in which the elements of a topic unfold step by step. Along the way, you have to answer questions that check your understanding. The answer is always available, so you can quickly tell whether you're on the right track or not.

One principle involved in this approach to teaching is active learning. This means that the work you do in attacking problems and thinking about your answers helps you to understand and remember, making it easier the next time you come across another problem in which the same logic and ideas are important. To just read the problems and answers or glance ahead at the solutions to the problems defeats the process, so don't take any shortcuts. Put in the time, do the work, and think about what you are doing. Treat the review questions at the end of each chapter as your quality control check. Do all the problems before glancing at the answers.

Most every frame contains a dashed line. Take an index card and place it on the frame you're working on so that it covers everything below the dashed line. (That's where the answer is.) Read the information in the frame and work on the question or questions in the space provided. When you're done, check your answer. If you are correct, move on to the next frame. If not, try the problem again before consulting the solution. You may uncover your error and be able to do it right the second time. If you still can't get it, you may want to go back a few frames to see if there is something you missed. If all else fails, look at the solution. You may also want to consult the solution if you struggled to get the right answer and wondered if a more direct approach is available.

We are assuming that you can perform the usual arithmetic operations of addition, subtraction, multiplication, and division and that you can solve simple algebraic and exponential equations. Outlines of certain techniques are provided in case you have gotten rusty and need to review. Working through the problems that are provided will usually get you back on track.

Some topics will be easy for you and you'll zoom through the text. Difficult spots will take more time; in any case, stick with it and take as much time as you need to work through the material until you understand it and can handle the problems. Happy problem solving!

J. L. Z.
M. G. T.

An errata sheet with corrections to *Pharmaceutical Calculations, Fourth Edition* can be found and downloaded at the following webpage:

ftp://ftp.wiley.com/public/sci_tech_med/pharmaceutical_calculations/

GENERAL PRINCIPLES OF CALCULATIONS

***Learning Objectives:** After completing this chapter the student should be able to:*

1. Perform mathematical operations on units.
2. Apply dimensional analysis to problems involving conversion of units.
3. Compare two like quantities (ratio).
4. Apply ratio and proportion in problem solving.
5. Estimate results using rounding and power-of-10 notation.
6. Express tolerances in terms of amounts and percentages.
7. State the accepted tolerances for prescription compounding and manufacturing calculations.

Calculations are part of the daily practice of pharmacists, nurses, doctors, and most health-related professionals. Pharmacists, determine quantities of materials required to fill prescriptions, compound formulas, calculate doses, etc. The dosage of each medication that is dispensed must be finally checked by the pharmacist, who is legally accountable for an overdose. The fact that most pharmaceuticals are prefabricated and not prepared inside the pharmacy does not lessen the pharmacist's responsibility.

Modern drugs are effective, potent, and therefore potentially toxic. An overdose may be fatal. Knowing "how to" calculate the amount of each drug and "how to" combine them is not sufficient. Of course, dispensing a sub-potent dose is not satisfactory either. The drug(s) given will probably not elicit the desired therapeutic effect and will therefore be of no benefit to the patient. Clearly, the only satisfactory approach is one that is completely free of error. Absolute accuracy is our goal. Since our goal when performing calculations is the correct answer, it is logical to suppose that any rational approach to a problem that results in the correct answer is acceptable. While this is true, some approaches are more rational than others. Try to use a method that requires as few steps as possible and that you feel comfortable with. The simplest, most direct pathway to the solution allows less opportunity for error in computation than does one that is roundabout.

In this chapter, we will go into some manipulative techniques basic to all types of calculations. We are going to assume that you can add, subtract, multiply, and divide; that you can work with decimals and fractions; and that you can solve simple algebraic expressions. You will probably find that you are already familiar with all or most of the techniques covered. When that is so, you will make rapid progress through the program. But if you need further review or instruction, they will be provided.

Pharmaceutical Calculations, Fourth Edition, By Joel L. Zatz and Maria Glaucia Teixeira
ISBN 0-471-67623-3

We will see how units participate in arithmetic operations and how we can take advantage of this property in our calculations. We will review dimensional analysis, ratio, proportion, estimation, rounding, and power-of-10 notation.

UNITS

1. In compounding prescriptions, the pharmacist deals with measured quantities. The magnitude of each such quantity is expressed as the product of a number and a unit. The unit name specifies the scale of measurement. My pencil, for example, is 5 inches (in.) long. I may also say that it is 12.7 centimeters (cm) long. Changing the scale of measurement (unit) causes the multiplying number to change as well. Thus, in describing any measured quantity, it is necessary to specify the unit used. The unit is an integral part of the designation of a value that is either measured directly or calculated from measured data. *Units must not be permitted to drop off or fade away during calculation.*

It is sometimes found that the units in which a measured quantity is expressed are not convenient for the user. For example, an American traveling in France wants to buy 2 pounds of beef at the neighborhood butchery. Before going to the store around the corner, the traveler may wish to convert this amount into the equivalent metric unit (grams or kilograms), used in that country. To perform the conversion, it is necessary to know that

$$1 \text{ lb} = 454 \text{ g}$$

---

2. One advantage of performing a calculation involving units is that they may be multiplied and divided in much the same way as numbers or algebraic symbols. If the same unit appears in both the numerator and denominator, they will cancel each other. For example,

$$2 \text{ lb} \times \frac{454 \text{ g}}{1 \text{ lb}} = 908 \text{ g}$$

The traveler can go to the butcher and ask for 900 grams of beef to get approximately the 2 pounds needed.

Computations involving units will always require some knowledge of different systems of measurement and intersystem conversions (Chapter 2 and Appendix 1)

Perform the operations indicated:

A. $3 \text{ dL} \times \frac{100 \text{ mL}^*}{1 \text{ dL}} =$

B. $154 \text{ lb} \times \frac{1 \text{ kg}}{2.2 \text{ lb}} =$

* The abbreviation for liter (capital L) and all other metric volume denominations follow the USP standard expression rather than official abbreviation recommended in SI units.

C. $250 \text{ mg} \times \frac{1 \text{ tablet}}{50 \text{ mg}} =$

Solutions.

A. 300 mL
B. 70 kg
C. 5 tablets

---

3. How many fluid ounces are there in 1/2 qt of Scotch whiskey? (1 qt = 32 fluidounces)

Solution. 16 fluidounces

CALCULATIONS

$$0.5 \text{ qt} \times \frac{32 \text{ fluidounces}}{1 \text{ qt}} = 16 \text{ fluidounces}$$

---

4. Sometimes, the relationship between the units given and the units desired is not known. Say, for example, that we wish to convert 17 ft to meters. Although we do not know the number of feet in 1 m, we do know that 1 m = 39.4 in. and 1 ft = 12 in. We may therefore first convert feet to inches and then inches to meters. But rather than treat our problem as two separate parts, we may set it up as follows:

$$17 \text{ ft} \times \frac{12 \text{ in.}}{1 \text{ ft}} \times \frac{1 \text{ m}}{39.4 \text{ in.}} = 5.18 \text{ m}$$

The first fraction converts feet to inches; the second converts inches to meters. Notice that all units except for meters cancel out. There is no change in the value of the length represented by "17 ft." This technique may be extended to any number of successive conversions.

How many fluidounces are there in 1.75 liters (L)? (1 L = 1000 mL; 1 fluidounce = 29.6 mL)

Solution. 59.1 fluidounces

CALCULATIONS

$$1.75\,\text{L} \times \frac{1000\text{ mL}}{1\text{ L}} \times \frac{1\text{ fluidounce}}{29.6\text{ mL}} = 59.1\text{ fluidounces}$$

---

5. If a mercury barometer reads 30.3 in., what is the pressure in atmospheres (atm)? (1 in. = 2.54 cm; 1 atm = 76 cm)

____

Solution. 1.01 atm

CALCULATIONS

$$30.3\text{ in.} \times \frac{2.54\text{ cm}}{1\text{ in.}} \times \frac{1\text{ atm}}{76\text{ cm}} = 1.01\text{ atm}$$

DIMENSIONAL ANALYSIS

6. Calculations in pharmacy may be performed by dimensional analysis, a method that involves the understanding of placing the ratios of the data and the equivalence between the units in a way that like units will cancel out and only the desired terms will be left. This method provides one single expression that takes the place of multiple calculation steps, reducing the opportunity for error.

A clear view of this method is shown through the following example.

A pharmacist wants to know how many inhalers should be dispensed to a patient to provide a 60 day supply of beclomethasone. The recommended dose is 168 μg twice daily. The commercial inhaler delivers 42 μg per metered dose and contains 200 inhalations.

$$2 \times \frac{168\,\mu\text{g}}{1\text{ day}} \times \frac{1\text{ dose}}{42\,\mu\text{g}} \times \frac{\text{inhaler}}{200\text{ doses}} \times 60\text{ days} = 2.4\text{ inhalers}$$

Thus, 3 inhalers must be dispensed.

Now, try to do the next problem using dimensional analysis.

A drug is administered as a single daily dose of 10 mg/kg. How many milliliters of a 10 mL vial containing 100 mg per milliliter would be administered to a patient weighing 154 lb? (1 kg = 2.2 lb)

____

Solution. 7 milliliters

CALCULATIONS

$$\frac{10\text{ mg}}{1\text{ kg}} \times \frac{\text{kg}}{2.2\text{ lb}} \times \frac{154\text{ lb}}{\text{patient}} \times \frac{\text{mL}}{100\text{ mg}} = 7\text{ mL}$$

---

7. Practice your dimensional analysis skills with the following problems.

A. If the adult dose of a solution is 0.2 mL/kg of body weight to be administered once daily, how many teaspoonfuls (tsp) should be administered to a person weighing 220 lb? (1 kg = 2.2 lb, 1 tsp = 5 mL)

B. An antibiotic provides 25,000 units of activity in each 250 mg tablet. How many total units would a patient receive by taking four tablets a day for 10 days?

C. A medication order calls for 500 mL of D5W/NS solution to be infused over 6 hours using an administration set that delivers 15 drops per milliliter. How many drops per minute should be delivered to the patient?

Solutions.

A. 4 tsp

B. 1,000,000 units

C. 21 drops/min

RATIO

8. It is very common to use ratio and its practical application (proportion) in dosage calculations. Ratio provides a comparison between two like quantities and may be expressed in several different ways (quotient, fraction, percentage, decimal). Traditionally, for example, if a comparison is made between 2 and 10, the ratio between these numbers would be expressed as 2 : 10 or $\frac{2}{10}$, and would be read as 2 to 10.

Because a ratio is a quotient, it is regulated by the same rules used for common fractions, for example:

The two terms may be multiplied or divided by the same number, e.g., the ratio 2: 10 (or $\frac{2}{20}$) will not change if both terms are multiplied by 2. The ratio will become 4:20 or $\frac{4}{20} = \frac{2}{10} = 0.2$.

Two ratios with the same value and their cross products are equivalent, e.g., $\frac{2}{5} = \frac{4}{10}$ and $2 \times 10 = 5 \times 4$.

PROPORTION

9. Proportion represents the equality between two ratios. You are probably familiar with this method, and the problems that follow should present no difficulty. A brief review is provided in case you have forgotten. Remember to write all units and to make certain that the expressions on opposite sides of the equal signs have the same units.

Try to solve the following problems, using proportion, before verifying your answers.

A. $\dfrac{1\text{ kg}}{\$3.50} = \dfrac{j}{\$2.00}$ $\quad j =$ ______kg

B. If 127 paper clips weigh 1.5 oz, how many paper clips will weigh $\frac{1}{2}$ lb? (1 lb = 16 oz)

C. An analytical instrument that is in constant use needs a new battery every 73 days. How many batteries will be required for a year?

Solutions.

A. 0.571 kg

B. 677 paper clips

C. 5 batteries

---

10. For a review of proportion, analyze the next examples and try to solve the practice problems that follow. If you don't need the review, skip ahead to frame 14.

Proportions are useful in those situations where two properties are directly related to each other. For example, if a drug costs 5¢ per gram, 2 g will cost 10¢. The two properties, cost and amount of drug, are directly related to each other. If the quantity of drug is increased five times, the cost will increase five times. If the amount of drug is cut in half the cost will be halved also. If we wanted to know the cost of 12.5 g of this drug, we could write

$$\frac{1\text{ g}}{5¢} = \frac{12.5\text{ g}}{j}$$

This equation states, "If 1 g of a drug costs 5¢, then 12.5 g will cost j." Notice that the same units are found on both sides of the equality. The ratio on the left describes the known relationship between the related properties. The ratio on the right describes the unknown situation. The two ratios are equal to each other because there is a fixed relationship between cost and weight.

One sodium bicarbonate tablet contains 300 mg of the drug; we wish to find the number of tablets that will contain 1500 mg of sodium bicarbonate. Which of the following proportions will lead to the correct solution? Why are the others not correct?

A. $\dfrac{1\text{ tablet}}{300\text{ mg}} = \dfrac{1500\text{ mg}}{j}$

B. $\dfrac{1\text{ tablet}}{1500\text{ mg}} = \dfrac{j}{300\text{ mg}}$

C. $\dfrac{1\text{ tablet}}{300\text{ mg}} = \dfrac{j}{1500\text{ mg}}$

Solution. C is correct. The ratio on the left describes the known information; that on the right, the unknown situation. Both ratios have the same units. A is incorrect because the same units do not appear in both sides of the equality (tablets/mg do not equal mg/tablet). B is incorrect because the first ratio states that 1500 mg are found in each tablet (1500 mg and 300 mg are reversed). Although the units appear to be correct, the numbers have been jumbled.

---

11. To solve a proportion,

$$\frac{a}{b} = \frac{c}{d} \quad \text{or} \quad a:b = c:d$$

we make use of the fact that the product of the first and last terms or the extremes (a and d) is equal to that of the two central terms or the means (b and c). That is, $a \times d = c \times b$. To return to our problem,

$$\frac{1\text{ tablet}}{300\text{ mg}} = \frac{j}{1500\text{ mg}}$$

$$(1\text{ tablet}) \times (1500\text{ mg}) = j \times (300\text{ mg})$$

$$j = \frac{(1\text{ tablet})(1500\text{ mg})}{300\text{ mg}} = 5\text{ tablets}$$

A formula for 42 capsules (caps) calls for 300 mg of a drug. Using proportion, find how many milligrams of the drug would be needed to make 24 capsules.

Solution. 171 mg

CALCULATIONS

$$\frac{300\text{ mg}}{42\text{ caps}} = \frac{j}{24\text{ caps}}$$

$$j = 300\text{ mg} \times \frac{24\text{ caps}}{42\text{ caps}} = 171\text{ mg}$$

---

12. If 12.0 g of a powder occupy 7.00 mL, how many milliliters will be taken up by 150 g?

Solution. 87.5 mL

CALCULATIONS

$$\frac{12.0\text{ g}}{7.00\text{ mL}} = \frac{150\text{ g}}{j}$$

$$j = 7\text{ mL} \times \frac{150\text{ g}}{12\text{ g}} = 87.5\text{ mL}$$

---

13. If a chemical costs \$14 per kilogram, how many kilograms could be purchased for \$128?

Solution. 9.14 kg

CALCULATIONS

$$\frac{1\text{ kg}}{\$14} = \frac{j}{\$128}$$

$$j = 1\text{ kg} \times \frac{\$128}{\$14} = 9.14\text{ kg}$$

ESTIMATION: ROUNDING AND POWER-OF-10 NOTATION

14. Because of the importance of accuracy in performing calculations, it's a good idea to check all results. You might think that this is unnecessary, since calculators are in such widespread use. One problem with calculating machines is that we tend to take the results for granted, without thinking about them. An error in entering data is liable to go unnoticed just because we have so much confidence in the infallibility of these machines. For safety's sake, it is necessary to check every calculation in some way, to make sure that the result is reasonable.

One kind of check is particularly useful in preventing errors of large magnitude such as misplacement of the decimal point. The method to which we are referring is that of estimation, using rounded values. The first step in this process is to round all values to one figure. The figure is kept as it appears in the original number if the figure following it is 4 or less. The single figure is promoted to the next higher number if it is followed by a 5 or higher number. For example,

4.27 rounded to one figure is 4

0.37 rounded to one figure is 0.4

3508 rounded to one figure is 4000

0.00949 rounded to one figure is 0.009

Round the following to one figure:

A. 72

B. 0.08294

C. 0.452

D. 0.75

E. 820

Solutions.

A. 70

B. 0.08

C. 0.5

D. 0.8

E. 800

---

15. Before attempting to obtain the exact solution to a problem, estimate the answer. After solving the problem, compare the exact solution with the estimate. Unless they are reasonably close to each other, both should be recalculated. Unfortunately, it is necessary to know how to do the problem in order to come up with an estimate. It is therefore possible to "solve" a problem incorrectly and to have that wrong answer check against the estimate. Estimation is helpful in preventing errors and will give an idea of the order of magnitude of a calculated value but is not infallible. The estimated answer is found by rounding off the quantities involved in the calculation to one figure and then computing the result.

A formula for 42 capsules calls for 180 mg of sucrose. To estimate the amount of sucrose per capsule, round 42 capsules to 40 capsules and 180 mg to 200 mg:

$$\frac{200\text{ mg}}{40\text{ caps}} = \frac{5\text{ mg}}{\text{caps}}$$

(The exact answer is 4.28 mg per capsule.)

A certain tablet contains 32.5 mg of phenobarbital. Estimate the number of milligrams of phenobarbital in 24 tablets.

Solution. 600 mg

CALCULATIONS

$$\frac{30\text{ mg}}{\text{tablet}} \times 20\text{ tablets} = 600\text{ mg}$$

The exact answer is 780 mg. You may think that 600 mg is rather a poor estimate, but it is good enough to tell you that your answer is in the ballpark. Certainly, if you were to solve the problem and come up with an answer of 78 mg or 7800 mg, you would realize that an error had been made.

---

16. A liquid costs $3.27 per pint. Estimate the cost of 418 pints.

Solution. $1200.

CALCULAITONS

$$\frac{\$3}{\text{pt}} \times 400\text{ pt} = \$1200.$$ (The exact answer is $1366.86.)

---

17. It is often convenient to use *power-of-10 notation* in calculations. You should already be familiar with this way of writing numbers. In "standard notation" a number is expressed as the product of a multiplier between 1 and 10 inclusive, and a power of 10. The number in example B, below, is in standard notation. As a review, try the following examples.

A. $10^2 =$

B. $5.7 \times 10^{-3} =$

C. $60 \times 10^6 =$

D. $3 \times 10^1 =$

E. $70{,}000 = 7 \times 10^?$

F. $0.02 = 2 \times$

G. $20 = 2 \times$

H. $10^3 \times 10^2 =$

I. $\dfrac{10^3}{10^1} =$

J. $\dfrac{10^0 \times 10^4}{10^4} =$

K. $10^1 \times 10^{-3} =$

L. $(3 \times 10^2) \times (2 \times 10^3) =$

M. $\dfrac{(16\times10^{2})\times(2\times10^{-4})}{(4\times10^{-1})\times10^{1}} =$

N. $(3.83 \times 10^3) - (2.6 \times 10^2) =$

Solutions.

A. 100

B. 0.0057

C. 60,000,000

D. 30

E. 4

F. 10^{-2}

G. 10^{1}

H. 10^{5}

I. 10^{2}

J. $10^{0} = 1$

K. 10^{-2}

L. 6×10^{5}

M. 8×10^{-2}

N. 3.57×10^{3}

If you completed these successfully, you are familiar with power-of-10 expressions and may proceed to frame 26. If you had difficulty or feel a bit unsure of yourself, continue for a review of power-of-10 notation.

---

18. To change 10 raised to a power n, to a natural number, first write the figure "1" and then, if the exponent (power) is $+n$, move the decimal point to the right n places, but if the exponent is $-n$, move the decimal point n places to the left.

10^4 $(n = +4) = 10{,}000$

10^{-3} $(n = -3) = 0.001$

10^0 $(n = 0) = 1$

Complete the following:

A. $10^2 =$

B. $10^{-2} =$

C. $10^1 =$

D. $10^6 =$

E. $10^{-4} =$

F. $10^{-6} =$

G. $10^0 =$

H. $10^{-1} =$

Solutions.

A. 100
B. 0.01
C. 10
D. 1,000,000
E. 0.0001
F. 0.000001
G. 1
H. 0.1

---

19. To change the product of a power of 10 and a multiplying number to a single natural number, first write the multiplying number and then, if the exponent is $+n$, move the decimal point to the right n places; if the exponent is $-n$, move the decimal point n places to the left.

4.2×10^2 $(n = +2) = 420$
37.5×10^{-3} $(n = -3) = 0.0375$
0.29×10^1 $(n = +1) = 2.9$

Change the following expressions to a single natural number:

A. $5 \times 10^1 =$
B. $1.47 \times 10^4 =$
C. $1.2 \times 10^{-3} =$
D. $1.4 \times 10^{-2} =$
E. $5.7 \times 10^6 =$
F. $0.002 \times 10^3 =$

Solutions.

A. 50
B. 14,700
C. 0.0012
D. 0.014
E. 5,700,000
F. 2

---

20. To change a natural number to the product of a power of 10 and a multiplier, write the number and move the decimal point as many places as desired. If the decimal point is moved to the left, the exponent is a positive number equal to the number of places the decimal point is moved. If the decimal point is moved to the right, the exponent is a negative number equal to the number of places the decimal point is moved. Thus:

$300 = 300 \times 10^0 = 3 \times 10^2$

$50{,}000 = 50{,}000 \times 10^0 = 5 \times 10^4$

$0.087 = 0.087 \times 10^0 = 87 \times 10^{-3}$

$127 = 127 \times 10^0 = 1.27 \times 10^2$

$0.35 = 0.35 \times 10^0 = 3.5 \times 10^{-1}$

Notice that all of the examples have been written in standard notation (in which the decimal point follows the first number in the multiplier) except for example 3.

---

21. Fill in the proper *exponent* in the following expressions:

A. $480 = 4.8 \times 10$

B. $0.0095 = 9.5 \times 10$

C. $38 = 3.8 \times 10$

D. $0.013 = 1.3 \times 10$

E. $1000 = 1 \times 10$

F. $0.000001 = 1 \times 10$

G. $0.728 = 7.28 \times 10$

H. $270 = 2.7 \times 10$

Solutions.

A. 2

B. −3

C. 1

D. −2

E. 3

F. −6

G. −1

H. 2

---

22. When multiplying powers of 10, add exponents. When dividing, subtract exponents.

$10^2 \times 10^4 = 10^6$

$10^{-3} \times 10^2 = 10^{-1}$

$$\frac{10^4}{10^3} = 10^1$$

$$\frac{10^1}{10^4} = 10^{-3}$$

$$10^1 \times \frac{10^5}{10^2} = \frac{10^6}{10^2} = 10^4$$

Try these:

A. $10^6 \times 10^1 =$

B. $10^6 \times 10^{-3} =$

C. $\dfrac{10^2}{10^3} =$

D. $\dfrac{10^4}{10^7} \times 10^3 =$

E. $\dfrac{10^2 \times 10^{-1} \times 10^3}{10^3 \times 10^{-4}} =$

Solutions.

A. 10^7

B. 10^3

C. 10^{-1}

D. $10^0 = 1$

E. 10^5

---

23. When multiplying expressions containing powers of 10 add exponents and multiply the other numbers as usual. When performing division, subtract exponents and divide the other numbers as usual.

$(3 \times 10^1) \times (2 \times 10^2) = (3 \times 2) \times (10^1 \times 10^2) = 6 \times 10^3$

$$\frac{9 \times 10^2}{3 \times 10^4} = 3 \times 10^{-2}$$

$$\frac{(4 \times 10^1) \times (3 \times 10^{-2})}{(2 \times 10^{-2}) \times (1 \times 10^4)} = \frac{(4 \times 3) \times (10^1 \times 10^{-2})}{(2 \times 1) \times (10^{-2} \times 10^4)} = \frac{12 \times 10^{-1}}{2 \times 10^2} = 6 \times 10^{-3} = 0.006$$

Complete the following, expressing the answer in powers of 10:

A. $(3 \times 10^2) \times \dfrac{(4 \times 10^1)}{2 \times 10^{-1}} =$

B. $\dfrac{70{,}000 \times 0.8 \times 30}{20 \times 600 \times 0.02} =$

Solutions.

A. 6×10^4

B. 7×10^3

24. When expressions containing powers of 10 are to be added or subtracted, all expressions must contain the same power of 10.

$$(3.4 \times 10^2) + (5.2 \times 10^2) = 8.6 \times 10^2$$

$$(2.5 \times 10^2) + (8 \times 10^1) = ?$$

The latter operation cannot be performed unless one of the exponents is changed so that both are the same. We may change both to 10^1:

$$2.5 \times 10^2 = ? \times 10^1$$

Since the exponent will be reduced by 1, the decimal point must be moved one place to the right. Since the exponential portion of the term is divided by 10, the multiplier must be multiplied by 10 to keep the value of the number from changing. Thus,

$$2.5 \times 10^2 = 25 \times 10^1$$

Now we can add:

$$(25 \times 10^1) + (8 \times 10^1) = 33 \times 10^1 = 330$$

or both may be changed to contain 10^2:

$$8 \times 10^1 = ? \times 10^2$$

The exponent will be increased by 1, so the decimal point will move one place to the left:

$$8 \times 10^1 = 0.8 \times 10^2$$

$$(2.5 \times 10^2) + (0.8 \times 10^2) = 3.3 \times 10^2 = 330$$

Complete the following:

A. $(3.7 \times 10^1) - (2.5 \times 10^1) =$

B. $(12.4 \times 10^2) + (4.20 \times 10^3) =$

C. $(6.0 \times 10^{-1}) - (5 \times 10^{-2}) =$

Solutions.

A. 1.2×10^1

B. $5.44 \times 10^3 = 54.4 \times 10^2 = 5440$

C. 0.55

---

25. Power-of-10 notation makes it easy to keep track of the decimal point in a complex calculation. It also comes in handy when estimating. Consider this example:

$$\frac{387 \times 14}{82.2} = ?$$

To estimate the answer, round to one figure and write using power-of-10 notation:

A. 387 is rounded to 4×10^2

B. 14 is rounded to 1×10^1

C. 82.2 is rounded to 8×10^1

Thus

$$\frac{(4\times10^2)\times(1\times10^1)}{8\times10^1}=\frac{4\times10^3}{8\times10^1}=0.5\times10^2=50$$

Estimate the answer to these problems, using power-of-10 notation:

A. $\frac{2700\times0.008}{0.563}=$

B. $\frac{5070}{1000000}\times0.0132=$

Solutions.

A. $\frac{(3\times10^3)\times(8\times10^{-3})}{6\times10^{-1}}=40$

B. $0.00005=5\times10^{-5}$

---

26. In all of the problems that you will encounter from now on, estimate the result. Arrive at your estimate mentally, if possible. Use power-of-10 notation, when you need it, to keep track of the decimal point. Write down the estimate. Then perform the calculation using the estimate as a check. Use this procedure in doing the following problems:

A. A formula for vitamin B_{12} tablets calls for 0.020 mg of the vitamin per tablet. How many milligrams are required to make 350,000 tablets?

B. A pharmacist bought a 500-g bottle of a drug for $3.79. What is the cost of 33 g of that drug?

C. A chemical costs 3.3¢ per milligram. What is the cost of 8.8 g? (1 g = 1000 mg)

Solutions.

A. 7000 mg

B. 25¢

C. $290.40

CALCULATIONS

A. $\frac{0.020\text{ mg}}{\text{tablet}}\times350{,}000\text{ tablets}=7000\text{ mg}$

[Estimate: $(2\times10^{-2})\times(4\times10^5)=8\times10^3=8000$ mg]

B. $\dfrac{500\text{ g}}{\$3.79} = \dfrac{33\text{ g}}{j}$

$j = \$0.25 = 25¢$

[Estimate: $j = \dfrac{(3 \times 10^1) \times 4}{5 \times 10^2} = 2.4 \times 10^{-1} = \$0.24$]

C. $8.8\text{ g} \times \dfrac{1000\text{ mg}}{1\text{g}} = 8.800\text{ mg}$

$\dfrac{3.3¢}{1\text{ mg}} = \dfrac{j}{8800\text{ mg}}$

$j = \$290.40$

[Estimate: $j = 3 \times (9 \times 10^3) = 27{,}000¢$ (i.e., = \$270)]

TOLERANCE BASED ON AMOUNTS AND PERCENTAGE

27. Health professionals deal regularly with measurement and its accuracy. We need now to discuss the tolerances permitted and notations used in describing measured values.

We have just emphasized the necessity for accuracy in calculation, and explored some techniques to minimize *calculation* error. It is important to realize, though, that in the health field, coupled to nearly every calculation there is a physical *measurement.* Solid drug materials are usually handled in powdered form and are weighed on a balance. Liquids and solutions may also be weighed, but most often they are measured by volume in a device such as a graduate or pipette. All instruments have limitations. Some balances are intended to be more sensitive than others. Different volumetric instruments may not achieve the same level of accuracy. The particular instrument and technique chosen for a measurement depend on the degree of accuracy that is required.

All measurements are subject to error. When an instrument is faulty or when a reagent is incorrectly prepared, systematic errors (always positive or always negative) will be introduced and the results will be "biased". Errors of this type may be minimized by checking equipment for proper function, by using care in handling materials and running suitable checks where possible, and by insuring that proper techniques are utilized in performing the measurements required. Despite all precautions, errors in measurement will occur. Small random fluctuations cannot be eliminated. They are due to chance breezes, to local (minor) temperature changes, to limits in human vision, to accidental vibrations, and to whims of providence, among other causes. Since these deviations are random, they may be either positive or negative.

Which of the following errors would you consider as systematic? Which are random?

A. A balance is used to find the weight of some tablets. One of the 10-g weights has been chipped, and the recorded weight of the tablets is too high.

B. A pharmacist checks the rest point of a balance. The pointer indicates a reading of zero. A moment later, she checks it again. The pointer is slightly to the left. When she checks it again, the pointer is slightly to the right of zero.

C. A pharmacist uses a transfer pipette to measure out 1 mL of a liquid. Then he mistakenly blows through it to get out the last few drops. The volume delivered exceeds 1 mL.

Solution. B is a random error and probably cannot be eliminated. A and C are systematic. The error in A can be detected by checking the weights in the set against each other. The error in C can be eliminated by use of proper technique.

---

28. Each measurement is an estimate of the true value. However, our information is incomplete unless we have some notion of the amount of error involved. Without knowing what magnitude of deviation to anticipate, it is difficult to decide the degree of confidence to put in the measured quantity.

To adequately specify a measured quantity, the two elements required are *estimate of true value* and *indication of error*.

When working with large quantities of data, as does the sociologist interested in smokers' habits or the pharmacologist concerned about the effect of a drug on the reproductive capacity of rats, the results are analyzed statistically. The mean, or perhaps the median, becomes the estimate of "true value," and the standard deviation may be employed as a measure of dispersion or deviation. However, in compounding prescriptions or weighing materials for manufacturing, pharmacists usually perform only a single measurement on each material handled. They are thus unable to make use of these statistical tools and must designate experimental quantities in other ways.

If we know that determination of the weight of a sample of powder is subject to a maximum error of 10 mg, a 450-mg sample could be indicated as 450 mg ± 10 mg. We are therefore stating that the actual weight is approximately 450 mg and that it lies somewhere between 440 and 460 mg. Thus one way to indicate a measured quantity is to submit an estimate of true value and state the maximum deviation explicitly.

A. The volume of a sample of liquid is stated to be 27.0 ± 0.5 mL. Between what limits does the true volume fall?

B. A tablet is required to contain 45 to 55 mg of active ingredient. Express this requirement in terms of a desired weight and maximum error.

Solutions.

A. 26.5 to 27.5 mL

B. 50 ± 5 mg

---

29. A liquid product is required to have a density of 0.904 ± 0.012. Three batches of the liquid product are manufactured and their densities are:

Batch 1: 0.920 g/mL

Batch 2: 0.911 g/mL

Batch 3: 0.893 g/mL

Which batches fall within the density requirement? Which do not?

Solution. Batch 1 does not meet the standard because its density differs from 0.904 g/mL by a value greater than 0.012. Batches 2 and 3 pass this test.

---

30. Another way to indicate accuracy is to write the limit of error as a *percentage of estimated value*. Instead of writing 30 g ± 3 g, this quantity could be expressed as 30 g ± 10%, since 3 g are 10% of 30 g. Thus, 400 g ± 5% means 400 g ± 5% of 400 g.

What is 5% of 400 g?

Solution. 20 g

Considering that 5% of 400 g is 20 g, 400 g ± 5% means 400 g ± 20 g. The actual value falls between 380 and 420 g. The standards set by the USP for a particular type of drug might require that tablets contain 95 to 105% of the labeled amount of drug. This is another way of stating "labeled amount ± 5%," so that a tablet that is supposed to contain 200 mg of the drug may actually contain 200 mg ± 5%, or anywhere between 190 and 210 mg, and still be acceptable.

---

31. If you had difficulty with the previous example or would like to review percentage calculations, proceed to the following examples or check percentage in Chapter 5. If you don't need the review, skip ahead to frame 32.

The easiest way to handle percentage problems is first to convert percent to a decimal. This is accomplished by moving the decimal point two places to the left:

35% = 0.35

0.02% = 0.0002

To find the value of the percent of a quantity, change the percent to a decimal and multiply by the quantity.

13% of 40 tons = 0.13 × 40 tons = 5.2 tons.

1.5% of 12 g = 0.015 × 12 g = 0.18 g

To find the percent of a quantity represented by some component value, divide that value by the total quantity and move the decimal point two places to the right. Make sure that the units are the same. For example, 3 oz are what percent of 15 oz?

$$\frac{3\text{ oz}}{15\text{ oz}} = 0.2 = 20\%$$

What percent of 120 mg are 6 mg?

$$\frac{6\text{ mg}}{120\text{ mg}} = 0.05 = 5\%$$

ACCEPTED TOLERANCES: PRESCRIPTION AND MANUFACTURING CALCULATIONS

32. In many cases, measured weights and volumes are not written so as to indicate explicitly the maximum error incurred. However, the accuracy of the determination is implied by the number of figures used in its expression. The last figure written is always approximate. For example, the volume 70.8 mL implies that the "8" is uncertain. The true volume falls between 70.75 and 70.85 mL. In other words, 70.8 mL is accurate to the nearest 0.1 mL. Consider the following ways of writing the volume:

70.8 mL = 70.8 mL ± 0.05 mL
(accurate to nearest 0.1 mL)

70.80 mL = 70.80 mL ± 0.005 mL
(accurate to nearest 0.01 mL)

70.800 mL = 70.800 mL ± 0.0005 mL
(accurate to nearest 0.001 mL)

Here, 70.8 mL, 70.80 mL, and 70.800 mL all indicate the same estimate of volume, but with a different degree of accuracy implied in each case. By using this convention of *significant figures*, the way in which a measured quantity is written provides an indication of both true value and measurement accuracy.

A. If an object is said to weigh 37.38 g, between what limits is the actual weight expected to fall?

B. If a tablet's weight is recorded as 2.6 g, to how many grams is the measurement accurate?

Solutions.

A. 37.38 ± 0.005 g = 37.375 to 37.385 g

B. 2.6 g = 2.6 g ± 0.05 g, accurate to the nearest 0.1 g

---

33. A capsule is weighed on a triple beam balance, which is accurate to the nearest 0.1 g The balance riders indicate exactly twelve grams and the weight is recorded as 12.000 g. Is this designation correct?

Solution. No. Writing 12.000 g means that the measurement was accurate to the nearest 0.001 g. This implies a greater degree of accuracy than was achieved. The weight should have been recorded as 12.0 g.

If a tablet, placed on a balance accurate to the nearest 0.01 g, has a weight of exactly two grams, how should this value be written?

Solution. 2.00 g. More zeros would imply a higher degree of accuracy than the balance is capable of. Fewer zeros would imply less accuracy than was actually obtained.

---

34. The thickness of a section of frog skin is recorded as 0.014 cm. The device used to make the measurement must have been accurate to the nearest

A. cm

B. 0.1 cm

C. 0.01 cm

D. 0.001 cm

E. 0.0001 cm

F. It is impossible to tell.

Solution. D is correct. Writing 0.014 cm means that the actual value falls between 0.0135 and 0.0145 cm and is accurate to the nearest 0.001 cm.

---

35. Each digit that is part of an experimental value, including the single uncertain digit, is significant. Zeros are significant unless they are included only to locate the decimal point.

For each of the following measured quantities, determine the number of significant figures:

A. 4.73 g

B. 4.730 g

C. 0.0065 kg

D. 6500 kg

Solutions.

A. 3

B. 4. The final zero is not needed to fix the decimal point. It is included to indicate the degree of accuracy.

C. 2. The zeros locate the decimal point and are not really part of the measured value. The quantity could have been written 6.5 g, which clearly has two significant figures.

D. 2, 3, or 4. It is impossible to tell whether the zeros are significant or merely locate the decimal point. All ambiguity would be removed by using power-of-10 notation. Thus, if the weight were written 6.5×10^3 kg, we would see that it had two significant figures. If written 6.50×10^3 kg, the quantity would have three significant figures, since the zero is not needed to locate the decimal point.

---

36. One cannot take a relatively inaccurate weight or volume and make it more exact b performing some calculation or transformation. The only way to reduce uncertain is to use a more accurate instrument or technique. As a consequence, we must b careful that the result of a calculation is not represented as being more accurate th the measurement(s) on which the calculation is based. The result is permitted contain only one uncertain figure.

The following rules determine the way in which a *calculation based on measured qua tities* should be written.

(1) When measured quantities are added or subtracted, the result can have no mo *decimal places* than the measurement with the smallest number of decimal places.

(2) In multiplication or division involving experimental values, the final result can hav no more *significant figures* than does the measurement with the smallest number significant figures.

Consider the following problems:

A. 0.20 mL of oil is dissolved in enough alcohol to make 12.000 mL of a solution. Ca culate the amount of oil in each milliliter of the solution, paying attention to th number of significant figures.

B. A foot powder contains two ingredients; each was weighed on a different balance wi different accuracy. The powder contains 1.003 g of ingredient A and 35.4 g of ingr dient B. Estimate the total weight of the foot powder.

Solutions.

A. 0.017 mL. The experimental value with the smallest number of significant figure 0.20, has two figures. The result must also have two significant figures.

B. 36.4 g. In addition or subtraction involving experimental quantities, the result shou contain no more decimal places than the quantity with the fewest decimal places. Th simplest way to handle such a calculation is to carry it out to one decimal place mo than that of the quantity with fewest places and then round off.

---

37. All of the previous examples have dealt with situations in which calculations we performed using experimentally determined quantities. In those instances, the accu racy of the calculated result depended on the accuracy of the measurements made pr vious to the calculation.

In prescription compounding or pharmaceutical manufacturing, the quantity of a partic lar ingredient that is to be weighed or measured is specified by a formula or is calculate from it. In either case, the accuracy to which the measurement must be made is not dete mined by the number of significant figures that happen to appear in the formula quantit Its accuracy is determined by the nature of the end product and the manipulative processe that will be employed. Tolerances permitted in prescription compounding are generall larger than those allowed in industrial pharmacy.

A formula for morphine sulfate tablets directs that 5 g of morphine sulfate be weighed. Does this mean that a balance accurate to the nearest gram may be used, since the formula quantity has only one significant figure?

Solution. Absolutely not! The maximum permissible error in weighing is not determined by the number of significant figures in the formula.

---

38. As a general rule, the *maximum tolerable error in weighing or measuring* a sample *for prescription compounding is 5%*. If a prescription formula calls for 0.5 g of ammonium chloride the pharmacist is obliged to weigh the ammonium chloride so as to incur an error of no more than 5%. The balance used must be accurate enough so that the maximum error is 0.025 g.

A prescription requires 0.8 mL of glycerin and 30.00 mL of alcohol. What is the maximum error, in milliliters, that is permissible in measuring each of these liquids?

Solution. Glycerin: 5% of 0.8 mL = 0.04 mL
Alcohol: 5% of 30.00 mL = 1.5 mL

The tolerance in measurement does not depend on the number of significant figures in the formula.

---

39. In *pharmaceutical manufacturing*, tolerances in measurement are more stringent than in prescription compounding. Manufacturing involves larger quantities, more manipulation, and use of more complicated techniques than does compounding. Each operation provides an opportunity for error. Although the errors may each be small, they may be additive, leading to an unacceptable product. There is no generally accepted standard for measurement accuracy in industrial practice similar to the guideline for prescription compounding, but certainly the *maximum error in weighing or measuring should be less than 1%*.

A. The maximum percent error in measurement for prescription compounding is ________.

B. The maximum percent error in measurement for pharmaceutical manufacturing is ________.

Solutions.

A. 5%

B. Less than 1%

---

40. The instrument usually used by the pharmacist for *liquid volume measurement* is t graduate. Graduates may be cylindrical or conical in shape. The error incurred in usi a conical graduate depends on the volume measured. With a cylindrical graduate, t magnitude of the error is independent of the volume. But the percent error is not co stant. It does depend on the volume that is being measured in the graduate.

For example, let us say that we are using a 100-mL graduated cylinder in which the err in reading the graduations is 1.0 mL. If 100 mL are measured, the percent error may found by dividing the magnitude of the error by the actual volume.

$$\frac{1.0\ \text{mL}}{100\ \text{mL}} = 0.01 = 1\%$$

If 50 mL are measured, $\frac{1.0\ \text{mL}}{50\ \text{mL}} = 0.02 = 2\%$

What is the percent error if 20 mL are measured in the same graduate?

Solution. 5%

CALCULATIONS

$$\frac{1.0\ \text{mL}}{20\ \text{mL}} = 0.05 = 5\%$$

---

41. As the volume measured in a cylindrical graduate becomes smaller, the percent err

A. increases

B. decreases

C. remains the same

Solution. A is correct. This is apparent from our previous examples.

---

42. The pharmacist generally has an assortment of cylindrical graduates of various size To measure a given quantity of liquid, choose a graduate that will be filled as clo to capacity as possible, to minimize the percent error. This is true for conical grad ates as well, although in these graduates the error is a function of the volume of liqu measured. No graduate should be used to measure a volume that is less than 20% its capacity.

The smallest graduate commonly found in the pharmacy has a capacity of 10 mL. Wh is the smallest quantity that should be measured in this graduate?

Solution. 2.0 mL. If less than 2.0 mL is to be measured, the pharmacist must resort another type of instrument, such as a pipette or burette. It is also possible to use a ca brated dropper to measure small volumes (Appendix 2).

REVIEW PROBLEMS

A. A set of AA batteries lasts 3 hours (h) and 30 minutes in a radio under constant use. How many hours will the radio play on 14 sets of batteries?

B. If 4 chairs in an auditorium occupy 17 ft^2, how many square feet are needed to accommodate 304 chairs?

C. Five hundred penicillin tablets cost $43.09. What is the cost of 48 tablets?

D. A certain tablet is required to contain 90 to 110% of the labeled amount of active ingredient. Between what limits (in milligrams) must the amount of drug in a 200-mg tablet fall?

E. If the error in using a balance is 10 mg, what would the percent error be if 300 mg were weighed?

F. A tablet is required to contain 46 to 54 mg of active ingredient. Express this requirement as a desired quantity and maximum percent error.

G. How many significant figures are there in each of the quantities that follow?

(1) 0.035 g

(2) 0.350 g

(3) 427.2 kg

H. In the two examples that follow, an arithmetic operation is to be performed on quantities that have been measured. Make certain that your answers contain the correct number of figures.

(a) The amount of liquid in a bottle is measured in a graduated cylinder and found to be 172 mL. On separate occasions, using a pipette, the following quantities are withdrawn: 12.00 mL, 3.75 mL and 1.008 mL. Calculate the amount remaining in the bottle.

(b) Using a mold, 0.94 g of a drug is used to make six suppositories. How much of the drug is contained in each suppository? (A quantity that can easily be counted such as six suppositories, is taken to be an absolute, not an approximate, quantity and so is not subject to measurement error).

I. A formula for 100,000 tablets requires 400 mg of a potent drug. Between what limits must the actual weight of the drug fall if the maximum permissible error in weighing is 0.5%?

J. What is the smallest quantity that should be measured using a 120-mL conical graduate?

Solutions to Review Problems

A. 49 h

B. 1292 ft^2

C. $4.14

D. 180 to 220 mg

E. 3.3%

F. 50 mg ± 4 mg or 50 mg ± 8%

G. 2, 3 and 4

H. **(a)** 155 mL. The original measurement of the contents of the bottle was accurate to the nearest milliliter, so a sum or difference involving that quantity can only be accurate to the nearest milliliter.

(b) 0.16 g. When dividing measured quantities, the result can have no more significant figures than the component with the fewest significant figures.

I. 398 to 402 mg

J. 20% of 120 mL = 24 mL

UNITS, WEIGHING, AND MEASURING

LEARNING OBJECTIVES: *After completing this chapter the student should be able to:*

1. Use the metric units of weight, volume and length and state the relationships within each group of units.
2. Perform conversions using intersystem relationships.
3. Describe the use of a Class A prescription or torsion balance in small-scale compounding.
4. Calculate minimum weighable and measurable quantities for various instruments used by the pharmacist.
5. Use the aliquot method to achieve a desired degree of precision in a measurement beyond the capacity of the instrument available.
6. Use the density and specific gravity to convert liquid volume to weight and vice versa.

SYSTEMS OF MEASUREMENT

Although the metric system has not been generally adopted in the United States, all units of pharmaceutical measurement are defined in terms of metric standards. The *United States Pharmacopeia* (USP) and other compendia of drug specifications and product formulas use only the metric system, and essentially all prescription and hospital medication orders are written in terms of metric units. Two other systems require brief mention. The apothecary system, once widely used in prescription writing, is now largely of historical interest, but some units and symbols survive to this day. The avoirdupois system is still used in commerce in the United States. Typically, the capacity of containers for pharmaceutical products is described in terms of (avoirdupois) ounces.

After reviewing the metric system in this chapter and taking a quick look at the common systems in Appendix 1, we will explore the relationships that allow switching units from one system to another.

THE METRIC SYSTEM

1. The main advantage of the metric system is its decimal-notation base. Multiples of basic units receive a prefix (see Appendix 1) that indicates its relationship with the

Pharmaceutical Calculations, Fourth Edition, By Joel L. Zatz and Maria Glaucia Teixeira
ISBN 0-471-67623-3

basic unit. For example, a milligram (mg) represents 10^{-3} of a gram and a kilogra (kg) represents 10^3 grams. Since decimal fractions are always used with metric unit the USP recommends that all prescriptions and medication orders be written with *leading zero* (0.5 mg rather than .5 mg) to avoid overlooking the placement of th decimal point. Failure to notice the placement of the decimal point may lead to a error of one tenth or ten times the desired quantity. In contrast, a *trailing zero* shoul be avoided following a whole number, for the same reason. For example, if a pre scription is written as .2 mg/10 mL, it may be mistaken for 2 mg/10 mL and the patie will receive an overdose. Conversely, if written as 0.2 mg/10.0 mL, it could be rea as 0.2 mg/100 mL and an under-dose would be dispensed.

Computations in the metric system will frequently involve:

A. Conversions from one denomination to another, which may be performed by a simpl movement of the decimal point. For example, to convert gram (higher denomination to milligram (lower denomination), multiply by 1000 or move the decimal point thre places to the right. Similarly, to change milliliter (lower denomination) to liter (highe denomination), divide by 1000 or move the decimal point three places to the left.

B. Addition, subtraction, multiplication and division of different units, which must b converted all to the same denomination. For example, to add 0.2 g to 250 mg conve both to mg and then add: 200 mg + 250 mg = 450 mg.

MEASUREMENT OF WEIGHT

2. The basic unit of mass in the metric system is the *gram* (g). Relationships betwee pharmaceutically important metric units of mass are:

$$1 \text{ gram (g)} = 0.001 \text{ kilogram (kg)} = 10^{-3}\,\text{kg}$$

$$1\,\text{g} = 1000 \text{ milligrams (mg)} = 1{,}000{,}000 \text{ micrograms } (\mu\text{g})$$

$$= 1{,}000{,}000{,}000 \text{ nanograms (ng)}$$

$$10^{-3}\,\text{kg} = 1\,\text{g} = 10^3\,\text{mg} = 10^6\,\mu\text{g} = 10^9\,\text{ng}$$

$$1\,\text{mg} = 1000 \text{ micrograms } (\mu\text{g or mcg}) = 10^3\,\mu\text{g}$$

$$1\,\mu\text{g} = 1000 \text{ nanograms (ng)} = 10^3\,\text{ng}$$

$$1\,\text{kg} = 1000\,\text{g} = 10^3\,\text{g}$$

Review these relationships. When you are familiar with them, fill in the blanks.

A. 1 mg = ________g = ________kg = ________μg = ________ng

B. 1 kg = ________g = ________mg = ________μg

C. 1 g = ________μg = ________mg = ________ng = ________kg

Solutions.

A. 1 mg = 0.001 g (10^{-3} g) = 0.000001 kg (10^{-6} kg) = 1000 μg (10^{3} μg) = 1,000,000 ng (10^{6} ng)

B. 1 kg = 1000 g (10^{3} g) = 1,000,000 mg (10^{6} mg) = 1,000,000,000 μg (10^{9} μg)

C. 1 g = 1,000,000 μg (10^{6} μg) = 1000 mg (10^{3} mg) = 1,000,000,000 ng (10^{9} ng) = 0.001 kg (10^{-3} kg)

---

3. If you got these right, move on to frame 4; if not, review and try again.

Fill in the blanks:

A. 1275 mg = __________ g

B. 130 mg = __________ μg

C. 0.032 g = __________ mg

D. 455 mg = __________ kg

E. 0.0075 g = __________ μg

F. 0.030 mg = __________ g

G. 8.8×10^{5} μg = __________ kg

H. 0.094 kg = __________ g

I. 62 ng = __________ mg

Solutions.

A. 1.275 g

B. 130,000 μg (or 1.3×10^{5} μg)

C. 32 mg

D. 0.000455 kg (or 4.55×10^{-4} kg)

E. 7500 μg

F. 0.000030 g (or 3.0×10^{-5} g)

G. 8.8×10^{-4} kg

H. 94 g

I. 6.2×10^{-5} mg

If you had a perfect score, proceed. If you had trouble setting up your conversions, review Chapter 1. If you attempted the problems before *learning* the metric weight relationships, go back and study them. Check also Appendix 1. Do not go ahead in this chapter until you can do every problem presented above.

MEASUREMENT OF VOLUME

4. The standard unit of volume in the metric system is the *liter* (L). The following listing describes the metric units most frequently used by pharmacists in measuring volume and understanding clinical laboratory test values:

1 liter (L) = 1000 milliliters (mL) = 10^3 mL (cc)

1 mL = 1 cubic centimeter (cc)

1 mL = 1000 microliters (μL or mcL) = 10^3 μL

1 μL = 0.001 mL = 10^{-3} mL = 10^{-6} L

1 dL (deciliter) = 0.1 L = 100 mL

1 L = 10^3 mL = 10^6 μL

When you have learned the relationships above, fill in the blanks:

A. 1 mL = __________cc = __________L = __________μL

B. 1 L = __________μL = __________cc

C. 1 μL = __________mL = __________L

Solutions.

A. 1 mL = 1 cc = 0.001 L (10^{-3} L) = 1000 μL (10^3 μL)

B. 1 L = 1,000,000 μL (10^6 μL) = 1000 cc (10^3 cc)

C. 1 μL = 0.001 mL = 10^{-6} L

---

5. Now try these:

A. 2.6 mL = __________cc

B. 3.5 L = __________cc

C. 0.4 mL = __________μL

D. 0.1 μL = __________L

E. 6 mL = __________L

F. 2.37 L = __________mL

G. 0.072 cc = __________μL

Solutions.

A. 2.6 cc

B. 3500 cc = 3.5×10^3 cc

C. 400 μL

D. 0.0000001 L = 1×10^{-7} L

E. 0.006 L = 6×10^{-3} L

F. 2370 mL = 2.37×10^3 mL

G. 72 μL

If you had a perfect score, proceed to next frame. If you ran into trouble, review now to overcome any difficulties.

MEASUREMENT OF LENGTH

6. The *meter* (m) is the basic unit of length. The metric units of length most frequently encountered by pharmacists are as follows:

$$1 \text{ meter (m)} = 100 \text{ centimeters (cm)} = 10^2 \text{ cm}$$
$$1 \text{ m} = 1000 \text{ millimeters (mm)} = 10^3 \text{ mm}$$
$$1 \text{ mm} = 1000 \text{ micrometers } (\mu\text{m}) = 10^3 \mu\text{m}$$
$$1 \mu\text{m} = 1000 \text{ nanometers (nm)} = 10^3 \text{ nm}$$

$$\boxed{1 \text{ m} = 10^2 \text{ cm} = 10^3 \text{ mm} = 10^6 \mu\text{m} = 10^9 \text{ nm}}$$

When you are familiar with the above, do these problems:

A. 170 cm = __________mm

B. 12.5 μm = __________mm

C. 0.2 cm = __________m

D. 0.32 m = __________mm

E. 0.013 m = __________cm

F. 744 μm = __________cm

G. 6.19 mm = __________nm

H. 0.08 m = __________μm

Solutions.

A. 1700 mm

B. 0.0125 mm

C. 0.002 m = 2×10^{-3} m

D. 320 mm

E. 1.3 cm

F. 0.0744 cm

G. 6.19×10^6 nm

H. 8×10^4 μm

If you had a perfect score, proceed. If not, be sure you understand how to do each problem before going any further.

---

7. The following problems will review metric system conversions. If you were able to get all of the problems in the previous sections right on the first try, you may skip the review and go to frame 8.

A. 470 μL = __________mL

B. 0.095 g = __________mg

C. 1.5 L = __________cc

D. 340 mm = __________m

E. 2500 μg = __________ mg

F. 870 g = __________ kg

G. 3.7×10^4 mL = __________ L

H. 235 μm = __________ cm

I. 32 cc = __________ mL

J. 0.055 g = __________ μg

K. 57 mg = __________ kg

L. 42.2 μL = __________ L

M. 0.003 kg = __________ μg

N. 12.3 m = __________ cm

O. 958 μm = __________ mm

P. 78.3 μL = __________ cc

Q. 14 cm = __________ mm

R. 0.082 mg = __________ ng

Solutions.

A. 0.47 mL

B. 95 mg

C. 1500 cc

D. 0.34 m

E. 2.5 mg

F. 0.87 kg

G. 37 L

H. 0.0235 cm

I. 32 mL

J. 5.5×10^4 μg (55,000 μg)

K. 5.7×10^{-5} kg

L. 4.22×10^{-5} L

M. 3×10^6 μg

N. 1230 cm

O. 0.958 mm

P. 0.0783 cc

Q. 140 mm

R. 82,000 ng

INTERSYSTEM RELATIONSHIP: CONVERSIONS, APPROXIMATE EQUIVALENTS

8. Despite the widespread utilization of the metric system, there are occasions wh
is necessary to deal with quantities expressed in units from other systems of m

urement. For example, these units might be occasionally used to describe drug dosage (in grains, *gr*) or a finished quantity of a prescription product (in ounces, *oz* or fluidounces, *f*℥). In those cases, we will have to convert the values from one unit system to another (refer to the common systems in Appendix 1).

The various systems have very different origins, and the relationships between units are usually not in terms of whole numbers. They are approximations expressed in enough significant figures to satisfy the particular application.

The *United States Pharmacopeia* (USP) states, for example, that

1 gram (g) = 15.4324 gr (grain)

1 fluidounce (f℥) = 29.5729 mL (milliliter)

These relations, referred to as *exact equivalents*, are actually highly refined approximations. They are used when pharmaceutical formulas for commercially *manufactured products* are converted from one system to another. Conversions and other calculations for prescriptions do not require this high degree of accuracy. If a calculation is carried to three significant figures, the maximum error resulting from rounding is 0.5%. This is ample, considering that measurements for prescriptions are permitted a 5% tolerance. Consequently, for *prescription compounding*, the equivalents should be rounded to three significant figures:

1 g = 15.4 gr

1 f℥ = 29.6 mL

According to the USP, 1 oz = 28.350 g. What conversion relationship should be used for:

A. Prescription work?

B. Pharmaceutical manufacturing?

Solutions.

A. 1 oz = 28.4 g

B. 1 oz = 28.350 g

---

9. Following is a collection of mathematical statements that allow *conversion of weight* among the metric, apothecary, and avoirdupois systems. They are rounded to three significant figures and are intended for use in *prescription calculations*.

1 g = 15.4 gr

1 oz = 28.4 g

1 gr (apothecary) = 1 gr (avoirdupois)

1 gr = 64.8 mg

1 kg = 2.20 lb

1 lb = 454 g

Only the first two expressions may be memorized, since all of the others may be derived from them. However, it's easy enough to memorize the others, thus reducing computation and saving time.

When you think you are sufficiently acquainted with the relationships, fill in the blanks.

A. 1 gr (avoirdupois) = __________gr (apothecary)

B. 1 kg = __________lb

C. 1 g = __________gr

D. 1 gr = __________mg

E. 1 lb = __________g

F. 1 oz = __________g

Solutions.

A. 1 gr (apoth.)

B. 2.20 lb

C. 15.4 gr

D. 64.8 mg

E. 454 g

F. 28.4 g

---

10. Statements of equivalence that allow *conversion of volume* units include:

1 f℥ (fluid ounce) = 29.6 mL

1 pt (pint) = 473 mL

Accuracy is to three figures, so these values are suitable for prescription calculations.

When you know these statements, fill in the blanks.

A. 1 f℥ = __________cc

B. 1 pt = __________mL

C. 1 qt = __________L

Solutions.

A. 29.6 mL = 29.6 cc

B. 473 mL

C. 1 qt = 2 pt = 946 mL = 0.946 L

---

11. Here are some problems that require conversion from one system of measurement to another. Use them for practice.

A. How many milligrams of nitroglycerin are there in 60 tablets if each tablet contains 1/100 gr of nitroglycerin?

B. Convert 1 qt to microliters.

C. If an ounce of papaverine costs $3.75, how many dollars would 146 gr cost?

D. Convert 60 g to grains.

E. Convert 6 f℥ to milliliters.

F. If 1 lb of an ointment costs $6.30, what is the cost of 60 g of the ointment?

Solutions.

A. 38.9 mg

B. 9.46×10^5 μL

C. $1.25. The "ounce" was avoirdupois.

D. 924 gr

E. 178 mL

F. 83¢

---

12. Sometimes it will be necessary to convert several quantities to other units of a different system. This is a rare situation. Here is an example.

A syrup contains 12.0 gr of a pain reliever in each fluidounce. How many milliliters contain 650 mg?

Solution. 24.7 mL

CALCULATIONS

One way to attack this problem is first to find the number of milligrams per fluidounc and then use proportion to calculate the number of milliliters.

$$\frac{12.0\text{ gr}}{1\text{ fluidounce}} \times \frac{64.8\text{ mg}}{\text{gr}} = \frac{778\text{ mg}}{\text{fluidounce}}$$

$$\frac{778\text{ mg}}{29.6\text{ mL}} = \frac{650\text{ mg}}{j}$$

$$j = 24.7\text{ mL}$$

---

13. There is a table in the USP that lists *approximate equivalents*. The relationships give in that table are not to be used when calculating quantities of materials that will b weighed or measured for a prescription. However, when a prefabricated dosage for (one prepared by a pharmaceutical manufacturer) is prescribed in one system of uni but is available to the pharmacist in strengths given in a different system, it is pe missible to dispense the strength that is approximately equivalent to that prescribed

A pharmacist receives a prescription for tablets containing aminophylline, $1\frac{1}{2}$ gr (which i equivalent to 97.2 mg). He has on hand aminophylline, 100 mg. What should he do?

It is permissible to dispense aminophylline tablets, 100 mg, when tablets are written fc $1\frac{1}{2}$ gr.

Some examples of approximate equivalents:

$$1\text{ qt} = 1000\text{ mL}$$
$$1\text{ pt} \approx 500\text{ mL}$$
$$1\text{ f℥} \approx 30\text{ mL}$$
$$1\text{ grain (gr)} \approx 60\text{ mg}$$
$$1\tfrac{1}{2}\text{ gr} \approx 100\text{ mg}$$
$$1\text{ ounce (℥)} \approx 30\text{ g}$$

The following problems exemplify some practical uses of the metric system in pharmac practice. Use them to iron out any weak spots in your work.

A. How many milliliters of oil are needed to prepare 640 capsules if 15 capsules contai 137 μL of oil?

B. How many kilograms of sodium fluoride are needed to make 60,000 tablets of sodium fluoride tablets each containing 50 μg?

C. You have 15.8 g of tetracycline hydrochloride powder. How many 250-mg capsules can you make from this quantity of powder?

D. A pharmacist adds sufficient water to a vial containing 2 million units of penicillin to make a total volume of 5 mL of penicillin suspension. How many milliliters of the suspension should be injected into a child who is to receive a dose of 300,000 units?

E. A soft gelatin capsule contains 0.22 mL of an oil. How many liters of oil would be required for 4500 capsules?

F. How many tablets each containing 2.5 mg of amphetamine can be made from 0.620 kg of amphetamine?

G. How many micrograms of vitamin A are there in each tablet if 2.75 g of vitamin A are used to make 5000 tablets?

H. A manufacturer fills 490 bottles so that each bottle contains 30 mL of an oil. If he started with 15.5 L of the oil, how many milliliters are left after the filling operation is complete?

I. If 12.0 mg of a drug are present in 440 g of a powder, how many grams of drug would there be in 14.0 kg of powder?

Solutions.

A. 5.85 mL

B. 0.003 kg

C. 63 capsules

D. 0.75 mL

E. 0.99 L

F. 2.48×10^5 tablets

G. 550 μg

H. 800 mL

I. 0.382 g

CLASS A PRESCRIPTION OR TORSION BALANCE: MINIMUM WEIGHABLE AND MEASURABLE QUANTITIES

14. The balance used in weighing drugs for prescription compounding is known as a *Class A* or *torsion balance*. It is a fairly sensitive instrument. In a prescription balance that meets current standards, the *sensitivity requirement* (SR) is 6 mg. This means that a load of 6 mg causes a deflection of at least one scale division of the pointer. The balance is therefore capable of discriminating a minimum difference of 6 mg. This amount represents the maximum error in weighing that will be incurred using a prescription balance in proper working order. To calculate the percent error divide the sensitivity requirement by the amount to weigh and express the result as percent.

If 1500 mg are weighed, $\frac{6 \text{ mg}}{1500 \text{ mg}} = 0.004 = 0.4\%$

If 150 mg are weighed, $\frac{6 \text{ mg}}{150 \text{ mg}} = 0.04 = 4\%$

and so on.

Table 2.1 shows how the percent error changes depending on the amount that is weighed. Since the maximum percent error that is acceptable in weighing for prescription compounding is 5%, we see that it is all right to use the prescription balance to weigh 1500 mg or 150 mg but that the percent error involved in weighing 15 mg is much too great.

TABLE 2.1 Effect of amount weighed on percent error

Amount weighed (mg)	SR or error (mg)	% error
1500	6	0.4
150	6	4
15	6	40

Just as with a cylindrical graduate, the percent error increases as the amount weighed becomes smaller. From these data, we can tell that somewhere between 150 mg and 15 mg a particular quantity exists such that the error in weighing would be exactly 5%. That quantity is the smallest amount that can be weighed on a prescription balance with acceptable accuracy.

For example, the minimum weighable quantity for a prescription balance with a sensitivity requirement of 6 mg and 5% maximum acceptable percent error could be calculated by:

$$\text{(MWQ) Minimum Weighable Quantity} = \frac{\text{SR}}{\text{max. \% error}}$$

Let j equal minimum weighable quantity:

$$j = \frac{6\text{ mg}}{0.05} \text{ or } 0.05 = \frac{6\text{ mg}}{j}$$

$$j = 120\text{ mg}$$

This calculation shows that if a quantity of 120 mg or more is weighed on a prescription balance, the error will be 5% or less. The accuracy will therefore be acceptable. If less than 120 mg of a drug is required, direct weighing of that quantity on a prescription balance will lead to unacceptably large errors. Other techniques must be used. These are discussed below under *aliquot method.*

Now, try these problems.

A. What is the maximum percent error incurred if a balance with a sensitivity requirement of 15 mg is used to weigh 120 mg of a powder?

B. What is the minimum weighable quantity with a maximum error of 5% on a balance whose sensitivity requirement is 30 mg?

C. What is the smallest quantity that can be weighed with 1% maximum error on a balance whose sensitivity requirement is 1.0 mg?

Solutions.

A. 12.5%

B. 600 mg

C. 100 mg

CALCULATIONS

A. $\frac{15\text{ mg}}{120\text{ mg}} = 0.125 = 12.5\%$

B. Let j equal minimum weighable quantity:

$$0.05 = \frac{30\text{ mg}}{j}$$

$$j = 600\text{ mg}$$

C. Let j equal minimum weighable quantity:

$$0.01 = \frac{1\text{ mg}}{j}$$

$$j = 100\text{ mg}$$

ALIQUOT METHOD AND TRITURATIONS: SOLIDS AND LIQUIDS

15. According to the USP, weighing for prescription compounding must be performed i a Class A prescription balance with a sensitivity requirement of 6 mg and a maximu acceptable error of 5% (lower limit is 120 mg and upper limit is 120 g). As a conse quence, when quantities smaller than 120 mg or larger than 120 g are needed, the cannot be weighed directly on a Class A prescription balance. Some other means fo measuring the weight of a drug accurately, which are not beyond the limitations o the balance, must be found.

For quantities greater than the balance's capacity, the drug may be weighed in portions. The disadvantage of this method is that more errors could be introduced.

Greater accuracy can be achieved by using a high precision electronic analytical balance, when weighing very small quantities of drugs.

Another approach to the measurement of small quantities involves the use of triturations, which are mixtures of a drug with an inert diluent, usually lactose. For these techniques, a diluent will expand the weight (or volume) of the system, allowing convenient and accurate measurement of an *aliquot* (sample, part or portion). The aliquot method and triturations will allow us to obtain the necessary amount of the drug without compromising accuracy.

The use of the aliquot method includes:

1. Determination of the minimum weighable amount for the instrument in use;
2. Weighing of the minimum weighable amount (or any multiple) of the substance;
3. Mixing (diluting) with a determined amount of an inert substance to obtain a dilution or mixture.
4. Weighing of an aliquot (sample) of the mixture, which will provide the desired quantity of the substance.
5. All weighed quantities must be *equal or greater* than the minimum weighable amount.

The preparation of a trituration consists of three steps:

1. Determine the strength of the trituration.
2. Determine the quantities needed to make the trituration.
3. Determine the amount of trituration that contains the desired quantity of drug.

Let's try these approaches with the following example.

A prescription requires 30 mg of morphine sulfate. How should the pharmacist proceed?

It is not possible to weigh less than 120 mg with acceptable accuracy on a prescription balance, but the difficulty can be circumvented by using the aliquot method or by preparing a trituration. An inert substance such as lactose may be used as diluent. See the problem solved by 3 different approaches.

Determination of the minimum weighable amount:

$$0.05 = \frac{6\text{ mg}}{j}$$

$$j = 120\text{ mg}$$

1. Determination of an aliquot to contain 30 mg of drug:

$$\frac{120\text{ mg of morphine sulfate}}{480\text{ mg of mixture}} = \frac{30\text{ mg of morphine sulfate needed}}{j}$$

$$j = \frac{480 \times 30}{120} = 120\text{ mg of the mixture, which will provide 30 mg of morphine sulfate.}$$

Proof:
Since *4 times* the needed amount of morphine sulfate was weighed, an aliquot equal to one fourth of the 480 mg total mixture (= 120 mg), will contain 30 mg of morphine sulfate.

$$\tfrac{1}{4} \times 120\,\text{mg (drug weighed)} = 30\,\text{mg drug}$$

$$\tfrac{1}{4} \times 360\,\text{mg (diluent added)} = \underline{90\,\text{mg diluent}}$$

$$120\,\text{mg aliquot}$$

2. Consider the minimum weighable amount of 120 mg, from calculation above. Since only quantities ≥120 mg can be weighed with the desired accuracy in this balance, any multiple of the desired quantity that will yield 120 mg or greater can be used.

$$30\text{ mg (drug needed)} \times 4 = 120\text{ mg (drug to be weighed)}$$
$$120\text{ mg (aliquot containing drug)} \times 4 = 480\text{ mg (mixture to be prepared)}$$
$$480\text{ mg (mixture)} - 120\text{ mg (drug)} = 360\text{ mg of diluent}$$

The pharmacist should weigh 120 mg of morphine sulfate, combine with 360 mg of diluent to make 480 mg of mixture. If he weighs $\frac{1}{4}$th of the mixture ($\frac{1}{4} \times 480$ mg), the 120 mg aliquot will contain 30 mg of drug.

Proof:
The ratio of drug to mixture is 30 mg : 120 mg or 120 : 480 or 1 : 4. Considering the aliquot of 120 mg:

1 : 4 = 30 mg (drug): 30 mg + 90 mg (drug + diluent)

30 : 120 = 30 mg of morphine sulfate in each 120 mg of mixture

3. The pharmacist may prepare a trituration. A suitable quantity of morphine sulfate (at least 120 mg) must be combined with lactose. A portion of the trituration will then be used to obtain 30 mg of morphine sulfate.

A 1 : 10 trituration may be prepared. The designation 1:10 means that 10 units of the trituration contain 1 unit of drug (and 9 units of lactose).

$$\frac{30\text{ mg}}{\frac{1}{10}} = 300\text{ mg of trituration}$$

The pharmacist should use 300 mg of the trituration. This quantity can be weighed with acceptable accuracy using a prescription balance.

A 1 : 7 trituration may be prepared. The amount required would be

$$\frac{30\text{ mg}}{\frac{1}{7}} = 210\text{ mg trituration}$$

210 mg of a 1 : 7 trituration would be used. This is also acceptable since 210 mg exceeds the minimum weighable quantity for a prescription balance.

A 1 : 3 trituration may be prepared. The amount required would be

$$\frac{30\text{ mg}}{\frac{1}{3}} = 90\text{ mg trituration.}$$

90 mg of a 1:3 trituration are needed. Unfortunately, this is not an acceptable approach because 90 mg is not a weighable quantity. It is below the minimum weighable amount for the instrument.

Referring to the calculations above, does the quantity of trituration needed to deliver a fixed amount of drug have to be reduced, increased, or remain unchanged as the concentration of the trituration is increased?

Solution. Reduced (1:3 is a higher concentration than 1:7)

---

16. If the concentration of the trituration is made too high, the amount of trituration that contains the desired quantity of drug will be less than 120 mg and will therefore not be a weighable quantity.

To obtain the maximum permissible strength of the trituration, find the smallest whole number by which the required drug amount should be multiplied to give at least 120 mg. The trituration is the reciprocal of this number.

For example, let us say that 40 mg of a drug has to be weighed. 40 mg must be multiplied by 3 to equal 120 mg. A 1:3 trituration should be prepared. The amount of trituration necessary to supply 40 mg of the drug is 120 mg:

$$\frac{40\text{ mg}}{\frac{1}{3}} = 120\text{ mg}$$

As a second example, assume that 22 mg of a drug has to be weighed. The smallest whole number by which 22 mg must be multiplied to equal or just exceed 120 mg is 6. Therefore, a 1:6 trituration should be prepared. 132 mg of the trituration are required for the prescription:

$$\frac{22\text{ mg}}{\frac{1}{6}} = 132\text{ mg}$$

For the following situations, determine the trituration of maximum strength that could be used to supply the necessary amount of drug

A. 55 mg of drug are needed.

B. 14 mg of drug are needed.

C. 80 mg of drug are needed.

Solutions.

A. 1 : 3

B. 1 : 9

C. 1 : 2

---

17. Now let us consider a practical problem: A pharmacist has to weigh 25 mg of a dru for a prescription. This quantity is less than 120 mg and cannot be weighed direct on a prescription balance. We must resort to using a trituration.

We work these out as follows:

(1) By inspection, 25 mg must be multiplied by 5 to give a product of at least 120 m Use a 1 : 5 trituration.

(2) To make the 1 : 5 trituration, weigh 120 mg of the drug and mix intimately wi 480 mg of lactose (or another inert material, perhaps one included in the prescription

(3) Calculate the amount of trituration needed.

$$\frac{25\text{ mg}}{\frac{1}{5}} = 125\text{ mg}$$

125 mg of the trituration should be used for the prescription. If lactose is not part of th prescription formula, it is permissible for the pharmacist to add it to the formula becaus lactose is inert and will do no harm.

Note that a concentration weaker than 1 : 5 could have been used to solve th problem. Use of weaker dilutions will be just as accurate. But the use of large quantitie may lead to practical difficulties. If the drug is being incorporated into a powder mass th will be divided into capsules, the use of excessively dilute triturations may involve suc large quantities of material that the capsules will be too large to swallow.

Review the procedures for weighing quantities less than 120 mg. Then try this problem.

A pharmacist has a prescription for 6 capsules, each capsule to contain 8 mg of phen barbital. How should the phenobarbital be weighed?

Solution. The total needed is

$$\frac{8\text{ mg}}{\text{cap}} \times 6\text{ caps} = 48\text{ mg}$$

A trituration of phenobarbital must be prepared.

(1) Use a 1:3 trituration.

(2) To make the 1:3 trituration, combine 120 mg of phenobarbital with 240 mg of diluent.

(3) $\dfrac{48\text{ mg}}{\frac{1}{3}} = 144\text{ mg}$

Weigh 144 mg of the trituration. It will contain 48 mg of phenobarbital.

---

18. A pharmacist has to prepare 6 capsules, each containing 12 mg of codeine sulfate. How should the codeine sulfate be weighed?

Solution. The total needed is

$$\frac{12\text{ mg}}{\text{cap}} \times 6\text{ caps} = 72\text{ mg}$$

We must make up a trituration. Lactose may be used as the diluent.

(1) Use a 1:2 trituration.

(2) Prepare a 1:2 trituration by mixing 120 mg of codeine sulfate with 120 mg of lactose.

(3) $\dfrac{72\text{ mg}}{\frac{1}{2}} = 144\text{ mg}$

The trituration will contain 72 mg of codeine sulfate and 72 mg of lactose.

---

19. Less frequently, the aliquot method is applied to liquids. In pharmaceutical compounding, the selection of an instrument for measurement of volume is based on the level of precision required. Cylindrical graduates, micropipettes or syringes are the most common instruments used in pharmacy practice. As a rule of thumb, a graduate and a pipette are selected based on a capacity equal to or just exceeding the volume to be measured. A volume measured in a syringe should not be larger than 2/3 of its capacity.

In some situations, the compounding pharmacist may have limited tools for measurement of volumes. The use of the aliquot method will allow the measurement of the necessary volume of the drug without compromising accuracy.

The following exemplifies one of these situations.

A 150-mL prescription calls for 0.6 mL of a surfactant. The smallest measuring tool available in the pharmacy is a 10-milliliter graduate cylinder calibrated from 2–10 mL, in

1-mL units. How could this volume be measured? The surfactant is water-soluble, ar then water may be used as a diluent.

Solution. By the aliquot method.

Consider the minimum measurable amount of 2 mL and the maximum of 10 mL. Sinc only quantities ≥2 mL, in increments of 1 mL, can be measured with the desired accurac in this graduate, a multiple of the desired quantity that will yield 2 mL or greater can b used (remember that no more than 10 mL can be measured).

0.6 mL (surfactant needed) × 5 = 3 mL (surfactant to be measured)

2 mL (aliquot containing surfactant) × 5 = 10 mL (dilution to be prepared)

10 mL (mixture) − 3 mL (surfactant) = 7 mL of water

When 3 mL of surfactant is measured and 7 mL of water is added and mixed to make 1 mL of a diluted solution, 1/5th of the mixture (1/5 × 10 mL), or a 2 mL aliquot, will contai 0.6 mL of surfactant.

Proof:
The ratio of drug to mixture is 0.6 mL: 2 mL or 3: 10 or 1 : 3.33. Considering the aliqu of 2 mL:

1 : 3.33 = 0.6 mL (surfactant): 0.6 mg + 1.4 mL (surfactant + water)

3: 10 = 0.6 mL of surfactant in each 2 mL of diluted solution.

Note that no other different dilution could be prepared in this case. The final dilution pre pared must not exceed 10 mL, the limiting capacity of the graduate.

---

20. Now practice this concept with the following problems.

A. A prescription calls for 0.5 mL of a flavoring oil. Using a 10-mL graduate calibrate from 2–10 mL in 1-mL divisions and 95% alcohol as diluent, calculate how the desire quantity could be obtained.

B. A pharmacy receives a pharmaceutical formula for a syrup that calls for 1.25 mL of coloring solution. A 25-mL graduate calibrated in 1-mL units is the smallest capacit measuring tool available. Using water as the diluent, calculate how to obtain th desired quantity.

Solutions.

A. Measure 2 mL of oil, add 6 mL of alcohol and mix thoroughly. Take ¼ of the dilution prepared (= 2 mL) to obtain the desired amount of oil.

B. Prepare a 1 : 4 dilution. 5 mL of dilution will provide 1.25 mL of coloring solution.

CALCULATIONS

A. $0.5\,\text{mL} \times 4 = 2\,\text{mL}$ of oil

$2\,\text{mL (aliquot)} \times 4 = 8\,\text{mL (dilution prepared)}$

$8\,\text{mL} - 2\,\text{mL} = 6\,\text{mL}$ of alcohol

or: $= \dfrac{0.5\ \text{mL}}{2\ \text{mL}} = \dfrac{2\ \text{mL}}{j}$

$j = 8\,\text{mL}$ dilution to be prepared

6 mL of alcohol to prepare diluted solution.

or: By making a *trituration* (because this is a liquid mixture, it will be more appropriate to call it *dilution*)

By inspection, 0.5 mL must be multiplied by 4 to give a product of ≥2 mL.

$0.5 \times 4 = 2\,\text{mL}$

Prepare a 1 : 4 dilution by mixing 2 mL of oil with 6 mL of alcohol.

Volume of dilution needed to provide 0.5 mL of oil:

$$\frac{0.5\ \text{mL}}{\frac{1}{4}} = 2\ \text{mL}$$

B. $1.25 \times 4 = 5\,\text{mL}$

Prepare a 1 : 4 dilution by combining 5 mL of coloring solution and 15 mL of water.

$\dfrac{1.25\ \text{mL}}{\frac{1}{4}} = 5\ \text{mL}$ of dilution will provide 1.25 mL of coloring solution.

DENSITY AND SPECIFIC GRAVITY

21. Density is defined as the mass per unit-volume of a particular substance:

$$\text{Density} = \frac{\text{mass}}{\text{volume}}$$

The units of density depend, of course, on the units used to denote mass and volume. In the metric system, density is usually expressed in terms of grams per milliliter.

Since volume is a function of temperature, density is too. However, for purposes of calculation for prescription compounding, we ignore these temperature effects, since the error so introduced is small and does not materially increase the error of our final result. Thus, while the density of water at 25°C is actually 0.997 g/mL, we use a value of 1.00 g/mL in our calculations.

Certain pharmaceutical formulas list all the ingredients by weight, despite the fact that some may be liquids and are therefore more conveniently measured by volume.

For example, a formula calls for 2.75 g of methyl salicylate, a liquid. In order to convert this weight to a volume, we need a statement of equivalence. This is provided by the density. The density of methyl salicylate is 1.18 g/mL. In other words, 1 mL of methyl salicylate weighs 1.18 g, or

$$1.18\,\text{g methyl salicylate} = 1\,\text{mL methyl salicylate}$$

To convert 2.75 g of methyl salicylate to a volume:

$$2.75\,\text{g} \times \frac{1\,\text{mL}}{1.18\,\text{g}} = 2.33\,\text{mL}$$

Just in case this operation looks like sleight of hand, let us take another look at the fraction $\frac{1\,\text{mL}}{1.18\,\text{g}}$. If I had 1 mL of methyl salicylate in one vial and 1.18 g of methyl salicylate in another, the amounts in both vials would be identical since 1.18 g is the weight of 1 mL. Thus "1 mL" and "1.18 g" are merely different ways of describing the same quantity of methyl salicylate. The fraction $\frac{1\,\text{mL}}{1.18\,\text{g}}$ is therefore equal to unity, and multiplying by this fraction converts weight to volume.

The density of chloroform is 1.48 g/mL, and that of ether is 0.715 g/mL. For each of these liquids, write an equation that relates weight and volume.

Solution. Chloroform: 1.48 g = 1 mL

Ether: 0.715 g = 1 mL

---

22. A prescription for 60 g of an ointment contains 3 g of light mineral oil. The density of light mineral oil is 0.850 g/mL. The density of the finished ointment is 0.975 g/mL. Which of the following expressions will allow us to determine how many milliliters of light mineral oil should be used for this prescription? Why are the others incorrect?

A. $3.00\,\text{g} \times \frac{1\,\text{mL}}{0.850\,\text{g}} =$

B. $3.00\,\text{g} \times \frac{0.850\,\text{g}}{1\,\text{mL}} =$

C. $60.0\text{ g} \times \frac{1\text{ g}}{0.850\text{ g}} =$

D. $3.00\text{ g} \times \frac{1\text{ mL}}{0.975\text{ g}} =$

Solution.

A. Correct because 3.00 g is the amount of light mineral oil we must work with; the fraction 1 mL/0.850 g is equal to unity and the result will be expressed in milliliters.

B. Incorrect because the fraction, although equal to unity, is inverted, so that the result will be expressed as g^2/mL, not milliliters.

C. Incorrect because we do not have 60.0 g of light mineral oil, only 3.00 g.

D. Incorrect because of the fraction 1 mL/0.975 g. 0.975 g/mL is the density of the ointment not of light mineral oil. One mL of light mineral oil does not weigh 0.975 g, so that the fraction is not equal to unity. Consequently, this formula leads not only to an alteration of units but an alteration of quantity as well.

---

23. If 25.0 g of olive oil are required for a prescription, what volume should be used? (Density of olive oil = 0.910 g/mL.)

Solution. 27.5 mL

CALCULATIONS

$$25.0\text{ g} \times \frac{1\text{ mL}}{0.910\text{ g}} = 27.5\text{ mL}$$

---

24. How many grams would 2 fluidounces of peanut oil (density = 0.917 g/mL) weigh?

Solution. 54.3 g

CALCULATIONS

2 fluidounces = 59.2 mL

$$59.2 \text{ mL} \times \frac{0.917 \text{ g}}{\text{mL}} = 54.3 \text{ g}$$

---

25. From these examples it is evident that conversion from weight to volume, or from volume to weight, can be accomplished using the density of the material under consideration. Most often, however, reference sources will not list density; specific gravity will be given instead. Specific gravity (sp g) may be defined as the ratio of the density of a substance to that of some reference material. Water is used as the reference for liquids and solids:

$$\text{sp g} = \frac{\text{density of substance}}{\text{density of water}}$$

Both densities should be expressed in the same units, so that specific gravity is a *dimensionless quantity*. It should be noted that the specific gravity, as such, cannot be used in converting between weight and volume. We can work only with density and the units must be explicitly stated. However, we may use the definition of specific gravity to find density which can then be employed in conversion calculations.

The specific gravity of glycerin is 1.25. Let us calculate its density in the metric system:

$$\text{sp g} = \frac{\text{density of glycerin}}{\text{density of water}}$$

density of glycerin = sp g × density of water = 1.25 × 1.00 g/mL = 1.25 g/mL

The density, in g/mL, is numerically equal to the specific gravity. This is not true for other units. For example, the density of glycerin in the apothecary system is 569 grains per fluidounce. Only when density is expressed as g/mL are density and specific gravity numerically equal.

The specific gravity of white petrolatum is 0.850. What weight of this substance will fill an ointment jar whose capacity is 1/2 ounce? (*Hint*: Use the approximate equivalent 1 fluidounce = 30 mL for calculations involving container capacities. In this case, 1/2 ounce = 15 mL.)

____

Solution. 12.8 g

CALCULATIONS

With a sp g of 0.850, the density is 0.850 g/mL.

15 mL × 0.850 g/mL = 12.8 g

---

26. A formula calls for 0.623 kg of an oil whose specific gravity is 0.900. How many milliliters should be used?

Solution. 692 mL

CALCULATIONS

$$623\ \text{g} \times \frac{1\ \text{mL}}{0.900\ \text{g}} = 692\ \text{mL}$$

---

27. A prescription contains 88.8 mL of hydrochloric acid (sp g = 1.18), which costs $2.20 per pound. What is the cost of the quantity used in the prescription?

Solution. $0.51

CALCULATIONS

$88.8\ \text{mL} \times 1.18\ \text{g/mL} = 105\ \text{g}$ hydrochloric acid

$$\frac{454\ \text{g}}{\$2.20} = \frac{105\ \text{g}}{j}$$

$j = \$0.51$

---

28. If you want more practice, try these:

A. The specific gravity of an oil is 0.930. How much will 12.0 mL weigh?

B. If the specific gravity of a liquid is 1.10, how many milliliters would 27.0 kg occupy?

C. A formula calls for 1.32 g of an oil whose specific gravity is 0.886. How many microliters should be used?

Solutions.

A. 11.2 g
B. 24,500 mL
C. 1490 μL

REVIEW PROBLEMS

Do all of these review problems before verifying the answers. Work carefully using a
estimate as a check. Write all units down, and make certain that the units in your fin
solution are consistent with the quantities that enter into the calculation.

If any of your answers do not agree with the answer provided at the end, read th
problem(s) again and check your calculations. If you still have trouble, study the full solu
tions provided. To review the units of this system, go back to the beginning of this chapte
To review the techniques used in problem solving, go back to chapter 1. It is importa
that you clear things up now, before we leave the metric system and basic problem solvin
approaches.

A. A pharmacist, on separate occasions, dispenses 220 mg and 450 mg of a certain dru
How much of the drug remains if the bottle originally contained 1.5 g?

B. A patient receives 500 μg of estradiol benzoate by injection every day for 22 day
How many grams of estradiol benzoate does the patient receive altogether?

C. If 25 mL of oil are used in the manufacture of 125,000 capsules, how many micro
liters of oil are contained in each capsule?

D. A solution contains 125 mg of a drug in each milliliter. How many milliliters conta
2.5 g of the drug?

E. A 25 mL vial of an injectable solution has 4 mg of the drug. If the dose to be administered to a patient is 200 μg, how many milliliters of this solution should be used?

F. A patient is to be given capsules, each containing 22 mg of hydrocortisone. How many grams of hydrocortisone are required to make 60 such capsules?

G. If 0.5 L of a medicinal solution is dispensed to a patient who takes 1 tablespoonful of the solution four times a day for seven days, how many milliliters of the solution remain? (1 tablespoonful = 15 mL)

H. A capsule contains 4 μg of a potent drug. How many capsules could be made from 0.332 kg of the drug?

I. How many liters of a solution are needed to fill 250 bottles, each containing 45 cc of the solution?

J. If 3.17 kg of a drug are used to make 50,000 tablets, how many milligrams will 30 tablets contain?

K. A medicated disk has a thickness of 4.80 mm. What would be the height, in centimeters, of a stack of 24 medicated disks?

L. How many microliters of oil does each glass vial contain if 2.1 L of the oil were used to uniformly fill 6×10^4 glass vials?

M. If 0.625 kg of a drug is used to make 250,000 tablets, how many tablets contain 0.125 g?

N. What is the smallest quantity that can be weighed on a prescription balance (with a sensitivity requirement of 6 mg) with a maximum error of 2%?

O. What maximum percent error is anticipated if 450 mg of powder are weighed on a balance whose sensitivity requirement is 18 mg?

P. What is the smallest quantity that can be weighed with a maximum error of 1% on a balance whose sensitivity requirement is 0.1 g?

Q. What is the smallest volume that can be measured with acceptable accuracy for prescription compounding in a 2-mL pipette in which the maximum error is 0.005 mL, independent of the volume measured?

R. How many grams of lactose should be added to 120 mg of strychnine sulfate to make a 1 : 20 trituration?

S. How many milligrams of a 1 : 25 trituration of atropine should be used to prepare 40 capsules, each containing 0.4 mg of atropine?

T. A prescription for capsules calls for 18 mg of a drug. Calculate the maximum concentration of a trituration of simple ratio that may be used as a source of the drug.

U. ℞ Hyoscine hydrobromide 0.001 g
Benadryl hydrochloride 0.025 g
d.t.d. caps No. 12

How would you weigh the hyoscine hydrobromide for this prescription?

V. If a formula calls for 0.75 mL of peppermint oil and a pharmacist has available only a 25-mL graduate cylinder, calibrated in 1-mL units, how could the volume be accurately measured? Use alcohol as the diluent.

W. A prescription calls for 10.0 g of citric acid syrup, a liquid whose specific gravity is 1.17. What volume of citric acid syrup should be used?

X. The specific gravity of an oil is 0.812. How many cubic centimeters should be used to fill a prescription that calls for 9 g?

Y. A formula calls for 3.22 kg of an oil whose specific gravity is 0.922. How many milliliters should be used?

Z. A formula calls for 0.752 L of an oil whose specific gravity is 0.950. How many kilograms will the oil weigh?

Solutions to Review Problems

A. 830 mg or 0.83 g

B. 0.011 g

C. 0.2 μL

D. 20.0 mL

E. 1.25 mL

F. 1.32 g

G. 80 mL

H. 83×10^6 = 83,000,000 capsules

I. 11.3 L

J. 1900 mg

K. 11.5 cm

L. 35 μL

M. 50 tablets

N. 300 mg

O. 4%

P. 10 g

Q. 0.1 mL

R. 2.28 g

S. 400 mg

T. 1 : 7

U. Combine 120 mg of drug with 1.08 g of lactose and mix thoroughly. Use 120 mg of the mixture. (Other answers that give the same amount of drug in the final step are also correct.)

V. Prepare 12 mL dilution. An aliquot of 3 mL will contain 0.75 mL oil.

W. 8.55 mL

X. 11.1 cc

Y. 3490 mL

Z. 0.714 kg

Calculations for the Review Problems

A. Before adding or subtracting quantities, they must be expressed in the same units.
220 mg + 450 mg = 670 mg = 0.67 g (total dispensed)
1.5 g − 0.67 g = 0.83 g (remainder)

B. The total amount administered is equal to the product of the daily dose and the number of days.

$$\frac{500\ \mu g}{\text{day}} \times 22\ \text{days} = 11{,}000\ \mu g = 11\ \text{mg} = 0.011\ \text{g}$$

C. $$\frac{25\ \text{mL}}{125\,000\ \text{caps}} = 2 \times 10^{-4}\ \text{mL/cap} = 0.2\ \mu\text{L/cap}$$

D. 2.5 g = 2500 mg

$$\frac{125\ \text{mg}}{1\ \text{mL}} = \frac{2500\ \text{mg}}{j}$$

$j = 20.0$ mL

E. $$\frac{4000\ \mu g}{25\ \text{mL}} = \frac{200\ \mu g}{j}$$

$j = 1.25$ mL

Estimate by inspection, j is 1/20 of 25 mL or about 1 mL

F. $$22.0\ \text{mg/caps} \times 60\ \text{caps} \times \frac{1\ \text{g}}{1000\ \text{mg}} = 1.32\ \text{g}$$

G. The patient has taken a total of 28 tablespoonfuls. Since 1 tablespoonful = 15 mL, he has taken 28 tablespoonfuls × 15 mL/tablespoonful = 420 mL.

$500\,\text{mL} - 420\,\text{mL} = 80\,\text{mL}$

H. $0.332\,\text{kg} = 332\,\text{g} = 332 \times 10^6\,\mu\text{g}$

$$\frac{332 \times 10^6\ \mu\text{g}}{4\ \mu\text{g/cap}} = 83 \times 10^6\ \text{capsules}$$

I. $$\frac{45.0\ \text{cc}}{\text{bottle}} \times 250\ \text{bottles} = 11{,}300\ \text{cc} = 11.3\ \text{L}$$

J. $3.17\,\text{kg} = 3.17 \times 10^6\,\text{mg}$

$$\frac{3.17 \times 10^6\ \text{mg}}{5 \times 10^4\ \text{tabs}} = \frac{j}{30\ \text{tabs}}$$

$j = 1900\ \text{mg}$

(Only three significant figures should be written because the weight of the drug was accurate to three figures).

K. $4.80\,\text{mm} \times 24 = 115\,\text{mm} = 11.5\,\text{cm}$

L. $2.1\,\text{L} = 2.1 \times 10^6\,\mu\text{L}$

$$\frac{2.1 \times 10^6\ \mu\text{L}}{6 \times 10^4\ \text{vials}} = \frac{35\ \mu\text{L}}{\text{vial}}$$

M. $$\frac{625\ \text{g}}{2.5 \times 10^5\ \text{tabs}} = \frac{0.125\ \text{g}}{j}$$

$j = 50\ \text{tablets}$

N. $$0.02 = \frac{6\ \text{mg}}{j}$$

$j = 300\ \text{mg}$

O. $$j = \frac{18}{450} = 0.04 = 4\%$$

P. $$0.01 = \frac{100\ \text{mg}}{j}$$

$j = 10{,}000\ \text{mg} = 10\ \text{g}$

Q. 5% error is accepted for prescription compounding, thus:
$0.05 \times 2\,\text{mL} = 0.1\,\text{mL}$

R. $$\frac{120\ \text{mg}}{\frac{1}{20}} = 2400\ \text{mg trituration}$$

$2400\ \text{mg} - 120\ \text{mg} = 2280\ \text{mg} = 2.28\ \text{g}$

S. $\dfrac{0.4\text{ mg}}{\text{caps}} \times 40\text{ caps} = 16\text{ mg}$

$$\frac{16\text{ mg}}{\frac{1}{25}} = 400\text{ mg}$$

T. $18\,\text{mg} \times 7 = 126\,\text{mg}$
1 : 7 may be used.

U. $\dfrac{1\text{ mg}}{\text{cap}} \times 12\text{ cap} = 12\text{ mg}$

$12\,\text{mg} \times 10 = 120\,\text{mg}$

Make a 1 : 10 trituration by combining 120 mg of drug with 1080 mg of a diluent such as lactose.

$$\frac{12\text{ mg}}{\frac{1}{10}} = 120\text{ mg}$$

V. $\dfrac{0.75\text{ mL}_{\text{oil}}}{3\text{ mL}_{\text{mix}}} = \dfrac{3\text{ mL}_{\text{oil}}}{j_{\text{mix}}}$

$J = 12\,\text{mL}$ dilution (mix) has to be prepared by mixing 3 mL oil + 9 mL alcohol. An aliquot of 3 mL will contain 0.75 mL oil.

Proof: $0.75\,\text{mL}_{\text{oil}} : 3\,\text{mL}_{\text{mix}} = 1:4 = 0.75$ mL oil in each 3 mL dilution

W. $10.0\text{ g} \times \dfrac{1\text{ mL}}{1.17\text{ g}} = 8.55\text{ mL}$

X. $9\text{ g} \times \dfrac{1\text{ mL}}{0.812\text{ g}} = 11.1\text{ cc}$

Y. $3.22\text{ kg} \times \dfrac{1000\text{ g}}{1\text{ kg}} \times \dfrac{1\text{ mL}}{0.922\text{ g}} = 3490\text{ mL}$

Z. $752\text{ mL} \times \dfrac{0.950\text{ g}}{1\text{ mL}} \times \dfrac{1\text{ kg}}{1000\text{ g}} = 0.714\text{ kg}$

CHAPTER 3

PRESCRIPTION AND MEDICATION ORDERS

LEARNING OBJECTIVES: *After completing this chapter the student should be able to:*

1. Identify the basic components of prescriptions and medication orders.
2. Interpret the information provided in prescriptions and medication orders.
3. Recognize the meaning of common abbreviations used in prescription writing
4. Calculate the quantities needed for prescriptions written in terms of the tot quantity desired or as a single dosage unit.
5. Enlarge and reduce formulas.
6. Rewrite a parts formula in terms of quantities to be measured.

In this chapter, we will explore the form and content of prescriptions and medicati orders, given that many of our calculations will be concerned with prescripti compounding. We will also work with USP formulas and how to reduce and enlar them.

PRESCRIBING AUTHORITY

1. In the USA, several licensed health-care providers have traditionally been given th authority by state statutes to prescribe drug products. This authority is restricted prescribing within the scope of their practices. These include among others, oste pathic doctors, podiatrists, dentists and veterinarians. More recently, several stat have allowed other properly licensed health care practitioners, such as optometrist pharmacists, nurse practitioners and physician assistants, to prescribe medication with restrictions (certification exams, prescribing only under the supervision of licensed physician, or under a established protocol, etc). When writing a prescriptio some of these licensed health practitioners will sign the prescriber's name and degre followed by a slash and their name and degree, using v.o., for voice order, or t.o., f telephone order.

Amoxicillin 250 mg/5 mL susp., 150 ml

t.o. M. Drew, MD / G. Thompson, Pharm.D

Pharmaceutical Calculations, Fourth Edition, By Joel L. Zatz and Maria Glaucia Teixeira
ISBN 0-471-67623-3

TABLE 3.1 Abbreviations for prescribers

Abbreviation	Meaning
ARNP	Advanced Registered Nurse Practitioner
CNM	Certified Nurse Midwife
DDS	Doctor of Dental Surgery
DO	Doctor of Osteopathy
DPM	Doctor of Podiatric Medicine
DVM	Doctor of Veterinary Medicine
MD	Medical Doctor
ND	Doctor of Naturopathy
OD	Doctor of Optometry
PA	Physician's Assistant or "Medex"
Pharm.D	Doctor of Pharmacy
RPh	Registered Pharmacist

A pharmacist received the prescription below.

John M. Cooks, DDS
999 S. 3rd Street
Happyland, YZ
Name: James F. Murray Date: 8/05/04
Address: 1010 Bradley Street, Evanstown, XZ
℞ Naprosyn 250 mg tabs
20
Sig. i tab q 6 h, not to exceed 1000 mg/day
Refills : no t.o. John M. Cooks, DDS/ M.S. Neely, RPh
Substitution: yes DEA No. AB 0494168

(a) What is the profession of the prescriber?

(b) Only 250 mg tablets of Naproxen sodium (generic for Naprosyn®) are available. Would it be correct for the pharmacist to fill the prescription without consulting the prescriber?

(c) How was this prescription ordered (prescription form, telephone order, voice order or computer order)?

Solutions.

(a) A dentist: Doctor of Dental Surgery

(b) Yes, substitution was authorized by the prescriber

(c) The prescription was written by the registered pharmacist of a dentist telephone order.

PRESCRIPTIONS AND MEDICATION ORDERS

Components

2. A prescription is a lawful order of a practitioner for a *drug* or *device* for a specific patient. In a more specific sense, it means a written request for the preparation and administration of any medication for an outpatient. A prescription may order a manufactured drug product or a compounded drug product. It may order a wheelchair or a walker. A medication order, drug order or physician's order is a request for a drug product for inpatients in institutional settings.

Preprinted forms containing information about the prescriber are generally used by practitioners to write prescriptions. Additionally, pharmacists frequently receive prescription orders over the phone, through voice message, through the internet, or by direct communication. Typical medication orders (for inpatient) use different forms, which may be handwritten in ink by the physician, typed or just sent through the institution's database system.

Writing a Prescription

3. Prevention of medication errors is of utmost importance, and it must start at the time a prescription is written and transcribed/read. Some general recommendations by the USP and NABP to prevent medication errors:

- Written prescriptions/medication orders must be legible.
- Prescribers should avoid the use of certain abbreviations (e.g., abbreviations of drug names or combinations of drugs; OD for once daily since OD also stands for right eye; U for units, which may be mistaken for a zero when poorly handwritten).
- Prescriptions/medication orders should always be written using the metric system.
- Prescriptions/medication orders should include the drug name, metric weight or concentration, and dosage form. In addition, when transcribing verbal orders or typing medication names keep *one space apart* each the drug, the units, and the weights.
- A leading zero should always precede a decimal point in quantities less than one (e.g. 0.5 mg, not .5 mg) and a trailing zero should never be used after a decimal point (1 mg, not 1.0 mg).
- Numbers above 999 should have properly placed commas (e.g. 500,000 and not 500000).
- Prescribers should provide the age and, when appropriate, weight of the patient.
- Prescriptions/medication orders should include, when possible, a notation of purpose of the medication.
- Prescribers should not use imprecise instructions such as "Take as directed" or "Take as needed". Orally transmitted directions may be forgotten or misinterpreted. There is also the legal aspect for controlled substances (overdose, mishandling).
- Phoned prescriptions (and voice orders) should be reduced to writing and/or computer data entry immediately. In addition, the dose and drug strength should be checked by individual numbers as they may sound alike (e.g. "one-five" for fifteen and "five-zero" for fifty).

4. A handwritten medication order received by the hospital pharmacist contained the following:

(1) Penicillin G sodium 100000 U IM q 8 h

(2) Clonazepam .75mg bid

Point out what is wrong in each order.

Solutions.

(1) 100,000 (with comma) and *units* should be spelled.

(2) The dosage of clonazepam should have a leading zero (0.75). If the decimal point was not seen, a 100 fold overdose would be used. One *space* is needed between the number and its unit (0.75 mg).

5. A prescription order for a compounded drug product is:

℞	Indomethacin	0.025 g
	Fattibase	q.s.
	M. & Ft. supposit.	#6
	Sig. Take as directed.	

What is incorrectly written or imprecise in this prescription?

Solution. Mistakes can be made when reading numbers less than 1 with decimals: should be written 25 mg.

Abbreviation for suppository is not standard (correct is supp.)

Imprecise instruction in the signa.

Examples of Prescriptions and Medication Orders

6. The illustrations that follow will show some examples of prescriptions and medication orders with their basic components. According to the National Association of Boards of Pharmacy, the *minimal recommended legal requirements for outpatient prescriptions* are:

1. Prescriber information: name, degree, address, phone number, signature, DEA* registration number.

*DEA = Drug Enforcement Administration, a branch of the Department of Justice.

2. Patient information: name, address, other.
3. Date: date of prescription or date of issue.
4. Superscription: ℞ symbol = "take thou", "you take", "recipe".
5. Inscription: the drug product prescribed = name, strength, dosage form of drug product prescribed (manufactured product), or name and quantity of each ingredient (compounded product).
6. Subscription: dispensing instructions to the pharmacist.
7. Signa (Sig.): directions for use by the patient.
8. Special instructions: refills, generic product substitution. For refills, the common interpretation for absence of information is that zero refills are authorized.

Prescription order for a *manufactured* drug product (outpatient)

JAMES D. BLACK, M.D. **(1)**
100 N. Main Street, Suite 120
Laramie, WY 82072
Tel. (307) 745-5000

Name: *Mary J. Smith*(2) Date: *10/01/04* **(3)**
Address: 123 Grand Ave. Happytown, CE

(4) ℞
(5) *Bactrim DS tablets*
(6) *# 20 tabs.*
(7) *Sig. Take i tab bid for 10 days*

(8) Refills 1 2 3 4 5 6 *James Black* M.D. **(1)**
Substitution ✓ DEA No. ________

Prescription order for a *compounded* drug product (outpatient)

JAMES D. BLACK, M.D. **(1)**
100 N. Main Street, Suite 120
Laramie, WY 82072
Tel. (307) 745-5000

Name: *Mary J. Smith* **(2)** Date: *10/01/03* **(3)**
Address: 123 Grand Ave. Happytown, CE

(4) ℞
(5) Extract Belladonna 0.15 g
Magnesium oxide 6.5 g
(6) M. Div. caps. No. 10
(7) Sig. Cap i bid pc
(8) Refills 2 *James Black* M.D. **(1)**
Substitution ( ) yes (X) no DEA No. ____________

The components of medication orders are different from outpatient prescriptions and may vary in different institutions. Some items commonly included on completed *medication orders (inpatient)* are:

1. Patient identification: name, history number, age, weight, height
2. Room number
3. Date and time of order
4. Name, quantity, and dosage form of medications ordered
5. Route of administration
6. Dosage schedule
7. Physician's signature
8. Name or initials of person(s) who transcribed the order: nurse or pharmacist
9. Date of transcription

Typical inpatient medication order

St. Joseph Hospital
123 Hospital Ln.
90000 Citytown, US

[1]Patient	David R. Johnson	
[1]Age/Ht/Wt	63/ 5′4″/ 155 lb	
[1]History number	123456	
[2]Room number	1340	
Attending Physician	Kim J. Twain, MD	
Date	Time	Orders
[3]01/01/03	[3]1000	[4]5,000 units Heparin Sodium in 100 mL NS
		Infuse [5]IV over 4 hr[6] pre-op
	[3]1600	[4]Cefazolin [5]IM Inj. Give 100 mg stat[6]
		Then 50 mg q 6 hr[6]
		[7]K. J. Twain, MD
		[8]D. Green, RN 01/01/03[9]
[3]01/02/03	[3]1400	[4]Metoclopramide 10 mg [5]IVB over [6]2 min. and Morphine sulfate 0.5 mg/mL in 100 mL D5W. IV[5] infusion through PCA. [6]Stat post-op.
		[7]J. D. Peterson (resident MD)
		[8]J Osborn, RPh 01/02/03[9]

7. In a typical prescription, the Inscription defines ____________ ____________ the Subscription shows ____________ ____________ and the Signa displays ____________ ____________.

Solutions. Inscription defines the drug product prescribed.

Subscription shows the dispensing instructions to the pharmacist.

Signa displays the directions for use by the patient.

8. In the following prescriptions for outpatients, identify the basic parts of minimal recommended legal requirements that are missing or are incomplete.

(a)

Dr. Ron Barrow Tel: (123) 456-7890
123 Grand Ave.
Thomasville, KM 30000

Margaret Murdoch
4560 N. 15th Street
Thomasville, KM 30000
Inderal 40
Sig. i tid

Substitution: yes
Ron Barrow, MD

(b)

Dr. Ron Barrow, MD

Larry Wright

℞ Augmentin 125 mg
Disp.120 mL

Refills: 2

Ron Barrow, MD
DEA No. AC4728169

Solutions.

(a) In superscription: ℞ symbol is missing; no prescription date; in inscription: metric units for drug strength and type of dosage form are missing; subscription is missing; signa: no instructions on how to handle refills; prescriber DEA number is missing.

(b) Prescriber complete address; patient complete address; no prescription date; in inscription: complete drug strength per unit of volume and type of dosage form; signa is missing; no instructions on how to handle substitutions.

ABBREVIATIONS

9. As you have seen in the examples provided above, certain abbreviations are routinely used by the prescriber in writing prescriptions. Despite the fact that abbreviations are a convenience, a time saver, a space saver and a way to avoid misspelled words, one should remember that occasionally a price has to be paid for their use. Sometimes abbreviations are misunderstood, misread, or misinterpreted, leading to a delay i

> patient care or even harm. Professionals in the health fields should *never* "create" abbreviations. There are many lists of abbreviations in common use by healthcare workers. Frequently, each health care establishment has its own standardized list, which is used by all personnel. As a rule, where uncertainty exists, the one who wrote the abbreviation must be contacted for clarification.

Abbreviations currently used in writing prescriptions may be related to the type of preparation to be compounded or dispensed (the dosage form), to the directions to the pharmacist, or in the prescription formula. Table 3.2 lists some currently used abbreviations, some of which are derived from the Latin, and their meaning. For a more comprehensive table of abbreviations currently used in the health field, check Appendix 3. Abbreviations may be used sometimes followed by periods or not, and may be written either in upper case or lower case letters.

The abbreviation "$\overline{\text{aa}}$," is used when two or more ingredients are present in the same amount. They are listed sequentially with the symbol placed next to the last item of the group to which it refers.

"$\overline{\text{aa}}$ qs ad" tells you to add more than one substance to achieve a specified total weight or volume. It is assumed that these substances will contribute equally. In other words, the missing weight or volume is divided equally between the ingredients identified.

TABLE 3.2 Some abbreviations currently used by health professionals

Abbreviation or term	Meaning
	Related to dosage forms
cap or caps	capsule
chart	divided powder; powder in a paper
elix	elixir
inj	injection
MDI	metered dose inhaler
pulv, pulvis	powder, bulk powder
sol or soln	solution
supp or PR	suppository
susp	suspension
syr	syrup
tab or tabs	tablet
ung or unguentum	ointment
	Used in directions to the pharmacist
disp	dispense
DTD, d.t.d.	give of such doses
f, ft	make
M	mix
No., #	number of units to be prepared or dispensed
S.A., secundum artem	according to art ("use your skill and judgment")
	Related to the prescription formula
$\overline{\text{aa}}$ or aa	of each
aq., aqua	water
aq. dest.	distilled water
aq. pur.	purified water
q.s.	a sufficient quantity
q.s. ad	a sufficient quantity to make
$\overline{\text{aa}}$ qs ad	a sufficient quantity of each to make

Now, try to identify the information that belongs to the superscription, inscription, su scription, signa and special instructions of the following prescription and interpret abbreviations.

Benjamin J. Anderson, MD
1678 Hospital Lane, Suite 106, tel. (123) 456-7890
Smithtown, AB

℞ Name: *Madeleine J. Stunt* Date: *20/03/(*

Address: 123 Grand Ave. Smithtown, AB

Lotensin® tabs 20 mg

Disp. 60 tabs

Sig. Tab i a.m. et h.s.

Refills ____02___ *Benjamin Anderson* M.D.

Substitution *yes* DEA No. ________________

Solutions. Superscription: ℞ symbol

Inscription: Lotensin tablets 20 mg

Subscription: Dispense 60 tablets

Signa: Take one tablet in the morning and at bedtime

Special instructions: 2 refills and substitution permitted.

---

10. Rewrite the following prescription omitting all abbreviations, so that it makes sens How much petrolatum should be weighed?

℞	Starch		
	Talc $\overline{aa}$		5.0 g
	Lanolin		10.0 g
	Petrolatum	qs ad	60.0 g

Solution.

℞	Starch	5.0 g
	Talc	5.0 g (5.0 g "of each" of starch and talc)
	Lanolin	10.0 g
	Petrolatum	40.0 g (60.0–20.0)

Make an ointment.

11. Prescriptions written in the metric system specify quantities using decimal notation. If units are omitted, and the prescription is written in decimal notation, it is understood that metric units are to be employed: solids are weighed in terms of grams, and liquids are measured in milliliters.

℞ Zinc oxide
Talc
Starch $\overline{aa}$ 5.0
Lanolin
Petrolatum $\overline{aa}$ qs ad 60.0

How much petrolatum should be used? (All ingredients are solids)

Solution. 22.5 g

Since the materials are all solids and the quantities are written using decimal notation, they represent weight in grams. 5.0 g of each of zinc oxide, talc and starch makes a total of 15.0 g. The expression "$\overline{aa}$ qs ad" directs that a sufficient quantity of each be used, or, in other words, that the ingredients under the jurisdiction of this abbreviation should be given an equal share in arriving at the final weight. The remaining 45.0 g are shared equally by the lanolin and petrolatum, so we need 22.5 g of each of these.

PRESCRIPTION WRITTEN IN TERMS OF TOTAL QUANTITY DESIRED

12. In compounding dosage forms such as capsules and powders in paper, the pharmacist weighs sufficient of each drug to make all of the required doses. The drugs are combined into a homogeneous mix that is then divided equally among the units being prepared.

℞ Calcium carbonate
Sodium bicarbonate $\overline{aa}$ 5.0
Charcoal 0.4
Div in chart No. x

This prescription contains the formula for 10 powders. The pharmacist is directed to weigh the quantities indicated, combine them, and then divide the mass into 10 papers (the abbreviation "div" means divide). The quantity of a drug that will be contained in each powder paper may be calculated by dividing the total weight of the drug by the number of dosage units.

For example, the amount of calcium carbonate in each powder is

$$\frac{5.00\ \text{g}}{10\ \text{units}} = 0.50\ \text{g/unit (or g/powder paper)}$$

Calculate the total weight of material that will be contained in each powder paper.

Solution. 1.04 g

CALCULATIONS

The total weight of powder in the formula is

Calcium carbonate:	5.0 g
Sodium bicarbonate:	5.0 g
Charcoal:	0.4 g
	10.4 g

$$\frac{10.4\ \text{g}}{10\ \text{powder papers}} = 1.04\ \text{g/powder paper}$$

PRESCRIPTION WRITTEN IN TERMS OF A SINGLE DOSAGE UNIT

13. Instead of giving the formula for the desired number of dosage units, the prescription may give the formula for a single unit. To calculate the amount of each drug to be measured, it is necessary to multiply the quantity per unit by the number of units.

℞			
	Aspirin		| 300
	Cocoa butter qs ad	2	|
	Ft suppositories, d.t.d. #6		

This is an old fashioned format, rarely in use these days. The vertical line represents the decimal point. The abbreviation "d.t.d." means "give such doses" and indicates the number of units that are to be compounded or dispensed. Whenever "d.t.d." is used, the formula is written for a single unit and must be enlarged by the pharmacist. Both materials are solids, so that the unit of measurement for both is the gram. Thus the amount of aspirin is 300 mg (0.3 g).

Calculate the quantity of each material that should be used to compound the prescription above.

Solution. Aspirin: 1.80 g

Cocoa butter: 10.2 g

CALCULATIONS

To calculate the total amount of aspirin needed, multiply the content of each unit (suppository) by the number of units:

$300\,mg \times 6 = 1800\,mg = 1.80\,g$

Total wt = 2 g per suppository, of which 300 mg is aspirin. Thus

2.00 g total − 0.30 g aspirin = 1.70 g cocoa butter

$1.70\,g \times 6 = 10.2\,g$

To check:

Sum of weights: 1.80 g + 10.2 g = 12.0 g

Weight of 6 suppositories: $2\,g \times 6 = 12\,g$

---

14. Calculate the quantity of each drug needed for the prescription that follows. All are solids.

℞		
	Phenobarbital	0.03
	Belladonna extract	0.015
	Sodium bicarbonate	0.6
	Ft. caps d.t.d. No. XXXVI	

Solution. Phenobarbital: 1.08 g

Belladonna extract: 0.540 g

Sodium bicarbonate: 21.6 g

CALCULATIONS

Phenobarbital: $\frac{0.03\text{ g}}{\text{caps}} \times 36\text{ caps} = 1.08\text{ g}$

Belladonna extract: $\frac{0.015\text{ g}}{\text{caps}} \times 36\text{ caps} = 0.540\text{ g}$

Sodium bicarbonate: $\frac{0.6\text{ g}}{\text{caps}} \times 36\text{ caps} = 21.6\text{ g}$

To check:

Weight of each capsule:

0.030 g
0.015 g
0.600 g

0.645 g

Weight of 36 caps: $\frac{0.645\text{ g}}{\text{caps}} \times 36\text{ caps} = 23.2\text{ g}$

Sum of weights:

1.08 g
0.54 g
21.6 g

23.2 g

FORMULAS: REDUCING AND ENLARGING

15. A number of formulas appear in the USP and *National Formulary (NF)* to provid standardization; the composition of products listed in these references should b uniform. These formulas are written for a fixed amount, and if larger or smaller quar tities are desired, the formulas have to be scaled accordingly. Depending on how prescription formula is represented, simple calculations allow us to determine th quantity of each component to be included.

Drugs can usually not be taken in pure form. Many of them are so potent that proper meas urement of the small doses required would be extremely difficult. A drug product, ofte called a *drug delivery system* or *dosage form*, is a mini-system that makes drug adminis tration accurate, convenient and practical. You are probably already somewhat familia with many types of drug delivery systems.

An example of a *bulk powder* is baby powder, which is applied to the skin directl from the container. Bulk powders are usually not administered internally because it is ver difficult for the patient to measure the dose accurately. The problem of controlling the dos is solved by putting a powdered drug or drug mixture into a dosage form that allows th patient to take a pre-measured quantity of the product. Capsules, tablets, and divide powders are examples. A *capsule* is a shell, usually made of gelatin, which contains th active ingredients. When a capsule is swallowed, the gelatin dissolves in the acid environment of the stomach, releasing the powdered material inside. Because each capsule contains a definite, accurately measured quantity of drug, proper dosage is determined by administering the correct number of capsules. Capsules can be prepared by the pharmacist for a prescription. Many commercial capsule products are also manufactured by pharmaceutical companies.

Several types of *tablets* can be prepared. The type most frequently encountered is the *compressed tablet*, which can be manufactured only on an industrial scale. Tablets are designed to break up into fragments when they enter the stomach fluid, allowing the drug to come into contact with the liquid environment. Many people use the word "pill" to describe a tablet. A pill is actually different. It is an older type of preparation, not commonly used anymore, with a round shape.

In the case of a *divided powder*, the drug or drug mixture is wrapped in a folded paper. The patient unfolds the paper and transfers the powder onto a tablespoon where it is mixed with a small amount of water. The concoction is then swallowed and followed with water or some other liquid to wash the powder down. Many drugs have bitter or otherwise unpleasant taste, and divided powders are not as popular as tablets or capsules, which are usually easier and more pleasant to take.

A *solution* contains a dissolved drug; solutions taken by mouth are usually administered by teaspoon or tablespoon and solutions for injection are administered through a syringe and needle. A *syrup* is a solution that is sweet and highly viscous. An *elixir* is a pleasant-tasting solution containing water and alcohol. *Spirits* are solutions in alcohol. *Tinctures* are also basically alcoholic solutions, although other solvents may also be present.

Ointments and *pastes* are semisolid preparations intended for application to a body surface like the skin. Pastes contain a high proportion of powdered material. Ointments and pastes are packaged in tubes or jars. *Suppositories* are firm semisolid units that are designed to be inserted into a particular body opening. Rectal and vaginal suppositories are the most common. After insertion, the suppository either melts or dissolves, and the drug comes in contact with the local fluids and membrane surfaces. Cocoa butter is a natural material frequently used as a base for suppositories prepared extemporaneously by the pharmacist.

These descriptions of dosage forms have been quite brief. They are intended to acquaint you, in a general way, with some characteristics of drug products that may appear in some of our examples. As you work through the rest of the book, refer to this text to refresh your memory of these dosage forms, if necessary.

Formulas for pharmaceutical products list their contents in various ways. A formula for aspirin tablets, for example, may contain the ingredients and quantities for a single tablet. If 500,000 tablets are to be manufactured, the quantities must be scaled up or enlarged. On the other hand, the formula may be written in terms of 500,000 tablets if this is the size of the usual batch. A change in batch size may necessitate enlargement or reduction of the formula. This adjustment of formula quantities is made so that each tablet has the desired composition, regardless of the number of tablets produced in that batch. Of course, the same considerations apply to other dosage forms, such as capsules and suppositories. These dosage forms are alike in that a particular number of units (such as 2 tablets, 1 suppository, etc.) are administered in order to supply a dose. They are called *unit dosage forms.*

Formulas for *bulk dosage forms*, such as ointments and solutions, generally list components in quantities that will yield a definite final weight or volume. Those in the USP are for 1000 g or mL. If some other quantity is desired, the formula must be reduced or enlarged. This adjustment is made so that each gram or milliliter of the finished product has the desired composition, regardless of the size of the batch.

A. When the formula for a unit dosage form (such as a tablet) is reduced or enlarge both the original and new formula, must have the same ________________ p ________________.

B. When the formula for a bulk dosage form (such as a solution) is reduced or enlarge both the original and new formula, must have the same ________________ p ________________.

Solutions.

A. composition, unit

B. composition, unit of weight or volume

---

16. A hospital uses the following formula for 2.5 mg prednisolone capsules:

Prednisolone: 2.5 mg

Lactose: 0.9 g

This is the formula for one capsule. We wish to rewrite the formula to yield 1500 capsule

Before going ahead with this problem, stop and look again at the formula. Th amount of lactose is 0.9 g. As written, this quantity has one significant figure. Does th mean that the result of a calculation based on this formula must necessarily have only or significant figure, too? The answer is an emphatic *no*! The rules dealing with significa figures that we covered in previous chapters apply to experimental quantities. In oth words, they apply to measurements that have been completed. Listing a quantity in formula in a particular way does not limit the accuracy of any subsequent calculation any measurement based on the calculation. The standards for accuracy depend on th nature of the product and the manipulative procedure to be used, not on the number figures in the formula.

In essence, we can consider the formula quantities to be exact and carry calculatior to the number of significant figures necessary for the particular application. When pe forming calculations for prescription work, the result should generally have three signi cant figures. Calculations that apply to large-scale manufacturing must be carried to least four significant figures.

A. With prescription calculations, the result should be accurate to ________ significa figures.

B. Results of calculations for large-scale manufacturing should be accurate to ________ significant figures.

Solutions.

A. three

B. four or more

---

17. Returning to our problem:

A hospital uses the following formula for 2.5 mg prednisolone capsules:

Prednisolone: 2.5 mg

Lactose: 0.9 g

This is the formula for one capsule. How many grams of each component should be used to make 1500 capsules?

Formulas are most conveniently reduced or enlarged by multiplying the amount of each ingredient by a suitable factor, f.

$$f = \frac{\text{yield of new formula}}{\text{yield of original formula}}$$

The yields must be given in the same units. The factor is dimensionless. In this problem,

$$f = \frac{1500 \text{ capsules}}{1 \text{ capsule}} = 1500$$

For the prednisolone,

$2.5\,\text{mg} \times 1500 = 3750\,\text{mg} = 3.75\,\text{g}$

---

18. Calculate the amount of lactose. Check your result by adding the amounts of prednisolone and lactose and comparing the sum with that of the total weight on one capsule (from the formula) multiplied by 1500.

Solution. Lactose: $0.9\,\text{g} \times 1500 = 1350\,\text{g}$

To check:

Sum of weights:

3.75	g
1350.	g
1353.75	g

Weight of 1500 capsules:

each capsule = $2.5\,\text{mg} + 900\,\text{mg} = 902.5\,\text{mg}$

$902.5\,\text{mg} \times 1500 = 1{,}353{,}750\,\text{mg} = 1353.75\,\text{g}$

---

19. Here is the formula for ergotamine tartrate and caffeine suppositories:

Ergotamine tartrate: 2 mg

Caffeine: 100 mg

Cocoa butter: 1.898 g

This formula is for one suppository. Your company wishes to manufacture 5000 suppositories. How many grams of each ingredient should be used?

Solution. Ergotamine tartrate: 10.00 g

Caffeine: 500.0 g

Cocoa butter: 9490 g

In essence, this type of calculation leads to a revised formula. By stating that the required amount of ergotamine tartrate is 10.00 g, I am emphasizing that the calculation was accurate to four figures. It would also be correct to write "10 g" for the amount of ergotamine tartrate since it is understood that the accuracy of measurement is not limited by the number of figures used in the revised formula.

CALCULATIONS

$$f = \frac{5000 \text{ suppositories}}{1 \text{ suppository}} = 5000$$

Ergotamine tartrate: $2\,\text{mg} \times 5000 = 10{,}000\,\text{mg} = 10\,\text{g}$

Caffeine: $0.1\,\text{g} \times 5000 = 500\,\text{g}$

Cocoa butter: $1.898\,\text{g} \times 5000 = 9490\,\text{g}$

We can check our calculations by comparing the sum of the weights of the ingredients with the calculated theoretical weight of 5000 suppositories:

Theoretical weight: $2.000\,\text{g} \times 5000 = 10{,}000\,\text{g}$

Sum of ingredients:

10 g
500 g
9,490 g
____
10,000 g

---

20. A pharmaceutical manufacturer utilizes the following formula to make 100 capsules:

Ephedrine sulfate: 30 g

Amaranth: 1 g

Lactose: 250 g

We wish to rewrite the formula (in kilograms) so as to yield 200,000 capsules. Calculate the value of the factor, *f*.

____

Solution.

$$f = \frac{2 \times 10^5 \text{ caps}}{100 \text{ caps}} = 2000$$

21. Now calculate the amounts (in kilograms) in the formula for 200,000 capsules.

Solution. Ephedrine sulfate: 60 kg
Amaranth: 2 kg
Lactose: 500 kg

CALCULATIONS

Ephedrine sulfate: 30 g × 2000 = 60,000 g = 60 kg
Amaranth: 1 g × 2000 = 2000 g = 2 kg
Lactose: 0.25 kg × 2000 = 500 kg

To check:
Sum of weights:
60 kg
2 kg
500 kg

---

562 kg

Weight of 200,000 caps:
total wt of 100 caps = 281 g
0.281 kg × 2000 = 562 kg

---

22. In the formula for cherry syrup presented below, notice that the exact quantity of water cannot be calculated, since we do not know the *volume occupied by the solids in solution*. The best we can do is to say "a sufficient quantity to make 1000 mL." It is generally true that liquid preparations are completed by adding sufficient diluent to reach the desired volume using an appropriate measuring device. On the other hand, the weight of each ingredient is known and should be indicated in formulas written on a weight basis.

The formula for cherry syrup is
Cherry juice: 475 mL
Sucrose: 800 g
Alcohol: 20 mL
Purified water: a sufficient quantity to make 1000 mL

A pharmacist wishes to prepare 120 mL of this syrup for a prescription. What value of f should be used in reducing the formula?

Solution.

$$f = \frac{\text{yield of reduced formula}}{\text{yield of original formula}} = \frac{120\ \text{mL}}{1000\ \text{mL}} = 0.12$$

---

23. What quantity of each ingredient should be used to prepare 120 mL of the syrup?

Solution. Cherry juice: 57 mL

Sucrose: 96 g

Alcohol: 2.4 mL

Purified water: a sufficient quantity to make 120 mL

CALCULATIONS

Cherry juice: 475 mL × 0.12 = 57 mL

Sucrose: 800 g × 0.12 = 96 g

Alcohol: 20 mL × 0.12 = 2.4 mL

---

24. Not all formulas indicate what the final yield will be. For example, the formula for an aromatic concentrate is

Camphor	20 g
Anise oil	5 mL
Alcohol	130 mL

If we desire to make 25.0 mL of this concentrate, we cannot arrive at precise quantities, because the final volume is not given by the formula. It is reasonable in such a case to make a little more than is required and discard the excess. Hopefully, this procedure will not be too expensive.

In this particular case, we can be certain of getting at least 25 mL of concentrate if we use 25 mL of alcohol. What quantities of the other ingredients should be used?

Solution. Camphor: 3.84 g
Anise oil: 0.96 mL

CALCULATIONS

This problem may be solved by proportion or by recognizing that reduction of the formula is accomplished by multiplying all quantities by the same dimensionless factor. Since 25 mL of alcohol will be used while the original formula is for 130 mL of alcohol,

$$f = \frac{25 \text{ mL}}{130 \text{ mL}} = 0.192$$

Camphor: 20 g × 0.192 = 3.84 g

Anise oil: 5 mL × 0.192 = 0.96 mL

---

25. A formula for baclofen injection is

Baclofen: 0.5 g

Sodium chloride: 0.9 g

Water for injection: sufficient to produce 100 mL

Your company wishes to make some of this injection. You have on hand 53 g of Baclofen, 0.75 kg of sodium chloride, and unlimited water for injection. How many liters of Baclofen injection can you manufacture?

Solution. 10.6 L

CALCULATIONS

Whatever quantity is made, each ingredient will be present in the same ratio to that quantity as in the formula. The quantity made is limited by the ingredient in shortest supply. Water for injection cannot be the limiting ingredient.

The amount of injection that can be made from 53 g of Baclofen may be found as follows:

$$\frac{0.5 \text{ g baclofen}}{100 \text{ mL}} = \frac{53 \text{ g baclofen}}{j}$$

$$j = 10{,}600 \text{ mL injection}$$

The amount of injection that can be made from 0.75 kg of sodium chloride is

$$\frac{0.9 \text{ g NaCl}}{100 \text{ mL injection}} = \frac{750 \text{ g NaCl}}{j}$$

$$j = 83{,}300 \text{ mL injection}$$

It is apparent that baclofen is the limiting ingredient. Thus 10,600 mL or 10.6 L of the injection can be manufactured.

26. In this next problem you will be given a formula to make 1000 mL of ferrous sulfa syrup. You will then be asked to calculate the quantities necessary to make 8 fl idounces of the syrup. Before going ahead with the calculation, stop and consider t use to which it will be put. There is no reason whatever to convert all quantities in apothecary units. Such a conversion unnecessarily complicates a simple calculati and introduces more opportunity for arithmetic error. The way to proceed is to use t approximate equivalent for the total volume required; since 1 fluidounce is approx mately equivalent to 30 mL, 8 fluidounces ≈ 240 mL.

It is not necessary to use a more exact conversion procedure (which would result in a val of 237 mL) for two reasons:

(1) Metric graduates have no 237-mL mark.

(2) The more important consideration is that the composition of each milliliter of the syru be the same as that in the original formula. Use of 240 mL for 8 f℥ does not contr dict our requirements for accuracy in prescription compounding, since the ratio of a components in the formula to the finished quantity will not be altered.

From the following formula calculate the quantities required to make 8 f℥ of ferro sulfate syrup:

Ferrous sulfate: 40 g

Citric acid, hydrous: 2.1 g

Peppermint spirit: 2 mL

Sucrose: 825 g

Purified water: a sufficient quantity to make 1000 mL

Solution. Ferrous sulfate: 9.60 g

Citric acid, hydrous: 0.504 g

Peppermint spirit: 0.480 mL

Sucrose: 198 g

Purified water: sufficient to make 240 mL

CALCULATIONS

$$f = \frac{240 \text{ mL}}{1000 \text{ mL}} = 0.24$$

Ferrous sulfate: 40 g × 0.24 = 9.60 g

Citric acid hydrous: 2.1 g × 0.24 = 0.504 g

Peppermint spirit: 2 mL × 0.24 = 0.480 mL

Sucrose: 825 g × 0.24 = 198 g

Purified water: sufficient to make 240 mL

---

27. The formula for iodine tincture USP is

Iodine: 20 g

NaI: 24 g

Alcohol: 500 mL

Purified water: a sufficient quantity to make 1000 mL

What quantity of each ingredient should be used to make one-half fluidounce of iodine tincture?

Solution. Iodine: 0.3 g

NaI: 0.36 g

Alcohol: 7.5 mL

Purified water: a sufficient quantity to make 15 mL

CALCULATIONS

$$f = \frac{15\text{ mL}}{1000\text{ mL}} = 0.015$$

Iodine: 20 g × 0.015 = 0.3 g

NaI: 24 g × 0.015 = 0.36 g

Alcohol: 500 mL × 0.015 = 7.5 mL

---

28. The formula for sodium phosphate solution is

Sodium phosphate: 755 g

Citric acid: 130 g

Glycerin: 150 mL

Purified water: a sufficient quantity to make 1000 mL

How much of each ingredient should be used to make 90.0 mL of sodium phosphate solution for a prescription?

Solution. Sodium phosphate: 68.0 g

Citric acid: 11.7 g

Glycerin: 13.5 mL

Purified water: a sufficient quantity to make 90.0 mL

PARTS FORMULAS

29. Some formulas are given in terms of *parts* rather than specific units of weight or volume. Such formulas indicate the ratio of ingredient quantities to each other. In a formula given in terms of parts by weight, any unit of weight may be used, but it must be applied to all of the components. In a formula given in parts by volume, any unit of volume may be used, provided that all components have the same units.

Here is the formula for zinc oxide paste written by parts:

Zinc oxide: 1 part

Starch: 1 part

White petrolatum: 2 parts

This formula could be written

Zinc oxide: 1 lb

Starch: 1 lb

White petrolatum: 2 lb

or

Zinc oxide: 1 kg

Starch: 1 kg

White petrolatum: 2 kg

The actual units applied are determined by the units describing the desired quantity.

How much of each ingredient is required to make 60 g of zinc oxide paste?

Solution. Zinc oxide: 15 g

Starch: 15 g

White petrolatum: 30 g

CALCULATIONS

To use the formula, we simply replace "parts" by "grams," since our target quantity is 60 g.

Zinc oxide:	1 g
Starch:	1 g
White petrolatum:	2 g
	4 g

$$f = \frac{60\text{ g}}{4\text{ g}} = 15$$

Zinc oxide: 1 g × 15 = 15 g

Starch: 1 g × 15 = 15 g

White petrolatum: 2 g × 15 = 30 g

The sum of all ingredients is 60 g, the desired quantity.

---

30. A foot powder has this formula:

Talc: 15 parts

Benzoic acid: 1 part

Bentonite: 3 parts

Rewrite the formula so that it will yield 25 kg of foot powder. (Your results should contain four significant figures.)

Solution. Talc: 19.74 kg

Benzoic acid: 1.316 kg

Bentonite: 3.948 kg

CALCULATIONS

The parts formula may be rewritten

Talc: 15 kg

Benzoic acid: 1 kg

Bentonite: 3 kg

This makes a total of 19 kg.

$$f = \frac{25}{19} = 1.316$$

Talc: 15 kg × 1.316 =	19.74 kg
Benzoic acid: 1 kg × 1.316 =	1.316 kg
Bentonite: 3 kg × 1.316 =	3.948 kg
	25.00 kg

---

31. The formula for an analgesic powder is:

Aspirin: 6 parts

Phenacetin: 3 parts

Caffeine: 1 part

All quantities are based on weight. How many grams of aspirin should be used to prepa
1.25 kg of the powder?

Solution. 750 g

CALCULATIONS

$$f = \frac{1250\text{ g}}{10\text{ g}} = 125$$

$$6\text{ g} \times 125 = 750\text{ g}$$

---

32. The formula for yellow ointment may be written

Yellow wax: 1 part

Petrolatum: 19 parts

How many kilograms of yellow ointment can be prepared from 660 g of yellow wax?

Solution. 13.2 kg

CALCULATIONS

$$\frac{1}{20} = \frac{0.660\text{ kg}}{j}$$

$$j = 13.2\text{ kg}$$

33. State how much sodium bicarbonate (a solid) and elixir phenobarbital (a liquid) should be used to prepare this solution:

℞			
℞	Sodium bicarbonate		0.4
	Elixir phenobarbital		2.5
	Aromatic elixir	qs ad	5.0
	M Ft sol 180 mL		

Solution. Sodium bicarbonate: 14.4 g
Elixir phenobarbital: 90.0 mL

CALCULATIONS

The formula must be enlarged to yield 180 mL

$$f = \frac{180\text{ mL}}{5\text{ mL}} = 36$$

Sodium bicarbonate: 0.4 g × 36 = 14.4 g
Elixir phenobarbital: 2.5 mL × 36 = 90.0 mL

34. Use these problems involving prescription formulas for further practice.

A. ℞ Sodium phenobarbital 15 mg
Tincture belladonna 0.5 mL
Aromatic elixir qs ad 5.0 mL
d.t.d. No. 48

What quantities should be used to prepare this prescription?

B. ℞ Hydrocortisone 0.3 g
Lactose
Ft caps Div in caps #40

How many micrograms of hydrocortisone are contained in each capsule?

C. ℞ Dilaudid 0.003
Aspirin 300
d.t.d. caps No. LXIV

How many tablets, each containing 4 mg of Dilaudid, should be used to supply the Dilaudid for this prescription?

D. ℞ Cinnamon water 1.5 mL
Elixir phenobarbital 2.5 mL
Aromatic elixir qs ad 5.0 mL
Dispense 120 mL

What quantity of each ingredient should be measured?

E. A formula for 12 bismuth subsalicylate suppositories calls for

Bismuth subsalicylate 1.4 g
Cocoa butter qs ad 24 g

How many kilograms of cocoa butter are needed for 150 suppositories?

Solutions

A. Sodium phenobarbital: 720 mg
Tincture belladonna: 24 mL
Aromatic elixir: qs ad 240 mL

B. 7500 μg

C. 48 tablets

D. Cinnamon water: 36 mL
Elixir phenobarbital: 60 mL
Aromatic elixir: qs ad 120 mL

E. 0.2825 kg

REVIEW PROBLEMS

A. A pharmacy located in a rural area received the prescription below.

℞ Rimadyl

#30

Sig. Use prn. for arthritis

v.o. Matt Stuart, DVM/JMB, Pharm.D

(a) What is the profession of the prescriber?

(b) How the prescription was ordered (prescription form, telephone order, voice order or computer order)?

B. Identify what is poorly written or imprecise in the following prescriptions:

(a) Penicillin G sodium 2000000 U q 6h IV

(b) Amphotericin B .1 mg q day

C. Interpret the following directions written by prescribers in Inscriptions and Signas:

(a) Ampicillin 250 mg cap #40

i cap po ac & hs

(b) HCTZ 25 mg

ii tab po qd

(c) Amoxicillin 125 mg/5 mL 100 mL

2.5 mL po tid × 10 d

(d) A prescription lists two different inhalers with the following instruction: i puff $\overline{aa}$ qid

(e) Vicodin tab
1 tab q6h prn pain

D. In the following prescriptions for outpatients, identify the basic parts of minimal recommended legal requirements that are missing or are incomplete.

(a)

David Neeley Tel: (123) 456-7890
123 Grand Ave.
Thomasville, KM 30000

Hugh West
4560 N. 15th Street
Thomasville, KM 30000
R Avandia 4
Sig. i bid

Substitution: yes
David Neeley

(b)

David Neeley, MD

5/11/04
Chris Welch

R Celebrex 100 mg
#10
Refills: none

David Neeley, MD
DEA No. AC4728169

E. ℞ Coal tar ointment
Disp. 30.0 g

The pharmacist must prepare the coal tar ointment for this prescription. What quantity of each ingredient should be used? The formula for coal tar ointment is

Coal tar: 10 g

Polysorbate 80: 5 g

Zinc oxide paste: a sufficient quantity to make 1000 g

F. The formula for green soap tincture is

Lavender oil: 20 mL

Green soap: 650 g

Alcohol: a sufficient quantity to make 1000 mL

How much lavender oil and green soap should be used to prepare 12.5 L of green soap tincture?

G. Here is the formula for placebo tablets:

Lactose: 100 g

Sucrose: 150 g

Starch, direct compressing formula: 250 g

Magnesium sulfate: 0.5 g

Yield: 1000 tablets

Rewrite the formula so as to yield 50,000 tablets. Express quantities in kilograms.

H. The formula for camphorated parachlorophenol is

Parachlorophenol: 7 parts

Camphor: 13 parts

What quantities should be used to make 150 g of camphorated parachlorophenol?

I. Compound undecylenic acid ointment contains

Undecylenic acid: 50 g

Zinc undecylenate: 200 g

Polyethylene glycol ointment: 750 g

How much compound undecylenic acid ointment could be made from 2.2 kg of undecylenic acid, 8.5 kg of zinc undecylenate, and 30 kg of polyethylene glycol ointment?

J. ℞ Prednisone 2.75 mg
Acetylsalicylic acid 0.60 g
Ft caps d.t.d. XXIX

Although you do not have any pure prednisone available, you do have 5-mg tablet How many tablets would you use in compounding this prescription?

K. ℞ Sodium salicylate 0.6 g
Sodium bicarbonate 0.15 g
Elixir lactated pepsin qs ad 15 mL
M. d.t.d. No. 8

What quantity of each ingredient should be used for this prescription?

L. ℞ Zinc oxide
Starch
Calamine $\overline{aa}$ 12 g
Sodium bicarbonate qs ad 44 g
Ft. pulv

How many grams of sodium bicarbonate should be used to prepare 12 g of thi mixture?

M. ℞ Calcium carbonate 0.5 g
Magnesium trisilicate 0.3 g
Bismuth subcarbonate qs ad 1.2 g
dtd chart #XLII

What quantity of each ingredient should be weighed for this prescription?

N. ℞ Calcium gluconate 48 g
Thiamin 8 g
Div. in caps Ft. No. 60

How many milligrams of thiamin will each capsule contain?

O. The formula for zinc gelatin is:

Zinc oxide: 100 g
Gelatin: 150 g
Glycerin: 400 g
Purified water: 350 g

How much of each ingredient should be used to make 240 g of zinc gelatin?

P. The formula for nitromersol solution is:

Nitromersol: 2 g
Sodium hydroxide: 0.4 g
Monohydrated sodium carbonate: 4.25 g
Purified water: a sufficient quantity to make 1000 mL

What quantities should be used to make 90 mL of nitromersol solution for a prescription?

Q. A formula for a dusting powder contains

Salicylic acid: 1 part
Benzoic acid: 3 parts
Starch: 20 parts

How many grams of each ingredient should be used to make 12 kg of dusting powder?

R. ℞ Dexamethasone 0.045 g
Lactose qs
Ft caps. Div in caps No. 68

How many tablets of dexamethasone, each containing 0.75 mg, should be used provide the drug for this prescription?

S. ℞ Sodium bromide 1.2 g
Syrup tolu 2.0 mL
Syrup wild cherry qs ad 5.0 mL
d.t.d. #24

How many grams of sodium bromide should be used in filling this prescription?

T. The formula for Ringer's solution is:

Sodium chloride: 8.60 g
Potassium chloride: 0.30 g
Calcium chloride: 0.33 g
Water for injection: a sufficient quantity to make 1000 mL

How much of each ingredient is necessary to make 90 mL of Ringer's solution?

U. ℞ Dextroamphetamine sulfate 7.5 mg
Lactose qs ad 300 mg
d.t.d. caps No. 36

How many tablets, each containing 10 mg of dextroamphetamine sulfate, should b used for this prescription?

V. Calculate the amounts necessary to compound the following prescription:

℞	
Sodium bromide	0.6 g
Syrup	2.5 mL
Aq dest qs ad	5.0 mL
Ft sol d.t.d. No. 24	

W. ℞ Humulin R 100 units/mL
#10 mL vial
Sig. 10 units bid sc

How many days will the medication last?

X. How many grams of each ingredient would the pharmacist need to weigh to prepare the following product?

℞ Calcium carbonate
Magnesium oxide 5 : 1 : 4 : 2
Sodium bicarbonate
Bismuth subcarbonate
Disp. 60 g

Sig. i tsp in a glassful of water for indigestion and heartburn.

Y. The USP formula to prepare hydrophilic petrolatum is:

Cholesterol		
Stearyl alcohol	$\overline{aa}$	3 parts
White wax		8 parts
White petrolatum		86 parts

How many grams of each ingredient would be needed to prepare $\frac{1}{2}$ kg of the product?

Z. ℞ Flagyl 250 mg

Disp. _______ tablets

Sig. tab ii stat, then 5 mg/kg bid × 5 d thereafter

How many tablets does the pharmacist need to dispense to a patient weighin
143 lb?

Solutions to the Review Problems

A. **(a)** A veterinarian: Doctor of Veterinary Medicine

(b) The order was a voice order from the veterinarian and was transcribed by th doctor in pharmacy (Pharm.D).

B. **(a)** 2,000,000 *units*

(b) 0.1 mg. Dosage form and route not specified.

C. **(a)** Ampicillin 250 mg capsules

Dispense 40 capsules

One capsule by mouth 4 times a day, before meals and at bed time.

(b) Hydrochlorothiazide 25 mg tablets

Two tablets by mouth every day

(c) Amoxicillin suspension 125 mg/5 mL

Dispense 100 mL

Half teaspoon by mouth, three times a day for 10 days

(d) One puff (inhalation) of each four times a day.

(e) Vicodin tablets

One tablet every 6 hours as needed for pain

D. **(a)** No prescription date; metric unit for drug strength and type of dosage form ar missing; number of doses to dispense is missing; Signa: missing the type o dosage form; No instructions on how to handle refills; Prescriber DEA numbe is missing.

(b) Prescriber complete address; Patient complete address; In inscription: type o dosage form; Signa is missing; No instructions on how to handle substitutions

E. Coal tar: 0.300 g

Polysorbate 80: 0.150 g

Zinc oxide paste: 29.6 g

F. Lavender oil: 250 mL

Green soap: 8125 g

G. Lactose: 5 kg

Sucrose: 7.5 kg

Starch : 12.5 kg

Magnesium sulfate: 0.025 kg

H. Parachlorophenol: 52.5 g
Camphor: 97.5 g

I. 40 kg

J. 16 tablets

K. Sodium salicylate: 4.80 g
Sodium bicarbonate: 1.20 g
Elixir lactated pepsin: sufficient quantity to make 120 mL

L. 2.18 g

M. Calcium carbonate: 21.0 g
Magnesium trisilicate: 12.6 g
Bismuth subcarbonate: 16.8 g

N. 133 mg

O. Zinc oxide: 24 g
Gelatin: 36 g
Glycerin: 96 g
Purified water: 84 g

P. Nitromersol: 0.18 g
Sodium hydroxide: 0.036 g
Monohydrated sodium carbonate: 0.383 g
Purified water: sufficient to make 90 mL

Q. Salicylic acid: 500 g
Benzoic acid: 1,500 g
Starch: 10,000 g

R. 60 tablets

S. 28.8 g

T. Sodium chloride: 0.774 g
Potassium chloride: 0.0270 g
Calcium chloride: 0.0297 g
Water for injection: a sufficient quantity to make 90.0 mL

U. 27 tablets

V. Sodium bromide: 14.4 g
Syrup: 60.0 mL
Distilled water: sufficient to make 120 mL of solution

W. 50 days

X. Calcium carbonate: 25 g
Magnesium oxide: 5 g
Sodium bicarbonate: 20 g
Bismuth subcarbonate: 10 g

Y. 15 g Cholesterol
15 g Stearyl alcohol

40 g White wax

430 g White petrolatum

Z. 15 tablets

Calculations for the Review Problems

A. See answers above.

B. **(a)** 2,000,000 (with commas) and *units* should be written.

(b) The dosage of Amphotericin B should have a leading zero (0.1). If the decimal point was not seen, a 10 fold overdose would be used. One *space* is needed between the number and its unit (0.1 mg). The type of dosage form is not mentioned, nor the route of administration.

C. See answers above.

D. See answers above.

E. $f = \frac{30\ \text{g}}{1000\ \text{g}} = 0.03$

Coal tar: $10\,\text{g} \times 0.03 = 0.300\,\text{g}$

Polysorbate 80: $5\,\text{g} \times 0.03 = 0.150\,\text{g}$

Zinc oxide paste: $985\,\text{g} \times 0.03 = 29.6\,\text{g}$

Total weight $= 30.0\,\text{g}$

F. $f = \frac{12.5\ \text{L}}{1\ \text{L}} = 12.5$

Lavender oil: $20\,\text{mL} \times 12.5 = 250\,\text{mL}$

Green soap: $650\,\text{g} \times 12.5 = 8125\,\text{g}$

G. $f = \frac{50{,}000\ \text{tablets}}{1000\ \text{tablets}} = 50$

Lactose: $0.1\,\text{kg} \times 50 = 5\,\text{kg}$

Sucrose: $0.15\,\text{kg} \times 50 = 7.5\,\text{kg}$

Starch: $0.25\,\text{kg} \times 50 = 12.5\,\text{kg}$

Magnesium sulfate: $0.0005\,\text{kg} \times 50 = 0.025\,\text{kg}$

H. The total number of parts is 20. This may be equated to 20 g:

$f = \frac{150\ \text{g}}{20\ \text{g}} = 7.5$

Parachlorophenol: $7\,\text{g} \times 7.5 = 52.5\,\text{g}$

Camphor: $13\,\text{g} \times 7.5 = 97.5\,\text{g}$

I. The formula makes 1000 g

Undecylenic acid:

$\frac{50\ \text{g}}{1000\ \text{g}} = \frac{2.2\ \text{kg}}{j}$

$j = 44\,\text{kg}$ compound undecylenic acid ointment

Zinc undecylenate:

$\frac{200\ \text{g}}{1000\ \text{g}} = \frac{8.5\ \text{kg}}{j}$

$j = 42.5\,\text{kg}$ compound undecylenic acid ointment

Polyethylene glycol ointment:

$$\frac{750\text{ g}}{1000\text{ g}} = \frac{30\text{ kg}}{j}$$

$j = 40\,\text{kg}$ compound undecylenic acid ointment

Only 40 kg of compound undecylenic acid ointment can be made.

J. $\frac{2.75\text{ mg}}{\text{caps}} \times 29\text{ caps} \times \frac{1\text{ tablet}}{5\text{ mg}} = 16\text{ tablets}$

K. Sodium salicylate: $0.6\,\text{g} \times 8 = 4.80\,\text{g}$

Sodium bicarbonate: $0.15\,\text{g} \times 8 = 1.20\,\text{g}$

Total volume: $15.0\,\text{mL} \times 8 = 120\,\text{mL}$

L. Since 12 g of each of the other components are used to make 44 g of the mixture, the amount of sodium bicarbonate is

$$44\text{ g} - (3 \times 12\text{ g}) = 8\text{ g}$$

$$\frac{8\text{ g}}{44\text{ g}} = \frac{j}{12\text{ g}}$$

$j = 2.18\text{ g}$

M. Calcium carbonate: $0.5\,\text{g} \times 42 = 21.0\,\text{g}$

Magnesium trisilicate: $0.3\,\text{g} \times 42 = 12.6\,\text{g}$

Bismuth subcarbonate: $0.4\,\text{g} \times 42 = 16.8\,\text{g}$

N. $\frac{8000\text{ mg}}{60\text{ caps}} = \frac{133\text{ mg}}{\text{caps}}$

O. $f = \frac{240\text{ g}}{1000\text{ g}} = 0.24$

Zinc oxide: $100\,\text{g} \times 0.24 = 24\,\text{g}$

Gelatin: $150\,\text{g} \times 0.24 = 36\,\text{g}$

Glycerin: $400\,\text{g} \times 0.24 = 96\,\text{g}$

Purified water: $350\,\text{g} \times 0.24 = 84\,\text{g}$

P. $f = \frac{90\text{ g}}{1000\text{ g}} = 0.09$

Nitromersol: $2\,\text{g} \times 0.09 = 0.18\,\text{g}$

Sodium hydroxide: $0.4\,\text{g} \times 0.09 = 0.036\,\text{g}$

Monohydrated sodium carbonate: $4.25 \times 0.09 = 0.383\,\text{g}$

Q. $f = \frac{12{,}000\text{ g}}{24\text{ g}} = 500$

Salicylic acid: $1\,\text{g} \times 500 = 500\,\text{g}$

Benzoic acid: $3\,\text{g} \times 500 = 1500\,\text{g}$

Starch: $20\,\text{g} \times 500 = 10{,}000\,\text{g}$

R. $45\text{ mg} \times \frac{1\text{ tablet}}{0.75\text{ mg}} = 60\text{ tablets}$

S. $1.2\,\text{g} \times 24 = 28.8\,\text{g}$

T. $f = \frac{90.0\text{ mL}}{1000\text{ mL}} = 0.09$

Sodium chloride: 8.60 g × 0.09 = 0.774 g

Potassium chloride: 0.30 g × 0.09 = 0.0270 g

Calcium chloride: 0.33 g × 0.09 = 0.0297 g

Water for injection: a sufficient quantity to make 90.0 mL

U. $\frac{7.5\text{ mg}}{\text{tab}} \times 36\text{ tab} = 270\text{ mg needed}$

1 tab/10 mg × 270 mg = 27 tablets

V. Sodium bromide: 0.6 g × 24 = 14.4 g

Syrup: 2.5 mL × 24 = 60.0 mL

Total volume: 5.0 mL × 24 = 120 mL

W. $\frac{1\text{ day}}{2\text{ doses}} \times \frac{\text{dose}}{10\text{ units}} \times \frac{10\text{ units}}{\text{mL}} \times 10\text{ mL} = 50\text{ days}$

X. 5 : 1 : 4 : 2 = 12 parts total

$$\frac{60\text{ g}}{12\text{ parts}} = \frac{5\text{ g}}{\text{part}}$$

$$5\text{ part} \times \frac{5\text{ g}}{\text{p}} = 25\text{ g}$$

$$1\text{ part} \times \frac{5\text{ g}}{\text{p}} = 5\text{ g}$$

$$4\text{ part} \times \frac{5\text{ g}}{\text{p}} = 20\text{ g}$$

$$2\text{ parts} \times \frac{5\text{ g}}{\text{p}} = 10\text{ g}$$

Y. 3 + 3 + 8 + 86 = 100 parts total

$$\frac{500\text{ g}}{100\text{ parts}} = \frac{5\text{ g}}{\text{part}}$$

3 × 5 = 15 g cholesterol and stearyl alcohol

8 × 5 = 40 g white wax

86 × 5 = 430 g white petrolatum

Proof: 15 + 15 + 40 + 430 = 500 g product

Z. 2 tablets immediately

$$143\text{ lb} \times \frac{1\text{ kg}}{2.2\text{ lb}} \times \frac{5\text{ mg}}{\text{kg}} \times \frac{2\text{ tab}}{\text{day}} \times 5\text{ days} \times \frac{1\text{ tab}}{250\text{ mg}} = 13\text{ tablets}$$

Dispense 2 + 13 = 15 tablets.

CHAPTER 4

DOSAGE

***Learning Objectives:** After completing this chapter the student should be able to:*

1. Use dosing directions written in shorthand with usual abbreviations when performing dosage calculations.
2. Determine the dose size to be administered from a prescription formula.
3. Calculate the number of dosage units or amount of liquid to dispense based on the dose and dosing frequency.
4. Calculate the quantity of a specific ingredient present in the dose based on the total prescription.
5. Calculate the number of drops needed to supply a predetermined liquid dose from the number of drops in a fixed volume.
6. Consider special dosing for pediatric and geriatric patients.
7. Calculate doses using body weight or surface area.

In this chapter we will continue our study of the prescription, concentrating on calculations that relate to dosage. You will learn some more abbreviations used to indicate directions for use by the patient. The capacities of household measuring devices for taking medication will be reviewed. You will also calculate dosages based on body weight and surface area, based on a nomogram and will be introduced to the basic principles of pediatric and geriatric dosing. In general, dosage calculations can be carried to two significant figures and approximate equivalents are used when converting from one measurement system to another.

DOSING DIRECTIONS AND ABBREVIATIONS

1. It is the pharmacist's obligation to check the dosage of all drugs dispensed on prescription in order to prevent accidental overdose. The pharmacist must also make certain that the patient understands how the medication is to be taken. The directions to the patient that appear on the prescription label must be written in a way that is clear and unambiguous.

Included in the directions to the patient are, in general:

(1) The number of dosage units in each dose (e.g., 1 capsule, 2 teaspoonfuls, 15 drops);

(2) The frequency with which the medication is to be taken (e.g., three times a day, every four hours);

(3) Additional, clarifying instructions (e.g., after meals, with water).

Pharmaceutical Calculations, Fourth Edition, By Joel L. Zatz and Maria Glaucia Teixeira
ISBN 0-471-67623-3

The directions for use by the patient include:

A. ______________________

B. ______________________

C. ______________________

Solutions.

A. Number of dosage units in each dose

B. Frequency of administration

C. Clarifying instructions

---

2. The number of dosage units in each dose is usually specified in Arabic numerals pr ceded by a # sign but sometimes roman numerals are used. They may be written i either capital or small letters and precede or follow the name of the dosage form. F example,

5 tab = 5 tablets

caps ii = 2 capsules

i gtt = 1 drop

250 mg #40 tabs = 40 tablets of 250 mg

Latin abbreviations are often used to describe the frequency of administration. Some the commonly used abbreviations are listed in Table 4.1. Sometimes periods are omitte from the abbreviated terms.

TABLE 4.1 Some Latin abbreviations related to dosage frequency

Latin abbreviation	Meaning
ATC	around the clock
d. or d	day
h. or hr	hour(s)
q.	every
q. 4 h.	every 4 hours
q.d	every day
b.i.d.	twice a day
t.i.d.	three times a day
q.i.d.	four times a day (here, "q" does not stand for "every")
h.s.	at bedtime
stat	immediately; at once

As an example, let us consider these directions to the patient:

Sig: 1 tab q 3 h [Take one (1) tablet every three (3) hours]

Sig: Chart II h.s. [Take two (2) powders at bedtime]. Notice that in this example the number of doses is given in roman numerals and the number follows the designation for the type of product.

If you need to brush up on Roman numerals go to the next frame. Otherwise, skip to frame 4.

---

3. The number of dosage units in each dose is usually specified in Arabic numerals but sometimes Roman numerals are used; they may be written in either capital or small letters and precede or follow the name of the dosage form. The final "i" in a Roman numeral is sometimes replaced by a "j".

Some values for the Roman symbols of greatest interest to us:

Roman numeral	Arabic numeral
I or i	1
V or v	5
X or x	10
L	50
C	100
M	1000

Some rules that apply to Roman numerals:

(i) Repetition (2 or 3 times) of a numeral, duplicates or triplicates its value (ii = 2; XXX = 30).

(ii) Largest value numeral should always be used (XX not XVV = 20)

(iii) Roman numerals of lower value placed before a higher value numeral will be subtracted from the higher value (IV = 4).

(iv) Roman numeral of higher value followed by a lower value numeral have additive property (VI = 6, xvii = 17).

(v) For larger roman numerals, subtraction rule prevails over addition rule (XXIV = 10 + 10 + (5 − 1) = 24; MCMXCIX = 1999).

Examples:

i tab (1 tablet)

caps ii (2 capsules)

VIII gtt (8 drops)

What is meant by "chart III"?

Solution. Three powders in paper.

---

4. Translate each of the following signas into clear English:

A. Apply ung. b.i.d.

B. Tab II stat, tab I q 4 h

C. Soak feet qid

D. i tab qd hs

E. 1–2 caps q. 8 h.

F. Chart I t.i.d.

Solutions.

A. Apply ointment twice a day.
B. Take two (2) tablets at once; then take one (1) tablet every four (4) hours.
C. Soak feet four (4) times a day.
D. Take one (1) tablet every day at bedtime.
E. Take one (1) or two (2) capsules every eight (8) hours.
F. Take one (1) powder three (3) times a day.

---

5. Latin abbreviations often appear in clarifying phrases in the signa. These may indicate whether the medication is to be taken before or after meals, by mouth or some other route of administration, with or without liquids, and so on. Commonly used phrases are given in Table 4.2. Sometimes, periods are omitted from the abbreviated terms.

TABLE 4.2 Selected Latin abbreviations used in the signa

Latin abbreviation or term	Meaning
c.	food; meal(s)
a.c.	before meals
p.c.	after meals
c̄, cum	with
s̄, sine	without
rep.	repeat
p.o., per os	by mouth
e.m.p.	in the manner prescribed
p.r.n.	as needed, when needed

An example of a signa is:

Sig: i tab pc c̄ aq [Take one (1) tablet after meals with water.]

Translate each of the following signas into clear English.

A. 1 caps t.i.d. a.c. ut. dict.

B. 1 tab c̄ 2 aspirins q 4 h

C. One teaspoonful s̄ aqua; rep. q. 2 h.

D. Rub in prn pain

E. Caps II p.o. p.c. e.m.p.

F. Tab I q.i.d. p.c. & h.s.

Solutions.

A. Take one (1) capsule three (3) times a day before meals, as directed.

B. Take one (1) tablet with two (2) aspirin tablets, every four (4) hours.

C. Take one (1) teaspoonful without water. Repeat every two (2) hours.

D. Rub in as needed, to relieve pain.

E. Take two (2) capsules by mouth after meals, in the manner prescribed.

F. Take one (1) tablet four (4) times a day, after meals and at bedtime.

(If these Latin phrases are still Greek to you, review them, try the examples again or try some review problems.)

CALCULATIONS INVOLVING DOSE SIZE, NUMBER OF DOSES, AMOUNT DISPENSED, AND QUANTITY OF AN INGREDIENT

6. The size of the individual dose may be indicated by the prescriber in terms of the weight or volume of drug that the patient should obtain. However, directions put on the label for the patient must be written in such a way that they can be understood by the patient. Consider the prescription that follows:

℞ Tetracycline caps 250 mg
Disp. caps. No. XII
Sig: 250 mg q.i.d.

This prescription calls for 12 capsules, each containing 250 mg of tetracycline. If the directions for use were translated literally as "250 mg four (4) times a day," the patient would not know how many capsules to take. The label should read: "Take one (1) capsule four (4) times a day."

℞ Penicillin V caps. 250 mg

24

Sig: 250 mg q4h, 1h a.c.

How would you write the directions to the patient?

Solution. Take one (1) capsule every four (4) hours, one (1) hour before meals.

---

7. ℞ Sulfisoxazole tabs 0.5 g

Dispense #XLIV

Sig: 2.0 g stat, 1.0 g t.i.d.

A. How many tablets should be dispensed?

B. How should the directions to the patient be written?

Solutions.

A. 44

B. Take four (4) tablets immediately, then two (2) tablets three (3) times a day.

---

8. A prescription calls for penicillin G tablets, each containing 200,000 units. The signa reads "400,000 units q.i.d. for 10 days."

A. What directions should appear on the label?

B. How many tablets should be dispensed?

Solutions.

A. Take two (2) tablets four (4) times a day for ten (10) days.

B. 80 tablets (the patient will use 8 a day for 10 days.)

9. The prescription below is for Mr. Jones. Determine how many days the tablets will last him.

℞ Sulfadiazine tablets 0.5
d.t.d. No. L
Sig: Tabs II p.c. and h.s.

Solution. About 6 days

CALCULATION

Assuming that the patient eats 3 meals a day, he will take 8 tablets each day. Thus

$$\frac{50\ \text{tab}}{8\ \text{tab/day}} \approx 6\ \text{days}$$

---

10. To check the safety of a medication, the pharmacist must remember to calculate the dose of each drug being prescribed to the patient.

℞		
	Aspirin	300 mg
	Phenylpropanolamine hydrochloride	20 mg
	Lactose qs ad	600 mg
	d.t.d. caps #XXIV	
	Sig: II q 4 h	

This prescription is for an adult.

The usual dose of phenylpropanolamine hydrochloride is listed as "25 to 50 mg every 3 or 4 hours."

Calculate the dose of phenylpropanolamine hydrochloride in the prescription and compare it with the usual dose.

Solution. Each capsule contains 20 mg of phenylpropanolamine hydrochloride. The patient is therefore taking 40 mg of this drug every 4 hours, which appears to be a satisfactory dose.

11. Consider the following prescription.

℞	Ephedrine sulfate	0.48
	Aspirin	7.2
	Caffeine	0.24
	Div. in caps #XXIV	
	Sig: Caps II q. 6h. p.r.n.	

How much ephedrine sulfate will the patient be given in each dose, which consists of two capsules?

Solution. 40 mg

CALCULATION

The quantity of ephedrine sulfate in each capsule may be found by dividing the total quantity of this material by the number of capsules:

$$\frac{480\text{ mg}}{24\text{ caps}} = 20\text{ mg}$$

Thus two capsules will contain 40 mg.

12. Given the following, how many milligrams of hydrocortisone will the patient take each day?

℞	Hydrocortisone	0.6 g
	Sodium salicylate	30.0 g
	Div. in caps #LX	
	Sig: caps I q 8 h.	

Solution. 30 mg per day. (Each capsule contains 10 mg; 3 capsules are taken each day.)

13. Tablets, capsules, and suppositories are examples of dosage forms that comprise discrete units. When dealing with these unit dosage forms, indication of the size of a dose is a simple matter. The number of units is specified. However, such dosage forms as ointments and solutions are generally dispensed in bulk, and it becomes more difficult to measure dosage accurately.

The dosage of *bulk liquids* is usually given in terms of household measuring devices such as the *teaspoon or tablespoon*. There are no uniform, official standards to which house-

TABLE 4.3 Household devices used to measure liquid medication

Household measure	Nominal capacity (ml.)
1 teaspoonful (tsp)	5
1 tablespoonful (1/2 f℥) (tbsp)	15
1 glassful (8 f℥)	240

hold teaspoons are made, and a teaspoon found at home may be daintier or heftier than average. According to the official compendia, the average teaspoon holds about 5 mL, but teaspoons in use may contain 4.93 ± 0.24 mL.

In all of our calculations, we will assume that the capacities listed in Table 4.3 are correct. However, to ensure that the patient is actually receiving the intended volume, the pharmacist should give the patient an accurately calibrated measuring device each time a liquid medication is dispensed.

If a 10-mL dose is required, the patient should be directed to take 2 teaspoonfuls. Note that a fluidounce contains 2 tablespoonfuls or 6 teaspoonfuls.

Learn the table. Then fill in the blanks:

A. 2 f℥ = ________ teaspoonfuls.

B. 2 f℥ = ________ tablespoonfuls

C. 90 mL = __________ teaspoonfuls.

D. 120 mL = __________ tablespoonfuls.

E. 1 tablespoonful = _________ teaspoonfuls.

Solutions.

A. 12 teaspoonfuls

B. 4 tablespoonfuls

C. 18 teaspoonfuls

D. 8 tablespoonfuls

E. 3 teaspoonfuls

---

14. Translate each of the following signas into clear English:

A. I tsp c̄ aqua p.c.

B. 15 mL t.i.d. a.c.

C. 10 mL stat, 5 mL q 6 h

Solutions.

A. Take one (1) teaspoonful, with water, after meals.

B. Take one (1) tablespoonful three (3) times a day, before meals.

C. Take two (2) teaspoonfuls at once, then one (1) teaspoonful every six (6) hours.

---

15. How many milliliters of solution should be dispensed for a 30-day supply of the following prescription?
Tsp ii stat
Tsp i b.i.d. d 2 et 3
Tsp ii p.m. d 4 and thereafter.

Solution. 300 mL

CALCULATIONS

$$2\text{ tsp} \times \frac{5\text{ mL}}{\text{tsp}} = 10\text{ mL}$$

$$\frac{1\text{ tsp}}{\text{dose}} \times \frac{5\text{ mL}}{\text{tsp}} \times \frac{2\text{ doses}}{\text{day}} \times 2\text{ days} = 20\text{ mL}$$

$$\frac{1\text{ tsp}}{\text{dose}} \times \frac{5\text{ mL}}{\text{tsp}} \times \frac{2\text{ doses}}{\text{day}} \times 27\text{ days} = 270\text{ mL}$$

Total to be dispensed = 10 + 20 + 270 = 300 mL

---

16. Given the prescription,

℞ Tetracycline oral syrup
Sig: tsp $\overline{\text{ss}}$ q.i.d. for 6 days

How many milliliters of tetracycline oral syrup should be dispensed?

Solution. 60 mL

CALCULATIONS

The directions state: "One-half (1/2) teaspoonful four (4) times a day for six (6) days." The patient will therefore take

$$\frac{2\ \text{teaspoonfuls}}{\text{day}} \times 6\ \text{days} = 12\ \text{teaspoonfuls}$$

$$12\ \text{teaspoonfuls} \times \frac{5\ \text{mL}}{\text{teaspoonful}} = 60\ \text{mL}$$

---

17. According to the following prescription, how much potassium bromide will the patient receive each day?

℞		
Potassium bromide		5.4 g
Aqua		30.0 mL
Cherry syrup	qs ad	120.0 mL
Sig: 1 tsp qid		

Solution. 0.9 g

CALCULATIONS

A total of 4 teaspoonfuls, or 20 mL, of the syrup will be taken each day. The amount of potassium bromide in that quantity can be calculated in several ways. One approach is to set up a proportion, recognizing that the relative content of potassium bromide is the same in any volume of the syrup.

$$\frac{5.4\ \text{g}}{120\ \text{mL}} = \frac{j}{20\ \text{mL}}$$

$$j = 0.9\ \text{g}$$

---

18. The signa for a prescription for 90 mL reads
1 tsp tid

How many days will this medication last the patient?

Solutions. 6 days

$$90.0\ \text{mL} \times \frac{1\ \text{teaspoonful}}{5\ \text{mL}} = 18\ \text{teaspoonfuls in the bottle}$$

$$\frac{18\ \text{teaspoonfuls}}{3\ \text{teaspoonfuls/day}} = 6\ \text{days}$$

---

19. ℞ Chlorpheniramine maleate syrup 2 mg/5 mL
Disp 3-day supply
Sig: 4 mg b.i.d.

A. What directions for use should be put on the label?

B. How much chlorpheniramine maleate syrup should be dispensed?

Solutions.

A. Take two (2) teaspoonfuls twice a day.

B. 60 mL should be dispensed.

CALCULATIONS

B. Each 5 mL of syrup = 2 mg of drug. The patient takes 8 mg per day, or 24 mg during the three-day period.

$$24 \text{ mg} \times \frac{5 \text{ mL}}{2 \text{ mg}} = 60 \text{ mL of syrup}$$

---

20. ℞ Milk of bismuth 15.0
dtd #6
Sig: 15 mL ut dict

A. What are the directions to the patient?

B. What quantity of milk of bismuth should be dispensed?

Solutions.

A. Take one (1) tablespoonful as directed.

B. 15 mL × 6 = 90 mL

21. ℞ Acetaminophen drops 1.5 g/15 mL

Disp. 30 mL
Sig. 0.5 mL q.i.d.

A. How many milligrams are there in each prescribed dose?
B. How many milligrams will be taken per day?

Solutions.

A. 50 mg
B. 200 mg

CALCULATIONS

A. $\frac{1500\text{ mg}}{15\text{ mL}} = \frac{j}{0.5\text{ mL}}$

$j = 50\text{ mg}$

B. 50 mg × 4 doses = 200 mg

---

22. How many grams of codeine sulfate are required to make 120 ml solution, such that each teaspoonful will contain 25 mg of drug?

Solution. 0.6 g

CALCULATIONS

$$\frac{5\text{ mL}}{25\text{ mg}} = \frac{120\text{ mL}}{j}$$

$j = 600\text{ mg} = 0.6\text{ g}$

or,

$$\frac{1\text{ g}}{1000\text{ mg}} \times \frac{25\text{ mg}}{5\text{ mL}} \times 120\text{ ml} = 0.6\text{ g}$$

23. A cough syrup contains 4.8 grams of dextromethorphan in 8 fl.oz. How many milligrams of dextromethorphan are there in a 2-teaspoonful dose?

Solution. 200 mg

CALCULATIONS

1 fl.oz. = 30 mL = 6 tsp

8 fl.oz = 240 mL = 48 tsp

$$\frac{48\ \text{tsp}}{2} = 24\ \text{doses}$$

$$\frac{4800\ \text{mg}}{24\ \text{doses}} = 200\ \text{mg}$$

or,

$$\frac{48\ \text{tsp}}{4800\ \text{mg}} = \frac{2\ \text{tsp}}{j}$$

$$j = 200\ \text{mg}$$

or,

$$\frac{4800\ \text{mg}}{240\ \text{mL}} \times 10\ \text{mL} = 200\ \text{mg}$$

24. The dose of an aerosol inhalant is 250 μg three times a day. The commercial inhaler delivers 50 μg per metered dose and contains 250 doses. How many inhalers should be dispensed to a patient for a 30 day supply?

Solutions. 2 inhalers

$$250\ \mu\text{g} \times 3 = \frac{750\ \mu\text{g}}{\text{day}}$$

$$750\ \mu\text{g} \times \frac{1\ \text{dose}}{50\ \mu\text{g}} = 15\ \text{doses}$$

$$15\ \text{doses} \times \frac{1\ \text{inhal}}{250\ \text{doses}} = 0.06\ \text{inhal} \times 30\ \text{days} = 1.8 = 2\ \text{inhalers}$$

25. Here are some extra practice problems dealing with dosage calculations in prescriptions.

A. ℞ Elixir phenobarbital 30 mL
Lactated pepsin elix 30 mL
Syrup tolu qs ad 120 mL
Sig: 1 teaspoonful tid ac

How many doses will the patient get?

B. ℞ Donnagel PG
Sig: I tbsp t.i.d. p.c. c̄ aq.

(1) What directions would you put on the label?

(2) How many milliliters of Donnagel PG should be dispensed so as to last the patient 4 days?

C. ℞ Syrup Phenergan
Sig: tsp I qid, pc & hs

How many milliliters of syrup Phenergan should be dispensed if this prescription is to last the patient six days?

D. ℞ Ampicillin syrup (125 mg/teaspoonful)
Sig: 250 mg qid for 10 days

How many milliliters of this syrup will the patient take altogether?

E. ℞ Ammonium chloride 8 g
Benylin expectorant 40 mL
Water qs ad 120 mL

How many milligrams of ammonium chloride does each teaspoonful of the prescription contain?

F. ℞ Sodium bromide 3 g
Lactose qs ad 5 g
Div in caps #24
Sig: Caps II tid

Calculate the total daily dose of sodium bromide.

G. ℞ Ephedrine sulfate syrup 60.0
Chlortrimeton syrup 60.0
Mix.
Sig: tsp I q.i.d.

How many days will this quantity of medication last the patient?

H. ℞ Benadryl caps 50 mg
No. 30
Sig: Caps I a.c., caps II h.s.

How many days will this prescription last the patient?

I. ℞ Ferrous gluconate 1.5 g
Ascorbic acid 15.0 g
Div. in caps No. 96
Sig: caps I b.i.d.

In how many days will the patient have ingested 280 mg of ferrous gluconate?

J. ℞ PBZ expectorant 90 mL
Ephedrine syrup 90 mL
Sig: 1 tablespoonful tid

How many days will this prescription last the patient?

Solutions.

A. 24
B. **(1)** 1 tablespoonful 3 times a day after meals, with water
(2) 180 mL
C. 120 mL
D. 400 mL
E. 333 mg
F. 0.75 g
G. 6 days
H. 6 days
I. 9 days
J. 4 days

DOSAGE BASED ON DROPS

26. Liquid medication intended for use in the eye, ear, or nose, or for oral use by pediatric patients, is usually dispensed in a *dropper bottle*. The patient is directed to measure the dose by counting drops, based on dropper calibration (go to Appendix 2 and review dropper calibration). The Latin abbreviations listed in Table 4.4 are frequently used.

TABLE 4.4 Abbreviations for signa of liquids administered by drop

Latin abbreviation	Meaning
gtt	drop
o.d.	right eye
o.s.	left eye
o.u.	both eyes

Translate these signas into clear English:

A. 2 gtt o.s. q.d.

B. Gtt. i o.u. b.i.d.

Solutions.

A. Put two (2) drops in the left eye once a day.

B. Put one (1) drop in both eyes twice a day.

---

27. A prescription has 0.4 mL of peppermint oil as one of its ingredients. A dropper wa calibrated and found to deliver 2.0 mL of peppermint oil in 55 drops. How many drop of peppermint oil should be used?

Solution. 11 drops

CALCULATIONS

$$0.4\text{ mL} \times \frac{55\text{ drops}}{2.0\text{ mL}} = 11\text{ drops}$$

---

28. A pharmacist measured 35 drops of a medicated solution while filling a graduat cylinder to 5-mL mark. How many drops per milliliter did the dropper deliver?

Solution. 7 drops

CALCULATIONS

$$\frac{35 \text{ drops}}{5 \text{ mL}} \times 1 \text{ mL} = 7 \text{ drops}$$

---

29. If you need additional practice try these, or, skip to frame 30.

A. A dropper delivers 4.0 mL of caraway oil in 110 drops. How many drops will deliver 1.3 mL?

B. ℞

Opium tincture	0.3 mL
Cascara sagrada fluid extract	2.0 mL
Syrup tolu qs ad	5.0 mL
Ft sol dtd #12	

A dropper, calibrated with opium tincture, delivers 1.0 mL in 33 drops. How many drops of opium tincture should be used to fill this prescription?

C. ℞

Paregoric	1.5 mL
Milk of bismuth qs ad	5.0 mL
d.t.d. No. 6	
Sig: 20 drops a.c.	

If the dropper dispensed with this prescription delivers 2.0 mL in 26 drops, how many milliliters of paregoric are in each dose?

Solutions.

A. 36 drops
B. 119 drops
C. 0.462 mL

DOSAGES BASED ON AGE: PEDIATRIC AND GERIATRIC PATIENTS

30. The *usual adult dose* of a drug is that quantity which is expected to exert the desired effect in most adults. A valuable reference for the pharmacist in checking prescribed doses is the USP *DI volume I*. This publication contains a compilation of useful drug information directed primarily to health care professionals. Individual drug descriptions, called monographs, contain such information as the therapeutic category, pharmacology, precautions, and side effects in addition to dosing information.

Children (pediatric patients)—from birth to adolescence—and the elderly (geriatric patients)—over 65 years old—often require special consideration in the determination of drug dosage because their excretion and metabolism of certain drugs may differ significantly from the way these drugs are handled by the standard adult population. Immature or impaired kidney function and liver enzymes, for example, may reduce drug elimination, cause drug accumulation and prolongation of effects in the body and possible toxic levels and related adverse effects. Also, many adults vary in size from what is considered the norm and the dose may have to be adjusted for them.

Package inserts of many drugs express pediatric doses on the basis of body weight per kilogram. For more guidance with respect to dosing of children, the usual pediatric dose listed in the USP *DI* is a valuable reference. Another useful source of informa-tion is the *Pediatric Dosage Handbook* published by the American Pharmaceutical Association.

A pediatric patient should not be considered as a "miniature" adult and doses calculated as a fraction of the usual adult dose. Some potent drugs (e.g. digoxin) have pediatric doses determined by a combination of the patient's age and weight. However, for over-the-counter (OTC) medications, most labeling provides guidelines for safe and effective dosing, grouping pediatric patients by age (2–6 years old, 6–12 years old, and over 12 years old).

There are a number of ways for making dosage adjustments to suit individual requirements. The most accurate are based on data obtained from testing the patient's drug serum concentration (pharmacokinetic data). This is especially valuable (and cost effective) for highly potent drugs with a narrow margin of safety.

What is incorrect about pediatric patients?

(a) Children from birth to adolescence.

(b) May have immature or impaired kidney function and liver enzymes.

(c) Dosing may be calculated as a fraction of an adult dose.

(d) Sometimes are grouped by age for OTC medications.

Solution. **(c)** is incorrect. A child should never be considered a miniature adult.

DOSAGE BASED ON BODY WEIGHT

31. Sometimes dosage calculations and adjustment are accomplished by expressing the dose as a quantity of drug per unit weight of the patient. The dose, determined in this way, is usually calculated to two significant figures. It is also common to see *dosing tables* in some drug products as a way to facilitate dosing calculations by physicians and pharmacists.

The usual dose of diethylcarbamazine citrate is 2 mg/kg of body weight, three times a day for 1 to 3 weeks.

The dose for a particular person is determined by his or her weight in kilograms. For an 80-kg adult, the dose is

$$\frac{2\ \text{mg}}{\text{kg}} \times 80\ \text{kg} = 160\ \text{mg}$$

The daily dose of a drug is 12 µg per kilogram of body weight. How many milligrams should be administered each day to a woman who weighs 55 kg?

Solution. 0.66 mg

CALCULATIONS

$$\frac{12\ \mu\text{g}}{\text{kg}} \times 55\ \text{kg} = 660\ \mu\text{g} = 0.66\ \text{mg}$$

---

32. If the dose of a drug is 0.25 mg/kg, how many milligrams should be administered to a man weighing 175 lb?

Solution. 20 mg

CALCULATIONS

$$175\ \text{lb} \times \frac{1\ \text{kg}}{2.20\ \text{lb}} \times \frac{0.25\ \text{mg}}{\text{kg}} = 20\ \text{mg}$$

---

33. The usual dose of lucanthone hydrochloride is 5.0 mg per kilogram of body weight three times a day for 1 week. What should the total daily dose, in grams, be for a youth weighing 120 lb?

Solution. 0.81 g

CALCULATIONS

$$120\ \text{lb} \times \frac{1\ \text{kg}}{2.20\ \text{lb}} \times \frac{5\ \text{mg}}{\text{kg}} = 270\ \text{mg}$$

Since 3 doses are administered each day, the total will be 810 mg or 0.81 g.

BODY SURFACE AREA (BSA)

34. For some drugs, dosing is conventionally based on *body surface area* (BSA), considering the patient's height and weight. BSA in medication dosing is particularly used for chemotherapy drugs and occasionally in dosing for children.

Various tables, graphs, and equations that relate body surface area to the patient's weight and height have been published. Whereas a universal height-weight formula validated in infants, children, and adults (Haycock et al, 1978), has been proposed to calculate BSA, the most frequently used method is based on the following equation $S = W^{0.425} \times H^{0.725} \times 0.007184$ (S = body surface in m^2, W = weight in kg, and H = height in cm).*

The most popular methods to determine body surface area (BSA) are:

(1) Use of a general equation (Equation 4.1) that considers height in *centimeters* and weight in *kilograms*

$$BSA\,(m^2) = \frac{(\text{Height}_{(cm)} \times \text{Weight}_{(kg)})^{1/2}}{60} \tag{4.1}$$

(2) Use of a nomogram based on Dubois equation (Figs. 4.1 and 4.2):

Height	Body Surface Area	Weight

cm: 200, 195, 190, 185, 180, 175, 170, 165, 160, 155, 150, 145, 140, 135, 130, 125, 120, 115, 110, 105, 100

inch: 78, 76, 74, 72, 70, 68, 66, 64, 62, 60, 58, 56, 54, 53, 52, 51, 50, 49, 48, 47, 46, 45, 44, 43, 42, 41, 40, 39

m^2: 2.80, 2.70, 2.60, 2.50, 2.40, 2.30, 2.20, 2.10, 2.00, 1.90, 1.80, 1.70, 1.65, 1.60, 1.55, 1.50, 1.45, 1.40, 1.35, 1.30, 1.25, 1.20, 1.15, 1.10, 1.05, 1.00, 0.95, 0.90, 0.86

kg: 150, 140, 130, 125, 120, 115, 110, 105, 100, 95, 90, 85, 80, 75, 70, 65, 60, 55, 50, 45, 40, 35, 30

lb: 320, 300, 280, 260, 250, 240, 230, 220, 210, 200, 190, 180, 170, 160, 150, 140, 130, 120, 110, 105, 100, 95, 90, 85, 80, 75, 70, 66

Figure 4.1 Nomogram for determination of body surface area (BSA) in adults. $S = W^{0.425} \times H^{0.725} \times 0.007184$ (S = body surface in m^2, W = weight in kg, and H = height in cm).

*DuBois, *Arch. Intern. Med.*, **17**, 863–871 (1916).

Height	Body Surface Area	Weight

cm: 120, 115, 110, 105, 100, 95, 90, 85, 80, 75, 70, 65, 60, 55, 50, 45, 40, 35, 30, 25
inch: 46, 44, 42, 40, 38, 36, 34, 32, 30, 28, 26, 25, 24, 23, 22, 21, 20, 19, 18, 17, 16, 15, 14, 13, 12, 11, 10
m^2: 1.10, 1.00, 0.90, 0.80, 0.70, 0.60, 0.55, 0.50, 0.45, 0.40, 0.35, 0.30, 0.25, 0.20, 0.18, 0.16, 0.15, 0.14, 0.13, 0.12, 0.11, 0.10, 0.09, 0.08, 0.074
kg: 40., 30., 20., 18., 16., 14., 12., 10., 9.0, 8.0, 7.0, 6.0, 5.0, 4.5, 4.0, 3.5, 3.0, 2.5, 2.0, 1.5, 1.0
lb: 90., 80., 70., 60., 50., 45., 40., 35., 30., 25., 20., 15., 10., 9.0, 8.0, 7.0, 6.0, 5.0, 4.0, 3.0, 2.2

Figure 4.2 Nomogram for determination of body surface area (BSA) in children. $S = W^{0.425} \times H^{0.725} \times 0.007184$ (S = body surface in m^2, W = weight in kg, and H = height in cm).

There are different nomograms for determination of BSA of adult and children. In a nomogram, the BSA is calculated by joining the body weight and height of the patient through a straight line and reading the body surface area column where it is intersected.

Both methods are universally used by doctors, nurses, and pharmacists in clinical practice and provide similar values for BSA. However, as we proceed with examples, you will notice that child BSA values calculated by the equation and the nomogram are often approximately 10% different.

Some important considerations when calculating BSA:

(a) Intersystem conversions, such as 1 kg = 2.2 lb and 1 inch = 2.54 cm.

(b) BSA values used for adult dosages are frequently rounded to the nearest tenth of a square meter (1.4 m^2, not 1.42 m^2).

(c) BSA values for children are rounded to the nearest one-hundredth of a square meter (0.45 m^2, not 0.4 m^2).

(d) BSA for adults usually ranges from 1.4 m^2 to 2.4 m^2.

(e) Child BSAs (3 month old to 12 year old) are usually in the range 0.27–1.35 m^2, while newborns are $\leq$0.22 m^2.

Now we need to practice the calculation of BSA before using it to determine dosages. With the help of a ruler, go to the nomograms provided in Figs. 4.1 and 4.2 and join the body weight and height of the patient through a straight line and read the body surface area column where it is intersected. Consider the following examples:

1. An adult patient weighing 142 lb and measuring 62 in. has a BSA of 1.65 m^2
2. The BSA for a 10-kg child, 79 cm tall is 0.45 m^2.

Now, try these on your own.

Using the nomograms in Fig. 4.1 and 4.2, determine the BSA of the following patient

A. Adult weighing 65 kg and measuring 170 cm

B. Adult weighing 180 lb and measuring 68 in.

C. 40-lb child, 32 in. tall

Solutions.

A. 1.74 m^2

B. 1.95 m^2

C. 0.59 m^2

---

35. Based on previous problem, calculate the BSA of all three patients using equation 4.1 (1 kg = 2.2 lb; 1 in. = 2.54 cm)

Solutions.

A. 1.75 m^2

B. 1.98 m^2

C. 0.64 m^2

CALCULATIONS

A. $$\text{BSA}\,(\text{m}^2) = \frac{(\text{height}_{(\text{cm})} \times \text{weight}_{(\text{kg})})^{1/2}}{60}$$

$$\text{BSA} = \frac{(170 \times 65)^{1/2}}{60} = 1.75\ \text{m}^2$$

B. $$180\ \text{lb} \times \frac{1\ \text{kg}}{2.2\ \text{lb}} = 81.8\ \text{kg}$$

$$68\ \text{in.} \times \frac{2.54\ \text{cm}}{\text{in.}} = 172.7\ \text{cm}$$

$$\text{BSA} = \frac{(172.7 \times 81.8)^{1/2}}{60} = 1.98\ \text{m}^2$$

C. $\frac{40\ \text{lb}}{2.2\ \text{kg}} = 18.2\ \text{kg}$

$32\ \text{in.} \times 2.54 = 81.3\ \text{cm}$

$$\text{BSA} = \frac{(81.3 \times 18.2)^{1/2}}{60} = 0.64\ \text{m}^2$$

If we analyze all three BSAs calculated above, we may conclude that while the calculation of BSA for adults is quite similar using either the equation or the nomogram, for children the equation frequently generates a slightly larger BSA (within 10%). This difference is considered irrelevant in clinical practice.

If you feel you can handle BSA determinations from now on, you may skip to frame 37, otherwise, go to the next frame.

---

36. Using the nomogram and the equation provided, calculate the BSA of:

A. A child of 10 kg weight and 24 inches height

B. A child of 66 lb weight and 32 inches height

C. An adult weighing 140 lb and measuring 62 inches

D. An adult patient weighing 205 lb, 6′2″ height

Solutions.

A. $0.38\ \text{m}^2$ (nomogram) and $0.41\ \text{m}^2$ (equation)
B. $0.73\ \text{m}^2$ (nomogram) and $0.82\ \text{m}^2$ (equation)
C. $1.64\ \text{m}^2$ (nomogram) and $1.7\ \text{m}^2$ (equation)
D. $2.2\ \text{m}^2$ (nomogram) and $2.2\ \text{m}^2$ (equation)

CALCULATIONS

A. $$\text{BSA}(\text{m}^2) = \frac{(\text{height}_{(\text{cm})} \times \text{weight}_{(\text{kg})})^{1/2}}{60}$$

$$24 \text{ in.} \times \frac{2.54 \text{ cm}}{\text{in.}} = 60.96 \text{ cm}$$

$$\text{BSA} = \frac{(60.69 \times 10)^{1/2}}{60} = 0.41 \text{ m}^2$$

B. $$66 \text{ lb} \times \frac{1 \text{ kg}}{2.2 \text{ lb}} = 30 \text{ kg}$$

$$32 \text{ in.} \times \frac{2.54 \text{ cm}}{\text{in.}} = 81.28 \text{ cm}$$

$$\text{BSA} = \frac{(81.28 \times 30)^{1/2}}{60} = 0.82 \text{ m}^2$$

C. $$140 \text{ lb} \times \frac{1 \text{ kg}}{2.2 \text{ lb}} = 63.6 \text{ kg}$$

$$62 \text{ in.} \times \frac{2.54 \text{ cm}}{\text{in.}} = 157.5 \text{ cm}$$

$$\text{BSA} = \frac{(63.6 \times 157.5)^{1/2}}{60} = 1.67 = 1.7 \text{ m}^2$$

D. $$205 \text{ lb} \times \frac{1 \text{ kg}}{2.2 \text{ lb}} = 93.2 \text{ kg}$$

$$6'2'' = (6 \times 12) + 2 = 74 \text{ in.} \times \frac{2.54 \text{ cm}}{\text{in.}} = 188 \text{ cm}$$

$$\text{BSA} = \frac{(93.2 \times 188)^{1/2}}{60} = 2.2 \text{ m}^2$$

CALCULATION OF DOSAGE BASED ON BSA

37. When the BSA and the usual dose are known, dosage adjustments for both adults an children are occasionally performed using the equations below.

$$\text{Adult Dose} = \frac{\text{BSA of adult (m}^2)}{1.73 \text{ m}^2} \times \text{Usual adult dose}$$

$$\text{Approx. Child Dose} = \frac{\text{BSA of child (m}^2)}{1.73 \text{ m}^2} \times \text{Usual adult dose}$$

For children, this equation provides an *approximate dose* based on the adult dose.

What would be the dose of a drug for a child weighing 30 lb and measuring 24 inche in height, considering that the usual adult dose is 100 mg?

From the nomogram in Fig. 4.2, the child's BSA is 0.42 m^2.

$$\frac{0.42\ \text{m}^2}{1.73\ \text{m}^2} \times 100\ \text{mg} = 24.3\ \text{mg is the approximate dose for the child}$$

Using the equations above and the nomograms provided for adult and children, calculate the doses,

A. For an adult weighing 142 lb and measuring 62 inches, if the usual adult dose of a drug is 50 mg.

B. For a child weighing 20 kg and measuring 78 cm, if the usual adult dose of a drug is 75 mg.

Solutions.

A. 66.5 mg

B. 26 mg

CALCULATIONS

A. BSA = 2.3 m^2 (from nomogram)

$$\frac{2.3\ \text{m}^2}{1.73\ \text{m}^2} \times 50\ \text{mg} = 66.5\ \text{mg}$$

B. BSA = 0.60 m^2 (from nomogram)

$$\frac{0.60\ \text{m}^2}{1.73\ \text{m}^2} \times 75\ \text{mg} = 26\ \text{mg}$$

---

38. Sometimes the usual *adult or pediatric dose per square meter* is known; then the individual dose can be calculated through the patient's BSA.

The pediatric dose of a drug is 200 mg/m^2. What is the dose for a child whose body surface area is 0.8 m^2?

Solution. 160 mg

CALCULATIONS

$$0.8\ \text{m}^2 \times \frac{200\ \text{mg}}{\text{m}^2} = 160\ \text{mg}$$

---

39. The dose of a drug is $15\,mg/m^2$, twice a day, for 1 week. How many milligrams of the drug should the patient take altogether in the course of treatment? Assume that the patient is an average adult ($1.73\,m^2$).

Solution. 364 mg

CALCULATIONS

Each dose should be:

$$\frac{15.0\ mg}{m^2} \times 1.73\ m^2 = 26\ mg$$

Since the drug is given b.i.d. for 1 week, a total of 14 doses will be administered:

$26\ mg \times 14 = 364\ mg$

---

40. Now, for the sake of comparison, we will calculate the dose of a drug for a child weighing 50 lb and measuring 80 cm, with usual pediatric dose of $100\,mg/m^2$ and compare the equation and the nomogram methods to calculate the BSA.

$$BSA = \frac{(height_{(cm)} \times weight_{(kg)})^{1/2}}{60}$$

$50\ lb = 50/2.2 = 22.7\ kg$

$$BSA = \frac{(80 \times 22.7)^{1/2}}{60} = 0.71\ m^2$$

$$Dose = 0.71\ m^2 \times \frac{100\ mg}{m^2} = 71\ mg$$

BSA (from the nomogram) $= 0.65\ m^2$

Dose $= 0.65\ m^2 \times 100\ mg/m^2 = 65\ mg$

Here again you may notice the 10% difference in BSA calculated through the nomogram and the equation.

REVIEW PROBLEMS

A. Compose the following label instructions from the signas so that a patient would understand them:

(1) Tab I q 4 h prn

(2) 5.0 mL c̄ aq. p.c. h.s.

(3) Gtt. II o.s. b.i.d.

(4) Apply ung qd ut dict

(5) 1 tbsp s̄ aq. t.i.d.

(6) Gtt. X p.o. a.c.

(7) I cap q 6h ATC

B. Interpret the directions to the pharmacist and the signa of the following prescription of a compounded drug product:

℞		
	Acetaminophen	500 mg
	Pseudoephedrine hydrochloride	30 mg
	Chlorpheniramine maleate	2 mg
	Dextromethorphan hydrochloride	20 mg

M. ft. caps DTD No. 40
Sig. Caps ii q 6h x 5d

C. ℞ Amphetamine sulfate 90 mg
Lactose qs
Div in caps No XX

How many milligrams of amphetamine sulfate will each capsule contain?

D. ℞ Ephedrine sulfate syrup 60 mL
Dimetane syrup 120 mL
M.
Sig: 2 tsp t.i.d.

How many days will this prescription last the patient?

E. The preparation of 125 capsules of a drug used 25 mg of the drug. How many micrograms are represented in each capsule?

F. If a syrup contains 0.5 g of hydrocodone in 100 mL, how many milligrams are contained in each teaspoonful dose?

G. ℞ Cinnamon water 1.5 mL
Sodium phenobarbital 0.015 g
Aromatic elixir qs ad 5.0 mL
d.t.d. No. 24
Sig: 1 tsp q.i.d. a.c.

How much sodium phenobarbital should be used for this prescription? How many milligrams of sodium phenobarbital will the patient receive each day?

H. ℞ Penicillin V suspension 240 mL
(each teaspoonful contains 200,000 units)
Sig: 400,000 units t.i.d.

How many days will this prescription last the patient?

I. ℞ Hydrocortisone 0.45 g
Sodium salicylate 30 g
Div. in caps. No. LX
Sig: caps II q. 12 h. p.c.

How many milligrams of hydrocortisone will the patient take each day?

J. ℞ Papaverine hydrochloride 1.0 g
Aqua 30.0 mL
Syrup tolu q.s. ad 90.0 mL
Sig: i tsp t.i.d.

How many milligrams of papaverine will the patient receive each day?

K. ℞ Sodium pentobarbital 0.090 g
Cherry syrup qs ad 30 mL
Sig: 10 mg of sod. pentobarb. h.s.

If the dropper dispensed with the product delivers 2.0 mL in 48 drops, how many drops would contain 10 mg of sodium pentobarbital?

L. The dose of a drug is 250 μg/kg of body weight. How many milligrams should be administered to a 175-lb man?

M. The dose of a drug is 0.6 mg/m^2 of body surface area. How many micrograms should be administered to a child whose body surface area is 0.75 m^2?

N. The usual pediatric dose of a drug is 12 μg/m^2. Calculate the dose for a child weighing 35 pounds whose height is 32 inches.

O. Calculate the dose of a drug for a child weighing 40 lb and measuring 32 inches, with usual pediatric dose of 150 mg/m^2. Use both methods discussed in this text to calculate the BSA. (1 kg = 2.2 lb; 1 in. = 2.54 cm)

P. A child weighing 30 lb needs an anticonvulsant in a dose of 5 mg/kg daily. How many milliliters of a pediatric suspension of the drug, containing 20 mg/mL should the child receive?

Q. The pediatric dosage of digoxin is based on age and weight. Recommended doses are: full term newborn 20–30 μg/kg, 1–24 months 30–50 μg/kg, 2–5 years 25–35 μg/kg. Calculate the dosage range for:

(a) A 20-month-old baby weighing 16 lb.

(b) A 3-year-old child weighing 55 lb.

R. The usual pediatric dose of a drug is 50 mg/kg of body weight. What would be the dose for a child weighing 44 lb?

S. How many milliliters of acetaminophen elixir would you recommend for a child weighing 22 lb, if the usual dose is 10 mg/kg and the label reads 100 mg/half tsp?

T. ℞

Ammonium bromide	10 g
Syrup	20 mL
Peppermint water qs ad	120 mL

Sig: tsp i qid

How many milligrams of ammonium bromide does the patient take each day?

U. The average intravenous dose of doxorubicin for children is 30 mg/m^2. What would be the dose for a child weighing 48 lb and 115 cm in height?

V. A physician orders 5 mg of ampicillin sodium to be added to each milliliter of a 100 mL bag of D5W for intravenous injection.

(a) How many milligrams of ampicillin sodium should be added?

(b) If the dose of ampicillin sodium is 20 mg/kg and the 100 mL bag is the daily dose, how many pounds does the patient weigh?

W. A 13-year-old, 110-lb child is prescribed minocycline 4 mg/kg/day for 1 month for acne treatment. The daily dose is to be divided into two equal doses. The pharmacist carries 50- and 100-mg capsules. What strength, quantity and directions should be selected?

X. A pharmacist received a prescription that states:

Drug A:	30 days supply
Directions:	20 mg po bid day 1 and 2
	10 mg po bid day 3 and 4
	5 mg po bid day 5
	5 mg po qd thereafter

There are 5-mg, 10-mg, and 20-mg tablets of Drug A available in the pharmacy. All tablets are scored. Which strength, how many tablets, and which directions should be given to the patient?

Y. Calculate the following BSAs using a nomogram and using the equation.

(a) Adult patient weighing 180 lb, 5′7″ height

(b) Adult patient weighing 205 lb, 6′2″ height

(c) Pediatric patient weighing 66 lb, 3 ft height

Z. A 116-lb, 5′1″ patient is to receive capecitabine 1250 mg/m^2 PO twice daily for 14 days, repeated every 3 weeks. Using a nomogram to calculate the BSA,

(a) Calculate the daily dose for this patient.

(b) How many grams of capecitabine would the patient have received after each cycle of treatment (14 days)?

Solutions to Review Problems

A. **(1)** Take one (1) tablet every four (4) hours as needed.

(2) Take one (1) teaspoonful with water after meals and at bedtime.

(3) Put two (2) drops in the left eye twice a day.

(4) Apply this ointment every day as directed.

(5) Take one (1) tablespoonful without water three (3) times a day.

(6) Take ten (10) drops by mouth before meals.

(7) Take one capsule every 6 hours around the clock.

B. The prescription is written for one capsule and asks the pharmacist to prepare 40 doses by mixing all ingredients (needs previous calculation of amounts of each ingredient required for 40 capsules). The signa states that the patient will take two capsules every six hours for five days.

C. 4.5 mg

D. 6 days

E. 200 µg/cap

F. 25 mg

G. 360 mg should be used; 60 mg per day

H. 8 days

I. 30 mg/day

J. 167 mg

K. 80 drops

L. 20 mg

M. 450 µg

N. 7.2 µg

O. 96 mg; 90 mg

P. 3.4 mL

Q. (a) 218–364 µg

(b) 625–875 µg

R. 1000 mg

S. 0.5 tsp

T. 1667 or 1670 mg

U. 24.9 or 25 mg

V. **(a)** 500 mg

(b) 55 lb

W. Dispense 100 mg capsules # 60. Take one capsule two times a day.

X. 51 tablets of Drug A, 5-mg strength.

Take 4 tablets two times a day for two days, then take 2 tablets two times a day fc two days, then take 1 tablet two times a day for one day, then take 1 tablet every da thereafter.

Y. **(a)** 1.93 m^2 and 1.97 m^2

(b) 2.2 m^2 and 2.2 m^2

(c) 0.80 m^2 and 0.87 m^2

Z. **(a)** 3750 mg/day

(b) 52.5 g

Calculations for the Review Problems

A. See solutions above.

B. See solutions above.

C. $\frac{90 \text{ mg}}{20 \text{ caps}} = \frac{4.5 \text{ mg}}{\text{caps}}$

D. The total volume dispensed is 180 mL. Each day, the patient takes 6 teaspoonfuls, or 30 mL.

$$\frac{180 \text{ mL}}{30 \text{ mL/day}} = 6 \text{ days}$$

E. 25 mg = 25,000 μg

$$\frac{25{,}000 \ \mu\text{g}}{125 \text{ cap}} = \frac{j}{1 \text{ cap}}$$

$$j = 200 \ \mu\text{g/cap}$$

F. 0.5 g = 500 mg

$$5 \text{ ml} \times \frac{500 \text{ mg}}{100 \text{ ml}} = 25 \text{ mg}$$

G. sodium phenobarbital: $\frac{15 \text{ mg}}{\text{dose}} \times 24 \text{ doses} = 360 \text{ mg}$

daily dose: $\frac{15 \text{ mg}}{\text{tsp}} \times \frac{4 \text{ tsp}}{\text{day}} = 60 \text{ mg/day}$

H. 400,000 units = 2 teaspoonfuls

The patient takes 2 teaspoonfuls three times a day, for a total of 6 teaspoonfuls or 30 mL.

$$\frac{240 \text{ mL}}{30 \text{ mL/day}} = 8 \text{ days}$$

I. Four capsules are taken each day.

$$\frac{450 \text{ mg}}{60 \text{ caps}} \times \frac{4 \text{ caps}}{\text{day}} = 30 \text{ mg/day}$$

J. The total daily amount taken is 15 mL

$$\frac{1000 \text{ mg}}{90 \text{ mL}} = \frac{j}{15 \text{ mL}}$$

$$j = 167 \text{ mg}$$

K. $\dfrac{90\text{ mg}}{30\text{ mL}} = \dfrac{10\text{ mg}}{j}$

$j = 3.33\text{ mL}$

$3.33\text{ mL} \times \dfrac{48\text{ drops}}{2\text{ mL}} = 80\text{ drops}$

L. $175\text{ lb} \times \dfrac{1\text{ kg}}{2.20\text{ lb}} \times \dfrac{0.25\text{ mg}}{\text{kg}} = 20\text{ mg}$

M. $0.75\text{ m}^2 \times 600\,\mu\text{g/m}^2 = 450\,\mu\text{g}$

N. $\text{BSA} = \sqrt{\dfrac{\text{I} \times \text{P}}{3131}} = \sqrt{\dfrac{32 \times 35}{3131}} = 0.60\text{ m}^2$

Dose $= 0.6\text{ m}^2 \times 12\,\mu\text{g/m}^2 = 7.2\,\mu\text{g}$

O. $\text{BSA} = \dfrac{(\text{height}_{(\text{cm})} \times \text{weight}_{(\text{kg})})^{1/2}}{60}$

$40\text{ lb} \times \dfrac{1\text{ kg}}{2.2\text{ lb}} = 18.2\text{ kg}$

$32\text{ in.} \times 2.54\text{ cm/in.} = 81.3\text{ cm}$

$\text{BSA} = \dfrac{(81.3 \times 18.2)^{1/2}}{60} = 0.64\text{ m}^2$

Dose $= 0.64\text{ m}^2 \times 150\text{ mg/m}^2 = 96\text{ mg}$

BSA (nomogram) $= 0.60\text{ m}^2$

Dose $= 0.60\text{ m}^2 \times 150\text{ mg/m}^2 = 90\text{ mg}$

P. $30\text{ lb} \times \dfrac{1\text{ kg}}{2.2\text{ lb}} \times 5\text{ mg/kg} \times \text{mL}/20\text{ mg} = 3.4\text{ mL}$

Q. **(a)** $\dfrac{30\,\mu\text{g}}{1\text{ kg}} \times \dfrac{1\text{ kg}}{2.2\text{ lb}} \times 16\text{ lb} = 218.2\,\mu\text{g}$

$\dfrac{50\,\mu\text{g}}{1\text{ kg}} \times \dfrac{1\text{ kg}}{2.2\text{ lb}} \times 16\text{ lb} = 363.6\,\mu\text{g}$

(b) $\dfrac{25\,\mu\text{g}}{1\text{ kg}} \times \dfrac{1\text{ kg}}{2.2\text{ lb}} \times 55\text{ lb} = 625\,\mu\text{g}$

$\dfrac{35\,\mu\text{g}}{1\text{ kg}} \times \dfrac{1\text{ kg}}{2.2\text{ lb}} \times 55\text{ lb} = 875\,\mu\text{g}$

R. $\dfrac{50\text{ mg}}{1\text{ kg}} \times \dfrac{1\text{ kg}}{2.2\text{ lb}} \times 44\text{ lb} = 1000\text{ mg}$

S. $\dfrac{10\text{ mg}}{1\text{ kg}} \times \dfrac{1\text{ kg}}{2.2\text{ lb}} \times 22\text{ lb} = 100\text{ mg} = 0.5\text{ tsp}$

T. $\dfrac{10\text{ g}}{120\text{ mL}} \times \dfrac{1000\text{ mg}}{1\text{ g}} \times \dfrac{4\text{ tsp}}{1\text{ day}} \times \dfrac{5\text{ mL}}{1\text{ tsp}} = 1666.6\text{ mg/day}$

U. BSA $= 0.83\text{ m}^2$ (nomogram)

$30\text{ mg/m}^2 \times 0.83\text{ m}^2 = 24.9\text{ mg}$

V. **(a)** $5\text{ mg/mL} \times 100\text{ mL} = 500\text{ mg}$

(b) $\dfrac{2.2\text{ lb}}{1\text{ kg}} \times \dfrac{1\text{ kg}}{20\text{ mg}} \times 500\text{ mg} = 55\text{ lb}$

W. $\frac{4\text{ mg}}{\text{kg/day}} \times \frac{1\text{ day}}{2\text{ doses}} \times 110\text{ lb} \times \frac{1\text{ kg}}{2.2\text{ lb}} = 100\text{ mg bid}$

Dispense 100 mg capsules # 60

Take one capsule two times a day.

X. Tablets can be cut but it is a pain for the patient. The solution for this would be to us full dosage forms, or 5 mg tablets.

4 tab × 2/day × 2 days = 16 tabs

2 tab × 2/day × 2 days = 8 tablets

1 tab × 2/day × 1 day = 2 tablets

1 tab × 1/day × 25 days = 25 tablets

Fill prescription with 51 tablets of Drug A, 5-mg strength.

Y. **(a)** BSA (nomogram) = 1.93 m^2

$$\text{BSA}(\text{m}^2) = \frac{(\text{height}_{(\text{cm})} \times \text{weight}_{(\text{kg})})^{1/2}}{60}$$

$$67\text{ in.} \times \frac{2.54\text{ cm}}{1\text{ in.}} = 170.2\text{ cm}$$

$$180\text{ lb} \times \frac{1\text{ kg}}{2.2\text{ lb}} = 81.8\text{ kg}$$

$$\text{BSA} = \frac{(170.2 \times 81.8)^{1/2}}{60} = 1.97\text{ m}^2$$

(b) BSA (nomogram) = 2.2 m^2

$$6'2' = 74\text{ in.} \times \frac{2.54\text{ cm}}{1\text{ in.}} = 188\text{ cm}$$

$$205\text{ lb} \times \frac{1\text{ kg}}{2.2\text{ lb}} = 93.2\text{ kg}$$

$$\text{BSA} = \frac{(188 \times 93.2)^{1/2}}{60} = 2.2\text{ m}^2$$

(c) BSA (nomogram) = 0.80 m^2

$$3\text{ ft.} = 36\text{ in.} \times \frac{2.54\text{ cm}}{1\text{ in.}} = 91.4\text{ cm}$$

$$66\text{ lb} \times \frac{1\text{ kg}}{2.2\text{ lb}} = 30\text{ kg}$$

$$\text{BSA} = \frac{(91.4 \times 30)^{1/2}}{60} = 0.87\text{ m}^2$$

Z. **(a)** BSA = 1.5 m^2

$$1.5\text{ m}^2 \times \frac{1250\text{ mg}}{\text{m}^2} \times \frac{2\text{ doses}}{\text{day}} = 3750\text{ mg/day}$$

(b) $\frac{3750\text{ mg}}{\text{day}} \times 14\text{ days} = 52500\text{ mg} = 52.5\text{ g}$

PERCENTAGE STRENGTH

***Learning Objectives:** After completing this chapter the student should be able to:*

1. State the meaning of weight in weight (w/w), volume in volume (v/v) and weight in volume (w/v) systems.
2. Apply the *default rules* that decide which system type is indicated in the absence of a specific designation.
3. Calculate quantities needed to weigh or measure ingredients needed in prescriptions or other formulas written on the basis of percent.
4. Determine concentration from the quantities used in a preparation.
5. Determine the amount of solvent needed for a liquid preparation from weight designations or liquid volume and density.
6. Convert laboratory test results based on mg % or mg/dL to percentage strength.
7. Use drug product concentrations expressed in mg/mL and perform conversions from percentage strength and ratio strength.
8. Perform calculations in which concentration is specified as parts per million or parts per billion.

Up to this point we have dealt with prescriptions and formulas in which quantities of the ingredients have been expressed as weights or volumes. Some formulas specify the concentrations of the ingredients used, as a percent or ratio strength. Other expressions of concentration, such as parts per million and parts per billion are useful in describing very dilute solutions.

PERCENTAGE V/V, W/W, AND W/V

1. Expressions of concentration describe the amount of solute that will be contained in some definite quantity of the total preparation (a solution, in general). If sodium chloride is dissolved in water, the concentration of the solution is characterized by the weight of sodium chloride dissolved and the volume of solution (not solvent). We may define concentration thus:

$$\text{Concentration} = \frac{\text{quantity of solute}}{\text{quantity of preparation}}$$

Pharmaceutical Calculations, Fourth Edition, By Joel L. Zatz and Maria Glaucia Teixeira
ISBN 0-471-67623-3

"Preparation" is frequently a *solution* but may be an ointment, a suppository, a powder mixture, and so on.

---

2. If the solute and the preparation are expressed in the same units, the concentration is dimensionless. It indicates the portion of the preparation represented by solute, and is written as a decimal or fraction. For example, if 10.0 mL of alcohol were dissolved in a sufficient quantity of water to make 40.0 mL of solution, the concentration of alcohol would be

$$\frac{10.0 \text{ mL}}{40.0 \text{ mL}} = 0.250$$

Percent concentration is defined as the number of parts of solute in 100 units of solution. Concentration expressed as a decimal may be converted to percent by multiplying by 100:

$$0.250 \times 100 = 25.0\%$$

This concentration may be written 25.0% by volume or 25.0% volume in volume or 25.0% v/v to indicate that both the solute and solution were measured by volume.

If 12.0 mL of peppermint oil were dissolved in sufficient alcohol to make 80.0 mL of solution, what would be the concentration of peppermint oil in this solution expressed as

A. a decimal?

B. a percent?

Solutions.

A. 0.150 v/v

B. 15.0% v/v

CALCULATIONS

A. $\text{concentration} = \frac{12.0 \text{ mL}}{80.0 \text{ mL}} = 0.150 \text{ v/v}$

B. 0.150 v/v × 100 = 15.0% v/v

---

3. The formula for 1 liter of syrup contains 0.5 mL of orange oil. What is the percentage strength (v/v) of the oil in the syrup?

Solution. 0.05% v/v

CALCULATION

$$\frac{0.5\text{ mL}}{1000\text{ mL}} = 0.0005 \times 100\% = 0.05\%\,\text{v/v}$$

---

4. If the quantity of solute and of the preparation are expressed in the same units of weight, the concentration is dimensionless. If 10.0 g of charcoal are mixed with 65.0 g of another powder to make a total of 75.0 g, the charcoal concentration is

$$\frac{10.0\text{ g}}{75.0\text{ g}} = 0.133 \text{ by weight (or } 13.3\%\,\text{w/w)}$$

Note: In a preparation in which all concentrations are w/w, the sum of all contributions must be 100%.

If 12.0 g of lanolin are combined with 2.0 g of white wax and 16.0 g of petrolatum to make an ointment, what is the percentage concentration of lanolin in the ointment?

Solution. 40.0% w/w

CALCULATIONS

$$\frac{12.0\text{ g lanolin}}{30.0\text{ g ointment}} = 0.400 = 40.0\%\,\text{w/w}$$

---

5. If 250 g of dextrose are dissolved in 300 mL of water, what is the percentage strength of dextrose in the solution in a w/w basis?

Solution. 45% w/w

CALCULATIONS

300 mL water = 300 g

$$\frac{250\text{ g dextrose}}{550\text{ g solution}} = 0.45 = 45\%\,\text{w/w}$$

---

6. A pharmacist adds 5.25 g of hydrocortisone to 150 g of a 2.5% hydrocortisone ointment. What is the percentage (w/w) of hydrocortisone in the finished product?

Solution. 5.8% w/w

CALCULATIONS

$$150 \text{ g oint.} \times \frac{2.5 \text{ g hydroc.}}{100 \text{ g oint.}} = 3.75 \text{ g hydroc.}$$

$$3.75 \text{ g} + 5.25 \text{ g} = 9 \text{ g hydroc.}$$

$$150 \text{ g} + 5.25 \text{ g} = 155.25 \text{ g oint.}$$

$$\frac{9 \text{ g}}{155.25 \text{ g}} = 0.058 = 5.8\% \text{ w/w}$$

---

7. When the solute is measured by weight and the solution by volume, concentration i *not* dimensionless. If 1.25 g of sodium chloride is dissolved in sufficient water to mak 55.0 mL of solution, the concentration is

$$\frac{1.25 \text{ g}}{55.0 \text{ mL}} = 0.0227 \text{ g/mL}$$

Commonly, this concentration will be stated as a percent, w/v. The appendag "w/v" tells us that the solute is expressed in *grams* and that the quantity of solution i determined by volume in *milliliters*. These implied units must not be neglected i calculations.

By multiplying grams per milliliter by 100, we obtain the number of grams i 100 mL, which defines percent w/v:

$$0.0227 \text{ g/mL} = \frac{2.27 \text{ g}}{100 \text{ mL}} = 2.27\% \text{ w/v}$$

If 5.75 g of boric acid are dissolved in sufficient alcohol to make a total volume c 120 mL, what is the strength of boric acid in the solution in

A. g/mL?

B. percent w/v?

Solutions.

A. 0.0479 g/mL

B. 4.79% w/v

CALCULATIONS

A. $\frac{5.75 \text{ g}}{120 \text{ mL}} = 0.0479 \text{ g/mL}$

B. 0.0479 g/mL × 100% = 4.79% w/v

8. If 20 mL of a 3.5% (w/v) xylocaine solution are added to a 100 mL bag of D5W injection,

A. What is the percentage strength (w/v) of xylocaine in the final product?

B. Express the strength of final dilution in mg/mL.

Solutions.

A. 0.58%

B. 5.8 mg/mL

CALCULATIONS

A. $20\text{ mL} \times \dfrac{3.5\text{ g}}{100\text{ mL}} = 0.7\text{ g xylocaine}$

$$\frac{0.7\text{ g}}{120\text{ mL}} = 0.0058 = 0.58\%$$

B. $0.58\% = \dfrac{0.58\text{ g}}{100\text{ mL}} = 0.0058\,\text{g/mL} = 5.8\,\text{mg/mL}$

---

9. To review these definitions of percent, the concentration of substance in a solution or mixture is expressed in terms of the amount of substance and the finished preparation. If the concentration is p percent w/w, then 100 weight units of the preparation contain p weight units of substance. The same weight units must be used for both substance and preparation. If the concentration is q percent v/v, then 100 volume units of the preparation contain q volume units of substance. The same volume units must be used for both substance and preparation. If the concentration is r percent w/v, then 100 mL of the preparation contain r grams of substance.

In each case, determine the percent concentration of glycerin and indicate whether w/w, v/v, or w/v:

A. 4.00 g of glycerin are dissolved in sufficient alcohol to make 25.0 mL.

B. 1.50 g of phenol are dissolved in 8.00 g of glycerin.

C. 10.0 mL of glycerin are dissolved in sufficient water to make 38.0 mL of solution.

Solutions.

A. 16.0% w/v

B. 84.2% w/w

C. 26.3% v/v

CALCULATIONS

A. $\dfrac{4.00\text{ g}}{25.0\text{ mL}} = 0.160\,\text{g/mL} = 16.0\%\,\text{w/v}$

B. $\dfrac{8.00\text{ g}}{9.50\text{ g}} = 0.842\,\text{w/w} = 84.2\%\,\text{w/w}$

C. $\dfrac{10.0\text{ mL}}{38.0\text{ mL}} = 0.263\,\text{v/v} = 26.3\%\,\text{v/v}$

Default Rules if Percentage Type is not Indicated

10. Sometimes the concentration of a solution is indicated without stating whether it is t be w/w, w/v, or v/v. In such cases, solutions are prepared so that solids can be weighe and liquids measured by volume. When the solute and the solvent (and therefore th solution) are liquids, the solution strength is assumed to be percent v/v. If the solut is a solid and the solvent a liquid, solution strength is assumed to be percent w/v. I both the solute and solvent are solid (or in a mixture of solids) percent w/w is assume These *default rules* come into play only when the percentage type is not indicated.

Let us look at an example. Say that you make up a solution of peppermint oil in alcoh by dissolving 2.0 mL of oil in enough alcohol to make 100 mL of solution. The concen tration is

$$\frac{2.0\text{ mL}}{100\text{ mL}} = 0.02(\text{v/v}) = 2\%(\text{v/v})$$

If your product is labeled "Peppermint Oil Solution, 2.0%," the default rules apply. Sinc both peppermint oil and alcohol are liquids, it is assumed that both will be measured b volume. Therefore, "2.0%" means 2.0% v/v in this case, so that your label describes th product accurately.

However, if you were to make up a solution by dissolving 2.0 g of peppermint oi in sufficient alcohol to make up 100 mL of solution? Then "2.0%" would not accuratel describe the product. It would have to be labeled 2.0% w/v.

If 4.0 g of peppermint oil are dissolved in 96.0 g of alcohol, how would you describe th concentration of peppermint oil in this solution?

Solution. 4% w/w

CALCULATIONS

$$\frac{4\text{ g}}{100\text{ g}} = 0.04(\text{w/w}) = 4\%\ \text{w/w}$$

Peppermint oil and alcohol are liquids and would normally be measured by volume But in this example, the solution was prepared by weight. The "percent w/w" designatio transmits this information. If the concentration were written "4.0%," the default rules woul apply and we would be (incorrectly) led to believe that the solution was prepared by volume

---

11. If 500 mg of sodium bicarbonate are dissolved in water to make 50 mL of solution, what would be the percentage concentration of sodium bicarbonate in the solution?

Solution. 1% w/v

CALCULATIONS

$$\frac{500 \text{ mg}}{50 \text{ mL}} = \frac{0.5 \text{ g}}{50 \text{ mL}} = \frac{0.01 \text{ g}}{\text{mL}} \times 100 = 1\% \text{ w/v}$$

---

12. Decide whether each of the following systems is w/w, v/v, or w/v, using the default rules:

A. 1% solution of zinc sulfate (a solid) in distilled water

B. An ointment containing 3% sulfur in petrolatum (both solids)

C. 10% solution of sugar in alcohol

D. 15% solution of alcohol in distilled water

Solutions.

A. w/v

B. w/w

C. w/v

D. v/v

Prescriptions with Ingredients Listed as Percentage

13. In a moment we will work with some prescriptions in which the ingredients are listed by percent. Of course, in order to prepare the medication, it is necessary to convert the percentages to weights and volumes. In w/w prescriptions, we calculate the weight of each material, since each must be weighed individually. But in w/v and v/v solutions, the volume of the solvent is usually not determined explicitly. There are two reasons:

(a) The solution is usually completed in a graduate by adding sufficient solvent to make the desired final volume. It is thus not necessary to know the volume of solvent.

(b) It is often impossible to calculate exactly what the volume of solvent should be. In w/v solutions, the volume occupied by the dissolved solute is not known. In very dilute solutions, the volume of solute may be negligible. But in general, there is no way to tell how much solvent to use. In v/v solutions, there may be shrinkage of volume when certain liquids are mixed: 50 mL of alcohol plus 50 mL of water yield less than 100 mL of solution. However, when chemically similar liquids are combined (e.g., mixing one fixed oil with another or one hydrocarbon with another), the total volume is usually the sum of the volumes of the components.

℞ Zinc sulfate (a solid) 1/2 %
Aq. pur. qs ad 60.0 mL

In this prescription the drug content is specified on the basis of concentration. We mu calculate the quantities needed to prepare this solution. Zinc sulfate is a solid and the fin product is a liquid, so the concentration is taken to be w/v. Thus the zinc sulfate conce tration is 0.5 g/100 mL or 0.005 g/mL.

The most direct way of calculating the quantity of zinc sulfate is to rearrange th definition of concentration:

$$\text{concentration} = \frac{\text{quantity of solute}}{\text{quantity of solution}}$$

to quantity of solute = concentration × quantity of solution

$$0.005\ \text{g/mL} \times 60.0\ \text{mL} = 0.3\ \text{g}$$

Dissolve 0.3 g of zinc sulfate in sufficient purified water to make 60 mL.

Now consider this prescription:

℞ Eucalyptus oil 2.5% v/v
Mineral oil qs ad 30.0

What quantities should be used for this prescription?

Solution. Eucalyptus oil: 0.75 mL
Mineral oil: sufficient to make 30.0 mL of solution

CALCULATION

30.0 mL × 0.025 = 0.75 mL

---

14. ℞ Sulfur 10.0%
Benzoic acid 1.5%
Petrolatum qs ad 30.0
Ft ung

All ingredients are solids. What quantity of each should be used to make this ointment (Remember that all three ingredients must be weighed.)

Solution. Sulfur: 3.00 g
Benzoic acid: 0.450 g
Petrolatum: 26.6 g

CALCULATIONS

All ingredients are solids. According to the default rules, we treat the concentrations as percent w/w:

Sulfur: 30.0 g × 0.1 = 3.00 g

Benzoic acid: 30.0 g × 0.015 = 0.450 g

The sum of all percentages must be 100%. This is necessarily true only in a preparation in which all concentrations are w/w. From the prescription, we can see that 88.5% of the ointment is petrolatum.

Petrolatum: 30.0 g × 0.885 = 26.6 g

To check, verify that the sum of all contributions is 30.0 g.

---

15. ℞ Boric acid 2%
Camphor water 35%
Aqua pur qs ad 15.0 mL
Ft. sol

How much boric acid (a solid) and camphor water (a liquid) are needed for this prescription? (Here is a situation in which two concentration types appear in the same solution. The boric acid must be considered on a w/v basis, whereas the camphor water is handled on a v/v basis. Treat each substance separately in doing your calculations.)

Solution. Boric acid: 300 mg
Camphor water: 5.25 mL

CALCULATIONS

Boric acid: 15.0 mL × 0.02 g/mL = 0.300 g

Camphor water: 15.0 mL × 0.35 = 5.25 mL

---

16. Iodine tincture is a 2% w/v solution of iodine.

A. How many grams of iodine will 40.0 mL of the tincture contain?

B. How many milliliters of the tincture contain 0.600 g of iodine? (Use proportion or rearrange the concentration definition to solve for amount of preparation.)

Solutions.

A. 0.8 g

B. 30.0 mL

CALCULATIONS

A. $40.0\,\text{mL} \times 0.02\,\text{g/mL} = 0.8\,\text{g}$

B. $$\frac{2\text{ g}}{100\text{ mL}} = \frac{0.6\text{ g}}{j}$$

$$j = 30\text{ mL}$$

Alternatively,

$$\text{amount of preparation} = \frac{\text{amount of solute}}{\text{concentration}}$$

$$\frac{0.6\text{ g}}{2\text{ g}/100\text{ mL}} = 30\text{ mL}$$

---

17. How many liters of a 0.9% aqueous solution can be made from 20.0 g of sodium chloride?

Solution. 2.22 L

CALCULATIONS

$$\frac{0.9\text{ g}}{100\text{ mL}} = \frac{20.0\text{ g}}{j}$$

$$j = 2220\text{ mL} = 2.22\text{ L}$$

---

18. If you need more practice, try these. Otherwise, proceed to frame 20.

A. ℞ Menthol 0.8%
Alcohol qs ad 60.0 mL

Menthol is a solid. How many grams should be used to prepare this prescription?

B. ℞ Zinc oxide
Talc āā 15%
Lanolin
Petrolatum āā qs ad 60 g

All ingredients are solids. Calculate the weight of each ingredient needed.

C. What quantity of each ingredient in the following prescription should be used? Include units.

℞		
Ammonium chloride (solid)		5%
Syrup tolu (liquid)		35%
Syrup wild cherry	qs ad	120 mL

Solutions

A. 0.48 g
B. Zinc oxide: 9 g
Talc: 9 g
Lanolin: 21 g
Petrolatum: 21 g
C. Ammonium chloride: 6 g
Syrup tolu: 42 mL
Syrup wild cherry: qs to make 120 mL

---

19. Need still more practice? Try these.

A. Diluted hydrochloric acid is a 10% w/v solution of HCl in water. How many milligrams of HCl does each teaspoonful of diluted hydrochloric acid contain?

B. How many milliliters of a 6.70% v/v solution contain 850 μL of solute?

C. How many liters of a 2.50% w/v solution can be prepared using 42.5 g of solute?

D. A boric acid preparation contains:

Boric acid: 2 parts by weight

Liquid petrolatum: 1 part by weight

Petrolatum: 17 parts by weight

Calculate the percentage strength for each ingredient. Must the percentages total 100%?

Solutions

A. 500 mg

B. 12.7 mL

C. 1.7 L

D. Boric acid: 10%

Liquid petrolatum: 5%

Petrolatum: 85%

Since all of the contributions in a w/w preparation are additive, the percentages must total 100%.

---

20. Calculating percentage strength in the apothecary system is much more complex than working in the metric system. Fortunately, apothecary prescriptions are rarely encountered in present days. In the event that one should appear, the best procedure is to use the metric system anyway. Since percent w/v is defined specifically in terms of grams and milliliters, changing to another set of units introduces unnecessary complexity.

Start by converting the finished amount using the approximate equivalents:

$$1 ℥ \approx 30\,\text{g}$$

$$1\ \text{f}℥ \approx 30\,\text{mL}$$

Then calculate the needed quantities in the usual way.

Consider the prescription that follows:

℞	Menthol		1/2%
	Alcohol	qs ad	f℥ ii

The best procedure is to prepare 60 mL. The required amount of menthol is simply

$60\,\text{mL} \times 0.005\,\text{g/mL} = 0.3\,\text{g}$

21. Imagine that a solution is prepared by dissolving 1 g of sodium chloride in sufficient water to make a total solution volume of 12 mL. As was explained previously, the amount of water needed to make this solution cannot be calculated directly. This is because it is impossible to predict the amount of space occupied by the dissolved sodium chloride. The density of pure, crystalline sodium chloride cannot be used to calculate the volume occupied by the salt in solution because the volume of a dissolved solute is almost always different from that of the pure material if it is a solid or gas. When liquids are combined into a solution, it sometimes happens that the volume of each liquid in the solution is the same as that in the pure state. This is most likely with liquids that are chemically similar. For example, 2.0 mL of decyl alcohol may be dissolved in 10.0 mL of octyl alcohol to produce 12.0 mL of solution. However, if 3.0 mL of ethanol are dissolved in 7.0 mL of water, the volume of the resulting solution is less than 10.0 mL. The volumes, in this case, are not additive. We can therefore state that, except for certain liquid mixtures, the volume of solution is not equal to the sum of the volumes of pure solute and pure solvent.

True or False:

A. The volume occupied by 1 g of hydrogen chloride in water solution is the same as that of 1 g of gaseous hydrogen chloride.

B. The volume occupied by 1 g of potassium bromide in water solution is the same as that of 1 g of solid potassium bromide.

C. The volume occupied by 1 g of methanol in water solution is the same as that of 1 g of pure liquid methanol.

Solutions. All three statements are false.

Determination of Amount of Solvent Based on Density and Specific Gravity

22. If it should be necessary to determine the exact amount of solvent used in making a solution, we can make use of the fact that in contrast to volume, weight is always an additive property. That is,

$$\text{weight of solute} + \text{weight of solvent} = \text{weight of solution}$$

This equation is true for all solutions. Let us put it to use in the following example:

The specific gravity (review density and specific gravity in Ch. 2, if needed) of a 25% w/v solution of sodium acetate in water is 1.113. How many milliliters of water should be used to prepare 40.0 mL of this solution?

According to the equation, we can find the *weight* of the solvent by subtracting the weight of the solute from the weight of the solution. Then, by using the density of the solvent (in this case, water) we can find the *volume* of solvent to use. Begin by calculating the weight of solute in 40 mL of solution.

$$40.0\,\text{mL} \times 0.25\,\text{g/mL} = 10.0\,\text{g}$$

Next, using the density of the solution, calculate the weight of 40.0 mL of solution.

$$40.0\,\text{mL} \times 1.113\,\text{g/mL} = 44.5\,\text{g}$$

Now, determine the weight of solvent and then convert weight to volume by making us of its density.

$$44.5\ \text{g} - 10.0\ \text{g} = 34.5\ \text{g}$$

$$34.5\ \text{g} \times \frac{1.00\ \text{mL}}{1.00\ \text{g}} = 34.5\ \text{mL}$$

Here is another problem.

The formula for potassium iodide solution is:

Potassium iodide (KI): 1000 g

Purified water: a sufficient quantity to make 1000 mL

How many milliliters of water will be needed to make 120 mL of solution? (The specifi gravity of KI is 3.12; specific gravity of KI solution is 1.70.)

Solution. 84 mL of water are required.

If you ran into trouble, let us see if we can find what went wrong. Remember that th volume of solvent is found from its weight. Its weight is determined by subtracting th weight of solute from the weight of solution.

I hope you did not try to make use of the density of KI. The volume of a materia in solution is not the same as that of the pure substance. go back to the problem. Find th number of grams of KI in 120 mL of the solution. Calculate the weight of the solutio from its density. Determine the weight of water by difference and then the volume of wate

Since 1000 g of KI makes 1000 mL of solution, there will be 120 g of KI in 120 ml of solution. The weight of solution may be determined from its density:

$$120\ \text{mL} \times 1.70\ \text{g/mL} = 204\ \text{g}$$

$$204\ \text{g solution} - 120\ \text{g KI} = 84\ \text{g water}$$

Since the density of water is 1.00 g per milliliter, 84 mL of water are required.

---

23. A manufacturer wishes to prepare 2 L of sodium acetrizoate solution, 30.0% w/v. Th specific gravity of this solution is 1.195. How many milliliters of water will b required?

Solution. 1790 mL

CALCULATIONS

Weight of the solution: 2000 mL × 1.195 g/mL = 2390 g

Weight of the sodium acetrizoate: 2000 mL × 0.300 g/mL = 600 g

Weight of the water: 2390 g − 600 g = 1790 g

Volume of the water: $1790\ \text{g} \times \frac{1\ \text{mL}}{1.00\ \text{g}} = 1790\ \text{mL water}$

24. How many milliliters of absolute alcohol must be used to make 240 mL of a 10% w/v solution of a drug? (Sp g absolute alcohol = 0.798; Sp g 10% solution of drug in absolute alcohol = 0.851.)

Solution. 226 mL

CALCULATIONS

Weight of solution: 240 mL × 0.851 g/mL = 204 g

Weight of drug: 240 mL × 0.1 g/mL = 24 g

Weight of absolute alcohol: 204 g − 24 g = 180 g

$$\text{Volume: } 180\ \text{g} \times \frac{1\ \text{mL}}{0.798\ \text{g}} = 226\ \text{mL}$$

---

25. The USP describes the solubility of drugs by stating the number of milliliters of solvent needed to dissolve 1 g of the drug. This type of information is useful if you want to be able to prepare a solution using the smallest amount of solvent possible. Such a solution is saturated with the solute.

One gram of niacinamide dissolves in 1.5 mL of water. Calculate the volume of water needed to dissolve 12 g of niacinamide.

Solution. 18 mL

CALCULATIONS

$$\frac{1\ \text{g}}{1.5\ \text{mL}} = \frac{12\ \text{g}}{j}$$

$$j = 18\ \text{mL}$$

---

26. How many grams of niacinamide would be contained in 100 g of a saturated solution of niacinamide in water? (Recall that 1 g dissolves in 1.5 mL of water.)

Solution. 40 g

CALCULATIONS

When 1 g of the drug dissolves in 1.5 g of the solvent, 2.5 g of solution result. In other words, each 2.5 g of saturated solution contain 1 g of niacinamide.

$$\frac{1\text{ g}}{2.5\text{ g}} = \frac{j}{100\text{ g}}$$

$j = 40$ g

---

27. One gram of a drug dissolves in 12.0 mL of carbon tetrachloride (CCl_4). How many grams of the drug will be contained in 40.0 g of saturated solution? (Specific gravity of CCl_4 is 1.59.)

Solution. 2.00 g

CALCULATIONS

g drug + g CCl_4 = g solution

In order to find the weight of solution that contains 1 g of the drug, we must determine the weight of 12.0 mL of CCl_4:

$12.0\text{ mL} \times 1.59\text{ g/mL} = 19.1\text{ g}$

Therefore,

$1\text{ g drug} + 19.1\text{ g } CCl_4 = 20.1\text{ g solution}$

$$\frac{1\text{ g drug}}{20.1\text{ g solution}} = \frac{j}{40.0\text{ g solution}}$$

$j = 2.00$ g of drug

---

28. One gram of carbetapentane citrate dissolves in 6.5 mL of alcohol. What is the percentage strength of the solution w/w? (Sp g of alcohol is 0.814.)

Solution. 16% w/w

CALCULATONS

6.5 mL × 0.814 g/mL = 5.3 g

Each gram of carbetapentane citrate dissolves in 5.3 g of alcohol yielding 6.3 g of solution.

$$\frac{1\text{ g}}{6.3\text{ g}} = 0.16 = 16\%\text{ w/w}$$

MILLIGRAMS PERCENT (mg%) AND MILLIGRAMS PER DECILITER (mg/dL)

29. Health care professionals may occasionally see values of laboratory tests recorded as *mg%* or *mg/dL*. These expressions state the number of milligrams of a substance present in *100 mL* (or 1 dL) of a biologic fluid, usually plasma (liquid part of blood) or serum (liquid part of clotted blood). Blood chemistry results are used by the prescriber for diagnosis (cholesterol, creatinine) or to adjust the patient's needs of substances such as, glucose, calcium, sodium, etc, and several drugs after administration.

Blood alcohol concentration considered to diminish driving performance has been established to be 0.08%. If the blood level of alcohol of a driver was found to be 300 mg/dL, was he intoxicated?

Solution. Yes, he was highly intoxicated.

CALCULATIONS

300 mg/dL = 300 mg/100 mL = 0.3 g/100 mL = 0.3%

---

30. Creatinine is a metabolic breakdown product of muscle. For a patient with a serum creatinine of 1.2 mg/dL, express this concentration in mg/mL.

Solution. 0.012 mg/mL

CALCULATION

1.2 mg/dL = 1.2 mg/0.1 L = 1.2 mg/100 mL = 0.012 mg/mL

---

31. A blood glucometer shows that a patient's blood contains 175 mg% of glucose. Express this value as mg/mL

Solution. 1.75 mg/mL

CALCULATIONS

175 mg% = 175 mg/100 mL = 1.75 mg/mL

Milligrams per Milliliter (mg/mL)

32. It is also very common to see prescriptions written in a milligram per milliliter basis, if a health care practitioner is in a patient care facility, such as a hospital. Medication orders for intravenous infusions and many drugs for injection used to prepare parenteral admixtures are expressed as mg/mL. Some "quick conversion" must be performed if a product concentration is expressed as percentage, ratio, or milliequivalents (we will discuss milliequivalents (mEq), in Chapter 9).

Several "techniques" have been proposed. You may find your own shortcut for this kind of conversion, remembering that it has to be always reliable and never confusing. One simple way is to transform the units in the concentration on hand, in units of mg in the nominator and mL in the denominator. Dividing the numerical value of nominator by denominator will give the final mg/mL strength.

An ampul of calcium gluconate injection 10% will have in mg/mL:

$$10\% = \frac{10\text{ g}}{100\text{ mL}} = \frac{10000\text{ mg}}{100\text{ mL}} = 100\text{ mg/mL}$$

If a skin test for allergy includes an intradermal skin prick of 50 μL of a 1 : 100,000 (w/v) dilution of an allergen, how many micrograms will the patient receive? (1:100,000 w/v indicates a solution in which 100,000 mL contain 1 g.)

Solution. 0.5 μg

CALCULATIONS

$$50\ \mu\text{L} \times \frac{1\text{ mL}}{1000\ \mu\text{L}} \times \frac{1\text{ g}}{100000\text{ mL}} \times \frac{1000\text{ mg}}{1\text{ g}} \times \frac{1000\ \mu\text{g}}{1\text{ mg}} = 0.5\ \mu\text{g}$$

---

33. A physician ordered a 2-g vial of a drug to be added to a liter of normal saline and administered by intravenous infusion. How many milligrams of the drug will the patient receive in each milliliter of solution?

Solution. 2 mg/mL

CALCULATIONS

$$2\text{ g/Liter} = \frac{2000\text{ mg}}{1000\text{ mL}} = 2\text{ mg/mL}$$

PARTS PER MILLION (ppm) AND PARTS PER BILLION (ppb)

34. Using a percent is equivalent to stating the number of parts of solute in each 100 parts of solution. Parts per million (ppm) and parts per billion (ppb) are used to describe the concentration of very dilute solutions:

ppm represents the number of parts of solute in 10^6 parts of solution

ppb describes the number of parts of solute in 10^9 parts of solution.

When ppm or ppb are used as a designation for concentration, some systems are w/w, some are v/v, and some are w/v. Concentration is always a ratio or fraction in w/w and v/v situations. Weight by volume (w/v) concentrations are always defined in terms of grams and milliliters. The same default rules are followed as for percentage systems.

If 10^6 mg (in other words, 1 kg) of a preparation contain 3 mg of a drug, the drug concentration is 3 ppm w/w. If 10^6 mL of a solution contain 22 mL of alcohol, the alcohol concentration is 22 ppm v/v. Any units may be used in describing w/w and v/v systems. However, with w/v situations, weight is in grams and volume is in milliliters. A 1.5 ppm w/v solution of a drug is one in which 10^6 mL of the solution contain 1.5 g of the drug.

A powder mixture contains 12 ppm of penicillin (as a contaminant). That is, each 10^6 g of powder contain 12 g of penicillin, or more generally, each 10^6 parts of the powder mixture contain 12 parts of penicillin.

The concentration of penicillin is therefore,

$$\frac{12 \text{ parts}}{10^6 \text{ parts}} = 12 \times 10^{-6}$$

---

35. How many milligrams of penicillin are there in 5.6 kg of the powder mixture?

Solution. 67.2 mg

CALCULATIONS

quantity of solute = concentration × quantity of preparation

$12 \times 10^{-6} \times 5.6 \text{ kg} = 67.2 \times 10^{-6} \text{ kg} = 67.2 \text{ mg}$

An alternative solution makes use of the definition directly. Since the concentration of penicillin is 12 ppm, each 10^6 mg of the mixture contain 12 mg of penicillin. The amount of penicillin in 5.6 kg (= 5.6×10^6 mg) of the mixture can be found by proportion.

$$\frac{12 \text{ mg}}{10^6 \text{ mg}} = \frac{j}{5.6 \times 10^6}$$

$j = 67.2$ mg penicillin

36. If 14L of commercial ethyl alcohol are found to contain 0.010mL of butanol, what is the concentration of butanol, in ppm?

Solution. 0.71 ppm

CALCULATIONS

$$\text{concentration} = \frac{10 \times 10^{-3}\ \text{mL}}{14 \times 10^{3}\ \text{mL}} = \frac{0.71}{10^{6}} = 0.71\ \text{ppm}$$

or by proportion,

$$\frac{0.010\ \text{mL}}{14 \times 10^{3}\ \text{mL}} = \frac{V}{10^{6}\ \text{mL}}$$

$v = 0.71$ mL of butanol. From the definition for ppm, the concentration is 0.71 ppm.

37. If the content of sodium fluoride in drinking water is 2 ppm, this means that in each 10^6 mL of this solution are contained 2 g of sodium fluoride. How many milligrams of sodium fluoride would be found in 15 L of the drinking water described above? (*Hint*: Sodium fluoride is a solid, so the concentration defaults to w/v and is expressed in terms of grams of solute and milliliters of solution.)

Solution. 30 mg

CALCULATIONS

2 ppm in this case is defined as 2 g in 10^6 mL

$$\frac{2\ \text{g}}{10^{6}\ \text{mL}} = \frac{j}{15 \times 10^{3}\ \text{mL}}$$

$$j = 30 \times 10^{-3}\ \text{g} = 30\ \text{mg}$$

38. The content of parathion (a solid) in a particular patient's bloodstream is 0.085 ppm. How many micrograms of parathion does the patient's blood contain if the blood volume is 6.0 L?

Solution. 510 μg

CALCULATIONS

$$\frac{0.085\text{ g}}{10^6\text{ mL}} = \frac{j}{6000\text{ mL}}$$

$$j = 0.51 \times 10^{-3}\text{ g} = 510\ \mu\text{g}$$

---

39. A sample of a solution for injection is found to contain 1.4 ppm of lead chloride. How much of the solution will contain 50 μg of lead chloride?

Solution. 36 mL

CALCULATIONS

$$1.4\text{ ppm} = 1.4\text{ g in }10^6\text{ mL};\ 50\ \mu\text{g} = 50 \times 10^{-6}\text{ g}$$

$$\frac{1.4\text{ g}}{10^6\text{ mL}} = \frac{50 \times 10^{-6}\text{ g}}{j}$$

$$j = 36\text{ mL}$$

---

40. The drinking water in a remote desert area in the Middle East contains 0.34 ppb of selenium. How many micrograms of selenium will be ingested by a camel whose water intake is 110 L?

Solution. 37.4 μg

CALCULATIONS

$$\frac{0.34\text{ g}}{10^9\text{ mL}} \times (110 \times 10^3\text{ mL}) = 37.4 \times 10^{-6}\text{ g} = 37.4\ \mu\text{g}$$

41. Here are some problems that review ppm and ppb calculations.

A. If 250 mL of water contain 0.275 mg of lead ion, what is the concentration of lead ion in ppm? In percent?

B. How many liters of solution can be prepared from 42.3 mg of an antioxidant if the final concentration of the antioxidant is supposed to be 12.0 ppm w/v?

C. If the methyl mercury content in a certain lake is 3.30 ppm, how many grams of methyl mercury would 250 L of lake water contain?

D. The asbestos level of a talc deposit is 0.79 ppb. How many micrograms of asbestos would be found in 10 g of this talc?

Solutions.

A. 1.10 ppm; $1.10 \times 10^{-4}\%$

B. 3.53 L

C. 0.825 g

D. 0.0079 μg

CALCULATIONS

A. $$\frac{0.275 \times 10^{-3}\ \text{g}}{250\ \text{mL}} = \frac{j}{10^{6}\ \text{mL}}$$

$j = 1.10\,\text{g}$

That makes the concentration 1.10 ppm.

$$\frac{0.275 \times 10^{-3}\ \text{g}}{250\ \text{mL}} = \frac{j}{10^2\ \text{mL}}$$

$j = 1.10 \times 10^{-4}$ g

The concentration is 1.10×10^{-4} g per 100 mL or 1.10×10^{-4}%

B. $$\frac{12\ \text{g}}{10^6\ \text{mL}} = \frac{42.3 \times 10^{-3}\ \text{g}}{j}$$

$j = 3.53$ L

C. $$\frac{3.3\ \text{g}}{10^6\ \text{mL}} = \frac{j}{250 \times 10^3\ \text{mL}}$$

$j = 0.825$ g

D. $$\frac{0.79\ \text{g asbestos}}{10^9\ \text{g talc}} \times 10\ \text{g talc} = 7.9 \times 10^{-9}\ \text{g asbestos} = 0.0079\ \mu\text{g asbestos}$$

---

42. If you need more practice with ppm problems, try these.

A. What volume of solution can be prepared using 147 mg of sodium fluoride if the final concentration of sodium fluoride is to be 6.4 ppm?

B. Express the concentration of alcohol in parts per million if 25.0 mL of blood of a reckless driver contained 9.50 μL of alcohol.

C. If 125 kg of rauwolfia (a plant drug) contain 8.5 mg of lead, what is the concentration of lead in parts per million?

D. The concentration of a preservative in a solution is 37.5 ppm, w/v. How many micrograms of the preservative will each teaspoonful contain?

E. The concentration of a pesticide in animal feed is 12.5 ppm. How many kilograms of animal feed would contain 0.65 mg of pesticide?

F. If the lead content in a patient's teeth is 15.0 ppm, how many micrograms of lead are bound there if the teeth weigh 15.0 g?

Solutions.

A. 23 L

B. 380 ppm

C. 0.068 ppm

D. 188 μg

E. 0.052 kg

F. 225 μg

REVIEW PROBLEMS

This is a good time to stop and review. Answer all of these questions before verifying your results.

A. ℞

Salicylic acid	1.8 g
Benzoic acid	3.6 g
White ointment	54.6 g

What is the percentage strength of each of the three components?

B. How many liters of a 0.2% solution in alcohol can be prepared using 12.5 mL of spearmint oil?

C. ℞

Glycerin	12%
Resorcinol	3.5%
Aq dest	50%
Alcohol qs ad	180.0 mL

How much glycerin and resorcinol (a solid) are needed for this prescription?

D. ℞ Hydrochloric acid, 10% w/v
Prepare 60.0 mL solution
Sig: 15 drops in water

How many grams of HCl does the patient receive in each dose if the dropper dispensed with this medication delivers 32 drops/mL?

E. ℞

Boric acid	2.5%
Epinephrine solution	2%
Aqua dest qs ad	15 mL

How much boric acid (a solid) and epinephrine solution are required?

F. How many liters of a 1.50% w/v potassium nitrate solution can be prepared from 22.0 g of potassium nitrate?

G. ℞ Vitamin C 100 mg

Iron 15 mg

d.t.d. caps No. LX

How many grams of an iron choline citrate complex (containing 12% iron w/w should be used in compounding this prescription?

H. A pharmacist has to prepare 25.0 mL of a suspension containing 1/2% w/v of chloramphenicol. Instead of using pure chloramphenicol, she uses chloramphenicol palmitate, which contains 57.5% w/w chloramphenicol. How many milligrams of chloramphenicol palmitate should be used for this suspension?

I. How many milliliters of water are required to prepare 75.0 mL of a 20% w/v solution of a salt? (Sp g of the salt solution = 1.12; Sp g of the salt = 1.85; formula weight of the salt = 207.)

J. Lanolin contains 72.5% w/w wool fat. The remainder is water. How many milliliters of water are there in 12.0 g of lanolin?

K. Syrup is an 85% w/v solution of sucrose in water. It has a density of 1.313 g/mL. How many milliliters of water should be used to make 125 mL of syrup?

L. If 1 g of a salt dissolves in 3.5 mL of water to form a saturated solution whose specific gravity is 1.10, how many grams will dissolve in 420 mL of water?

M. A gram of a salt dissolves in 4.90 mL of water to form a saturated solution whose specific gravity is 1.08. How many grams of the salt will be contained in 100 mL of the saturated solution?

N. ℞ Elixir phenobarbital 120.0

Sig: 1 tsp. tid

Elixir phenobarbital contains 0.4% phenobarbital (w/v). How many milligrams of phenobarbital does the patient receive each day?

O. Phosphoric acid contains 86.5% w/w of H_3PO_4. How many grams of H_3PO_4 are contained in 55.0 mL of phosphoric acid? (Sp g of phosphoric acid = 1.71.)

P. If 1 g of a drug dissolves in 2.50 mL of glycerin (Sp g = 1.25), what is the percentage strength of the solution w/w?

Q. If a patient has a serum cholesterol level of 180 mg/dL

(a) What is the corresponding value in mg%?

(b) How many milligrams of cholesterol would be present in a 10 mL sample of this patient's serum?

R. A drug ointment is available in three strengths: 0.025%, 0.05% and 0.25%. Ho
would you express these concentrations in terms of milligrams of drug per gram
ointment?

S. A vial of an injection contains 5 µg of vitamin per 10 mL. What is the ratio streng
of the preparation?

T. What is the concentration of benzoyl peroxide in terms of mg/mL in a 5% (w/
benzoyl peroxide lotion?

U. If an injection contains 1:2000 (w/v) diltiazem hydrochloride, how many milligram
of drug will be in each milliliter of solution?

V. An IV solution contains 4500 µg of drug per deciliter. What is the correspondin
percent concentration?

W. The average blood alcohol concentration in fatal intoxication is 0.4%. Express thi
concentration as mg/dL.

X. Express a patient's blood glucose levels of 1.5 mg/mL in terms of mg%.

Y. The concentration of nonprotein nitrogen in the blood of a patient is 30 mg%. Express this concentration in terms of mg/mL.

Z. An alkaloid extracted from a South American plant is found to be contaminated with DDT, 4 ppm. How many grams of the alkaloid will contain 20 mg of DDT?

AA. If a source of drinking water contains 1.5 ppm of fluoride, how many micrograms of fluoride are present in one glass (240 mL) of water?

AB. Express the concentration of alcohol in parts per million if 925 mL of fermentation broth contain 12.7 μL of alcohol.

AC. How many micrograms of a pesticide are present in 22.0 kg of a plant drug if the concentration of pesticide is 27.3 ppb?

Solutions to Review Problems

A. Salicylic acid: 3.00% w/w
Benzoic acid: 6.00% w/w
White ointment: 91.0% w/w

B. 6.25 L

C. Glycerin: 21.6 mL
Resorcinol: 6.30 g

D. 0.047 g

E. Boric acid: 0.375 g
Epinephrine solution: 0.3 mL

F. 1.47 L

G. 7.50 g

H. 217 mg

I. 69.0 mL

J. 3.30 mL

K. 57.8 mL

L. 120 g

M. 18.3 g

N. 60 mg

O. 81.4 g

P. 24.2% w/w

Q. **(a)** 180 mg%
(b) 18 mg

R. 0.25 mg/g; 0.5 mg/g; 2.5 mg/g

S. 1 : 2,000,000

T. 50 mg/mL

U. 0.5 mg/mL

V. 0.0045%

W. 400 mg/dL

X. 150 mg%

Y. 0.3 mg/mL

Z. 5000 g

AA. 360 μg

AB. 13.7 ppm

AC. 601 μg

Calculations to Review Problems

A. The formula is for a total of 60.0 g.

Salicylic acid: $\frac{1.8\text{ g}}{60.0\text{ g}} = 0.0300 = 3.00\%\text{ w/w}$

Benzoic acid: $\frac{3.6\text{ g}}{60.0\text{ g}} = 0.0600 = 6.00\%\text{ w/w}$

White ointment: $\frac{54.6\text{ g}}{60.0\text{ g}} = 0.910 = 91.0\%\text{ w/w}$

B. $0.002 = \frac{12.5\text{ mL oil}}{j}$

$j = 6250\text{ mL} = 6.25\text{ L solution}$

C. Glycerin: $180\text{ mL} \times 0.12 = 21.6\text{ mL}$

Resorcinol: $180\text{ mL} \times 0.035\text{ g/mL} = 6.3\text{ g}$

D. $15\text{ drops} \times \frac{1\text{ mL}}{32\text{ drops}} = 0.47\text{ mL}$

$0.47\text{ mL} \times 0.1\text{ g/mL} = 0.047\text{ g}$

E. Boric acid: $15\text{ mL} \times 0.025\text{ g/mL} = 0.375\text{ g}$

Epinephrine: $15\text{ mL} \times 0.02\text{ g/mL} = 0.3\text{ mL}$

F. $0.0150\text{ g/mL} = \frac{22\text{ g}}{j}$

$j = 1470\text{ mL} = 1.47\text{ L}$

G. The total amount of iron needed is

$\frac{15\text{ mg}}{\text{caps}} \times 60\text{ caps} = 900\text{ mg}$

$0.12 = \frac{900\text{ mg}}{j}$

$j = 7500\text{ mg} = 7.50\text{ g}$

H. $25.0\text{ mL} \times 0.005\text{ g/mL} = 0.125\text{ g}$ chloramphenicol

$\frac{57.5\text{ g}}{100\text{ g}} = \frac{0.125\text{ g}}{j}$

$j = 0.217\text{ g} = 217\text{ mg}$

I. Weight of solution: $75.0\text{ mL} \times 1.12\text{ g/mL} = 84.0\text{ g}$

Weight of salt: $75.0\text{ mL} \times 0.2\text{ g/mL} = 15.0\text{ g}$

Weight of solvent: $84.0\text{ g} - 15.0\text{ g} = 69.0\text{ g} = 69.0\text{ mL}$

J. The concentration of water must be 27.5% w/w

$12.0\text{ g} \times 0.275 = 3.30\text{ g water} = 3.30\text{ mL}$

K. $125\text{ mL} \times 0.85\text{ g/mL} = 106.25\text{ g sucrose}$

$125\text{ mL} \times 1.313\text{ g/mL} = 164.1\text{ g syrup}$

$164.1\text{ g} - 106.3\text{ g} = 57.8\text{ g water} = 57.8\text{ mL}$

L. $\dfrac{1\text{ g}}{3.5\text{ mL}} = \dfrac{j}{420\text{ mL}}$

$j = 120\text{ g}$

M. 1 g salt + 4.90 g H_2O = 5.90 g solution

$5.90\text{ g} \times \dfrac{1\text{ mL}}{1.08\text{ g}} = 5.46\text{ mL}$

$\dfrac{1\text{ g salt}}{5.46\text{ mL solution}} = \dfrac{j}{100\text{ mL solution}}$

$j = 18.3\text{ g}$

N. The patient takes 15 mL per day:

15 mL × 0.004 g/mL = 0.060 g = 60 mg

O. Since we know only the w/w concentration, only the weight of phosphoric aci allows us to calculate the H_3PO_4 content.

weight of phosphoric acid: 55 0 mL × 1.71 g/mL = 94.1 g

94.1 g × 0.865 = 81.4 g

P. We must know the weight of the drug in a definite weight of solution.

2.50 mL × 1.25 g/mL = 3.13 g

One gram of drug dissolves in 3.13 g of glycerin to make 4.13 g of solution.

$\dfrac{1\text{ g}}{4.13\text{ g}} = 0.242 = 24.2\%\text{ w/w}$

Q. **(a)** 180 mg/dL = 180 mg/100 mL = 180 mg%

(b) $\dfrac{180\text{ mg}}{100\text{ mL}} = \dfrac{j}{10\text{ mL}} = 18\text{ mg}$

R. 0.025 g/100 g = 25 mg/100 g = 0.25 mg/g

0.05 g/100 g = 50 mg/100 g = 0.5 mg/g

0.25 g/100 g = 250 mg/100 g = 2.5 mg/g

S. $5\,\mu\text{g}/10\text{ mL} = \dfrac{5 \times 10^6\,\mu\text{g}}{10 \times 10^6\text{ mL}} = \dfrac{5\text{ g}}{10 \times 10^6\text{ mL}} = \dfrac{1\text{ g}}{2 \times 10^6\text{ mL}}$

$= 1:2{,}000{,}000$

T. 5 g/100 mL = 5000 mg/100 mL = 50 mg/mL

U. 1 g/2000 mL = 1000 mg/2000 mL = 0.5 mg/mL

V. 4500 μg/dL = 0.0045 g/100 mL = 0.0045%

W. 0.4 g/100 mL = 400 mg/100 mL = 400 mg/dL

X. 1.5 mg/mL = 150 mg/100 mL = 150 mg%

Y. 30 mg/100 mL = 0.3 mg/mL

Z. $\dfrac{4\text{ mg}}{10^6\text{ mg}} = \dfrac{20\text{ mg}}{j}$

$j = 5 \times 10^6\text{ mg} = 5 \times 10^3\text{ g}$

AA. $\dfrac{1.5\ \text{g}}{10^6\ \text{mL}} = \dfrac{j}{240\ \text{mL}}$

$j = 360 \times 10^{-6}\ \text{g} = 360\ \mu\text{g}$

AB. $\dfrac{12.7\ \mu\text{L}}{925 \times 10^3\ \mu\text{L}} = \dfrac{j}{10^6\ \mu\text{L}}$

$j = 13.7\ \mu\text{L}$

Therefore, the concentration is 13.7 ppm.

AC. $\dfrac{27.3}{10^9} = \dfrac{j}{22\ \text{kg}}$

$j = 6.01 \times 10^{-7}\ \text{kg} = 601\ \mu\text{g}$

CHAPTER 6

RATIO STRENGTH AND STOCK SOLUTIONS

Learning Objectives: *after completing this chapter the student should be able to:*

1. Use ratio concentrations and its conversion to other practical concentratio expressions in pharmaceutical calculations.
2. Calculate the quantity of drug in a given amount of a preparation whose con centration is expressed in terms of ratio strength.
3. Determine the amount of a preparation of given ratio strength that will contai a desired quantity of drug.
4. Determine the strength of a drug trituration to be prepared and the amoun needed to supply a desired quantity of drug.
5. Calculate the quantity of a stock solution needed to deliver a desired quantity o drug.

Ratio strength is another way of representing concentration. It may be considered as ar alternate to percentage. In this convention, concentration is denoted in terms of *one* uni of solute contained in the total amount of solution or mixture. Stock solutions and tritu rations provide a convenient means for handling a small quantity of potent drug and are used for mixtures containing a drug and an inert substance (diluent).

RATIO STRENGTH EXPRESSIONS

1. As the name implies, ratio strength describes drug concentration in terms of a ratio A 1 : 25 solution of cinnamon oil means that 1 mL of cinnamon oil is contained in each 25 mL of solution. Note that the second number in the ratio does not describe the quan- tity of solvent or diluent, but rather the *total quantity of the solution*, which in this case includes cinnamon oil.

In a ratio, the numbers can be read as *parts*:

$$1:25 = 1 \text{ part in } 25 \text{ parts}$$

You assign the units to the parts depending on whether you are dealing with w/w, w/v, or v/v preparations.

Pharmaceutical Calculations, Fourth Edition, By Joel L. Zatz and Maria Glaucia Teixeira
ISBN 0-471-67623-3

2. The very same *default rules* based on weighing of solids and volumetric measurement of liquids used for percentage strength applies here. Thus, the cinnamon oil solution is assumed to be v/v since no other indication is given. Both liquids must have the same units. We can express this concentration as a percentage by first converting the ratio to a decimal value.

$$1{:}25 = 1\text{ mL}{:}25\text{ mL} = \frac{1\text{ mL}}{25\text{ mL}} = \frac{1}{25} = 0.04\text{ v/v} = 4\%\text{ v/v}$$

What is the percentage strength of a 1 : 400 solution of an oil in alcohol?

Solution. 0.25% v/v

CALCULATIONS

1 : 400 = 1 mL oil in 400 mL solution

Using proportion,

$$\frac{1\text{ mL oil}}{400\text{ mL solution}} = \frac{j}{100\text{ mL}}$$

$j = 0.25$ mL oil

Since percent is defined as parts per 100, the strength of the solution is 0.25% v/v.

---

3. A viral vaccine is preserved with 1 : 10,000 solution of benzalkonium chloride (a solid). How would we express this concentration as percentage?

Solution. 0.01% w/v

CALCULATIONS

1 : 10,000 = 1 g benz. in 10,000 mL solution

$$\frac{1\text{ g}}{10000\text{ mL}} = \frac{j}{100\text{ mL}} = 0.01\text{ g in }100\text{ mL} = 0.01\%\text{ w/v}$$

---

4. A 1:40 dilution of atropine in a solid mixture means that 1 g of atropine is contained in 40 g of the mixture. In systems of this type, the other solid material usually has no medicinal action. It is called a *diluent*. This type of solid mixture, containing only an active drug and diluent, is called a drug *trituration*. Triturations provide a convenient means for handling materials used in very small quantities. The concentration of drug in the trituration is generally specified in terms of ratio strength.

The inert substance (or diluent) that is used most frequently in preparing triturations is lactose, otherwise known as milk sugar. Lactose has no drug action or ill effects and is pleasant tasting. In preparing a trituration, the drug and the diluent must be thoroughly mixed so that the resulting powder is completely uniform.

Notice that when both components are solids, they must have the same units. The concentration in this case is

$$1{:}40 = \frac{1\text{ g}}{40\text{ g}} = 0.025\text{ w/w} = 2.5\%\text{ w/w}$$

What is the percentage strength of a 1:50 w/w mixture?

Solution. 2% w/w

CALCULATIONS

$$\frac{1}{50} = \frac{j}{100}$$

$j = 2$; 2 parts per 100 defines a 2% mixture.

---

5. What is the percentage strength of a 1:500 zinc oxide ointment?

Solution. 0.2% (w/w)

CALCULATIONS

$$\frac{1}{500} = \frac{j}{100}$$

$j = 0.2 = 0.2\%$ w/w

6. Express each of the following as a percentage strength:

A. 1:400 (w/w)

B. 1:250 (w/w)

C. 1:6600 (w/w)

Solutions.

A. 0.25% (w/w)

B. 0.4% (w/w)

C. 0.015% (w/w)

---

7. Other expressions of concentration may be changed to ratio strength. The concentration is written as a fraction, reduced (so that the numerator has a value of "1"), and then converted to a ratio.

A doctor prescribed an ointment to contain 5 mg of hydrocortisone per gram of ointment. How would the pharmacist express this as a ratio strength?

Solution. 1:200 (w/w)

CALCULATIONS

$$\frac{5\text{ mg hydrocortisone}}{\text{g ointment}} = \frac{0.005\text{ g hydr.}}{\text{g oint}}$$

$$\frac{0.005\text{ g hydr.}}{\text{g oint}} = \frac{1\text{ g drug}}{j}$$

$j = 200$ g ointment; 1 part per 200 parts defines 1:200 ratio strength.

---

8. Express the following concentrations as ratio strength.

A. 5 mg drug in 50 mL of solution

B. 0.25 mL of drug in 10 mL of solution

C. 500 μg of drug per gram of solution

D. 40 mg of drug in 0.5 mL of solution

Solutions.

A. 1 : 10,000 (w/v)

B. 1 : 40 (v/v)

C. 1 : 2000 (w/w)

D. 1 : 12.5 (w/v)

---

9. A 1 : 1000 solution of thimerosal (a solid) means that 1 g of thimerosal is contained in each 1000 mL of the solution. As with percentage solutions, the ratio strength of w/v systems is defined in terms of grams (of solid) and milliliters (of liquid). The concentration is

$$\frac{1\text{ g}}{1000\text{ mL}} = \frac{0.001\text{ g}}{1\text{ mL}} = \frac{0.1\text{ mg}}{100\text{ mL}} \text{ or } 0.1\%\text{ w/v}$$

What is the percentage strength of a 1 : 2000 w/v solution?

Solution. 0.05% w/v

CALCULATIONS

$$\frac{1\text{ g}}{2000\text{ mL}} = \frac{j}{100\text{ mL}}$$

$j = 0.05$ g; thus the concentration is 0.05%

---

10. If an injection contains 5 mg of drug per milliliter, what would be its ratio strength?

Solution. 1 : 200 w/v

CALCULATIONS

$$5\text{ mg/mL} = \frac{500\text{ mg}}{100\text{ mL}} = \frac{0.5\text{ g}}{100\text{ mL}} = \frac{1\text{ g}}{j}$$

$j = 200$ mL; thus the ratio strength is 1 : 200

---

11. What is the ratio strength of a 1/4% (w/v) solution?

Solution. 1 : 400 w/v

CALCULATIONS

$$\frac{1}{4}\% = \frac{0.25\ \text{g}}{100\ \text{mL}} = \frac{1\ \text{g}}{j}$$

$j = 400$; thus the ratio strength is 1 : 400 w/v

---

12. If 150 mg of strychnine sulfate are intimately mixed with 7.35 g of lactose, an inert substance, what is the ratio strength of strychnine sulfate in the mixture?

Solution. 1:50 w/w

CALCULATIONS

Total weight of the mixture:

$$7.35\ \text{g} + 0.15\ \text{g} = 7.50\ \text{g}$$

$$\frac{0.15\ \text{g}}{7.5\ \text{g}} = \frac{1}{j}$$

$j = 50$; the ratio strength is 1:50

---

13. What is the ratio strength of a 0.01% w/v solution? (Remember that w/v systems are defined in terms of grams and milliliters.)

Solution. 1 : 10,000 w/v

CALCULATIONS

$$\frac{0.01\text{ g}}{100\text{ mL}} = \frac{1}{j}$$

$j = 10{,}000$; the ratio strength is 1:10,000.

---

14. What is the ratio strength of a 0.5% solution of eucalyptus oil in mineral oil?

Solution. 1 : 200 v/v

CALCULATIONS

$$\frac{0.5\text{ mL oil}}{100\text{ mL solution}} = \frac{1\text{ mL oil}}{j}$$

$j = 200$ mL of solution

The solution is therefore 1 : 200 v/v.

---

15. How many milligrams of mercury bichloride are needed to make 200 mL of 1 : 500 w/v solution?

Solution. 400 mg

CALCULATIONS

By definition, 500 mL of solution contain 1 g.

$$\frac{1\text{ g}}{500\text{ mL}} = \frac{j}{200\text{ mL}}$$

$j = 0.4\text{ g} = 400\text{ mg}$

---

16. A pharmacist has 3.0 mL of an oil. How many milliliters of a 1 : 25 solution in alcohc can she prepare?

Solution. 75 mL

CALCULATIONS

$$\frac{1}{25} = \frac{3.0 \text{ mL oil}}{j}$$

$j = 75$ mL solution

---

17. A 1:4 mixture containing codeine sulfate, with lactose as an inert diluent, is prepared. How many milligrams of lactose are present in 1.25 g of the mixture?

Solution. 938 mg

Since 4 parts of the mixture contain 1 part of drug, they must also contain 3 parts of lactose. The concentration of lactose is therefore

$$\frac{3 \text{ parts}}{4 \text{ parts}} = 0.75$$

$0.75 \times 1.25 \text{ g} = 0.938 \text{ g} = 938 \text{ mg}$

---

18. Here are some more practice problems involving ratio strength.

A. How many liters of a 1:1500 solution can be made by dissolving 4.80 g of cetylpyridinium chloride in water?

B. How many grams of quatricaine are needed to prepare 500 mL of a 1:800 w/v solution?

C. How many grams of lactose should be combined with 140 mg of a drug to make a 1:10 dilution?

D. What is the ratio strength of a drug if 260 mL of solution contain 0.65 μL of solute?

E. What is the concentration (expressed as ratio strength) of phenylmercuric nitrate if 27.0 mL of the solution contain 45.0 mg of phenylmercuric nitrate?

F. The concentration of butyl alcohol in a whiskey is 0.04%. Express this concentration (1) in ppm; (2) in terms of ratio strength.

Solutions.

A. 7.2 L

B. 0.625 g

C. 1.26 g

D. 1 : 400,000

E. 1 : 600

F. (1) 400 ppm; (2) 1 : 2500

STOCK SOLUTIONS

19. Prescriptions and industrial formulas sometimes call for small quantities of drugs that are difficult or inconvenient to measure using the usual instruments. A stock solution may be used to deliver the drug, as in the following example.

A prescription for a skin lotion that uses alcohol as a base contains 0.63 g of menthol. The pharmacist has a 12% *stock solution* of menthol in alcohol. This solution can be used to provide the menthol needed for the lotion. Since, by definition, 10 mL of the stock solution contain 12 g of menthol, the volume containing 0.63 g would be found from

$$\frac{12.0\text{ g}}{100\text{ mL}} = \frac{0.63\text{ g}}{j}$$

$$j = 5.25\text{ mL}$$

5.25 mL of the menthol solution should be used for this prescription.

℞		
Potassium permanganate		0.1 g
Aqua dest.	q.s. ad	60.0 mL

How many milliliters of a 2.5% solution will yield the desired amount of potassium permanganate for this prescription?

Solution. 4.00 mL

CALCULATIONS

$$\frac{2.5\ \text{g}}{100\ \text{mL}} = \frac{0.1\ \text{g}}{j}$$

$j = 4.00$ mL

---

20. ℞

Boric acid	300 mg
Camphor water	7.0 mL
Aqua dest. q.s. ad	15.0 mL

How many milliliters of a 5% solution of boric acid in water should be used for this prescription?

Solution. 6.00 mL

CALCULATIONS

$$\frac{5\ \text{g}}{100\ \text{mL}} = \frac{0.3\ \text{g}}{j}$$

$j = 6.00$ mL

---

21. Now, try your hand at these.

A. ℞

Menthol	300 mg
Mineral oil q.s. ad	30.0 mL

How many milliliters of a 15% stock solution of menthol in mineral oil should be used to fill this prescription?

B. A pharmacist needs 1.95 grams of a salt for a prescription. How many milliliters of a 32.5% solution of the salt are needed?

C. ℞

Cocaine HCl	90.0 mg
Boric acid solution	3.0 mL
Aqua pur. q.s. ad	7.5 mL

How many milliliters of a 1:40 solution of cocaine HCl in water should be used for this prescription?

Solutions.

A. 2.0 mL
B. 6.0 mL
C. 3.60 mL

CALCULATIONS

A. $\frac{15\text{ g}}{100\text{ mL}} = \frac{0.3\text{ g}}{j}$

$j = 2.0\text{ mL}$

B. $\frac{32.5\text{ g}}{100\text{ mL}} = \frac{1.95\text{ g}}{j}$

$j = 6.0\text{ mL}$

C. $\frac{1\text{ g}}{40\text{ mL}} = \frac{0.09\text{ g}}{j}$

$j = 3.60\text{ mL}$

TRITURATIONS

22. A drug *trituration* is also a stock preparation, one in which a solid drug is intimately dispersed with a solid, inert diluent. Since both components are solids, they are measured by weight so that *triturations are always w/w systems.*

What is the concentration, in ratio strength, of a trituration made by combining 120 mg of atropine sulfate and 3.48 g of lactose?

Solution. 1 : 30

CALCULATIONS

Remember that the ratio strength of a trituration is expressed as a ratio of active ingredient to total weight. In this case, the total weight is

$$0.120\ \text{g} + 3.48\ \text{g} = 3.60\ \text{g}$$

$$\frac{0.120\ \text{g}}{3.60\ \text{g}} = \frac{1}{j}$$

$$j = 30$$

The concentration is 1 : 30

---

23. ℞

Strychnine sulfate	45 mg
Belladonna extract	360 mg
Sucrose	36.0 g
Ft. chart; Div. in #XXXVI	

How many grams of a 1 : 80 trituration of strychnine sulfate should be used for this prescription?

Solution. 3.60 g

CALCULATIONS

We wish to know the quantity of trituration that contains 45 mg of strychnine sulfate. One approach is to use proportion:

$$1{:}80 = \frac{1}{80} = \frac{45\text{ mg}}{j}$$

$$j = 3600\text{ mg} = 3.60\text{ g}$$

Another possibility that leads to the same result is to rearrange the concentration equation

$$\text{concentration} = \frac{\text{amount of solute}}{\text{amount of preparation}}$$

to

$$\text{amount of preparation} = \frac{\text{amount of solute}}{\text{concentration}}$$

In the example,

$$\frac{45\text{ mg}}{\frac{1}{80}} = 3600\text{ mg} = 3.60\text{ g}$$

---

24. How many grams of a 1 : 25 trituration of saccharin sodium are needed to make 30 mL of a solution containing 2 mg of saccharin sodium in each milliliter?

Solution. 1.50 g

CALCULATIONS

Total saccharin sodium needed:

$$2\text{ mg/mL} \times 30\text{ mL} = 60\text{ mg}$$

$$\frac{60\text{ mg}}{\frac{1}{25}} = 1500\text{ mg} = 1.50\text{ g}$$

---

25. ℞

Atropine sulfate	0.4 mg
Sodium bicarbonate	0.4 g
Pepsin	0.2 g
Lactose q.s.	
d.t.d. chart No. 30	

How many milligrams of a 1:40 trituration of atropine sulfate (in lactose) should be weighed for this prescription?

Solution. 480 mg

Total quantity of atropine sulfate needed:

$$\frac{0.4\text{ mg}}{\text{chart}} \times 30\text{ chart} = 12\text{ mg}$$

$$\frac{12\text{ mg}}{\frac{1}{40}} = 480\text{ mg}$$

---

26. It is necessary to prepare a 1:15 trituration of emetine sulfate in sucrose. How much sucrose should be combined with 200 mg of emetine sulfate to make this trituration?

Solution. 2800 mg

CALCULATIONS

$$\frac{200\text{ mg}}{\frac{1}{15}} = 3000\text{ mg total trituration}$$

$$3000\text{ mg} - 200\text{ mg} = 2800\text{ mg}$$

---

27. A pharmacist wishes to prepare a 1:12 trituration of codeine phosphate, using lactose as the diluent. How many grams of lactose should be used to make the trituration if the pharmacist employs the smallest quantity of codeine phosphate that can be weighed on a prescription balance with acceptable accuracy?

Solution. 1.32 g

CALCULATIONS

The minimum weighable quantity on the prescription balance is 120 mg.

$$\frac{120 \text{ mg}}{\frac{1}{12}} = 1440 \text{ mg}$$

$$1440 \text{ mg} - 120 \text{ mg} = 1320 \text{ mg} = 1.32 \text{ g lactose}$$

---

28. Try these problems for review and practice.

A. A prescription calls for 16 mg of atropine. How many grams of a 1 : 30 dilution of atropine in lactose will supply the needed amount of atropine?

B. What is the ratio strength of a trituration made by combining 0.25 g of a drug with 2.75 g of lactose?

C. A pharmacist needs 5 mg of a drug. How many milligrams of a 1 : 30 dilution of the drug should be used?

Solutions.

A. 0.48 g

B. 1 : 12

C. 150 mg

REVIEW PROBLEMS

Do all of the following review problems before checking your answers.

A. How many milligrams of Zephiran chloride are required to make 1 L of a 1 : 750 solution in water?

B. A solution of potassium permanganate (1:2500) is used as a fungicide on a turtle's shell. If the shell is of such size as to require 1.2 mL of the solution for complete coverage, how many micrograms of potassium permanganate will be put on the shell?

C. How much kaolin should be used to prepare 2.00 g of a 1:30 dilution of hyoscine in kaolin? Both substances are solids.

D. What is the ratio strength of a solution in which 18.0 mg of solute is dissolved in sufficient water to make a total of 450 mL?

E. How many liters of a 1:800 solution can be made by dissolving 0.60 g of a drug in water?

F. The concentration of a sweetener in a solution is 1:400 w/v. How many milligrams of the sweetener does each teaspoonful contain?

G. The concentration of an antioxidant in an ointment is 1:250. How many milligrams of the antioxidant does each gram of the ointment contain?

H. How many liters of a 1:500 solution of cinnamon oil in alcohol can be made from 27.5 mL of cinnamon oil?

I. How many milliliters of a 4% stock solution of silver nitrate contain 150 mg of silver nitrate?

J. How many grams of lactose should be added to 120 mg of strychnine sulfate to make a 1:20 trituration?

K. How many milligrams of a 1:25 trituration of atropine should be used to prepare 40 capsules, each containing 0.4 mg of atropine?

L. A prescription for capsules calls for 18 mg of a drug. Calculate the maximum concentration of a trituration of simple ratio that may be used as a source of the drug.

M. The usual daily dose of a drug is 2.5 mg/kg/day divided into two equal doses. How many milliliters of a 12.5% (w/v) solution of drug should be added to a 100 mL bag of normal saline and administered to a patient weighing 110 lb to provide *one dose*?

N. ℞ Hydrocortisone

Hexachlorophene	aa	1/4 %
Coal tar solution (1:25, w/v)		10 ml
Hydrophilic ointment	ad	120 g

M. ft. 200 g

(a) How many tablets, each containing 20 mg of hydrocortisone, should be used in preparing this prescription?

(b) If you have on hand coal tar solution 20% (w/v), how many milliliters would you use to prepare the prescription?

O. ℞

Menthol		1 : 250 (w/w)
Hexachlorophene		1/8% (w/w)
Hydrophilic ointment	ad	60 g

Mix and divide into individual blisters containing 2 grams each

Sig. Apply one blister t.i.d. on affected area.

(a) How many milligrams of menthol will be present in each blister?

(b) How many milligrams of hexachlorophene will the patient receive per day?

P. A patient is to receive 45 mg of a drug. How many milliliters of a 1 : 20 (w/v) solution of the drug should be used?

Q. What is the ratio strength, expressed as w/w, of a solution prepared by dissolving 50 g of magnesium sulfate in 150 mL of water?

R. What is the resultant ratio strength of an ointment prepared by combining 10 g of 10% betamethasone valerate ointment, 40 g of 5% betamethasone valerate ointment and 10 g of ointment base?

S. A syrup is preserved with 0.05% (w/v) of methylparaben and 0.01% (w/v) sodium bisulfite. Express these concentrations as ratio strength.

T. A skin test involves the intradermal injection of 0.25 mL of a 1:1000 (w/v) dilution of an allergen. How many micrograms would be administered?

Solutions to Review Problems

A. 1330 mg

B. 480 µg

C. 1.93 g

D. 1:25,000

E. 0.48 L

F. 12.5 mg

G. 4 mg

H. 13.8 L

I. 3.75 mL

J. 2.28 g

K. 400 mg

L. 1:7

M. 0.5 mL

N. **(a)** 25 tablets
(b) 3.3 mL

O. **(a)** 8 mg/blister
(b) 7.5 mg/day

P. 0.9 mL

Q. 1:4 w/w

R. 1:20 w/w

S. 1:2,000 and 1:10,000

T. 250 μg

Calculations for Review Problems

A. $\frac{1\text{ g}}{750\text{ mL}} = \frac{j}{1000\text{ mL}}$

$j = 1.33\text{ g} = 1330\text{ mg}$

B. $\frac{1\text{ g}}{2500\text{ mL}} = \frac{j}{1.2\text{ mL}}$

$j = 4.8 \times 10^{-4}\text{ g} = 4.8 \times 10^{2}\ \mu\text{g}$

C. The concentration of kaolin in the mixture is 29/30 w/w:

$\frac{29}{30} = \frac{j}{2.00}$

$j = 1.93\text{ g}$

D. $\frac{0.018\text{ g}}{450\text{ mL}} = \frac{1\text{ g}}{j}$

$j = 25{,}000\text{ mL}$

concentration = 1:25,000

E. $\frac{1\text{ g}}{800\text{ mL}} = \frac{0.60\text{ g}}{j}$

$j = 480\text{ mL} = 0.48\text{ L}$

F. $\frac{1\text{ g}}{400\text{ mL}} = \frac{j}{5\text{ mL}}$

$j = 0.0125\text{ g} = 12.5\text{ mg}$

G. $\frac{1\text{ g antiox.}}{250\text{ g oint.}} = \frac{j}{1\text{ g oint.}}$

$j = 0.004\text{ g} = 4\text{ mg}$

H. $\frac{1\text{ mL oil}}{500\text{ mL solution}} = \frac{27.5\text{ mL oil}}{j}$

$j = 13750\text{ mL} = 13.8\text{ L}$

I. $\frac{0.150\text{ g}}{4\text{ g}/100\text{ mL}} = 3.75\text{ mL}$

J. $\frac{120\text{ mg}}{\frac{1}{20}} = 2400\text{ mg trituration}$

$2400\text{ mg} - 120\text{ mg} = 2280\text{ mg} = 2.28\text{ g}$

K. $0.4\text{ mg/caps} \times 40\text{ cap} = 16\text{ mg}$

$\frac{16\text{ mg}}{\frac{1}{25}} = 400\text{ mg}$

L. $18\text{ mg} \times 7 = 126\text{ mg}$

1:7 may be used.

M. $1.25\text{ mg/kg} \times \dfrac{1\text{ kg}}{2.2\text{ lb}} \times 110\text{ lb} \times \dfrac{100\text{ mL}}{12500\text{ mg}} = 0.5\text{ mL}$

N. **(a)** 1/4% in 120 g = 1/4% in 200 g

$$200\text{ g} \times \frac{0.25\text{ g}}{100\text{ g}} = 0.5\text{ g hydrocortisone} = 500\text{ mg}$$

$$500\text{ mg} \times \frac{\text{tab}}{20\text{ mg}} = 25\text{ tabs.}$$

(b) $\dfrac{1\text{ g}}{25\text{ mL}} \times 10\text{ mL} = 0.4\text{ g}$

$$\frac{0.4\text{ g hexac.}}{120\text{ g oint.}} \times 200\text{ g oint.} = 0.66\text{ g hexac. needed}$$

$$0.66\text{ g} \times \frac{100\text{ mL}}{20\text{ g}} = 3.3\text{ mL}$$

O. **(a)** $\dfrac{1\text{ g}}{250\text{ g}} \times 2\text{ g} = 0.008\text{ g} = 8\text{ mg}$

(b) $\dfrac{2\text{ g}}{\text{blister}} \times 3 = 6\text{ g/day}$

$$6\text{ g} \times \frac{0.125\text{ g}}{100\text{ g}} = 0.0075\text{ g} = 7.5\text{ mg}$$

P. $45\text{ mg} \times \dfrac{20\text{ mL}}{1000\text{ mg}} = 0.9\text{ mL}$

Q. $\dfrac{50\text{ g}}{200\text{ g}} = \dfrac{1\text{ g}}{j}$

$j = 4$, thus 1:4 w/w

R. 60 g = total ointment mixture

$10\text{ g} \times 0.10 = 1\text{ g}$ bet.

$40\text{ g} \times 0.05 = 2\text{ g}$ bet.

3 g = total betamethasone

$\dfrac{3\text{ g}}{60\text{ g}} = \dfrac{1\text{ g}}{j}$

$j = 20$, thus 1:20 w/w

S. $0.50\%\text{ w/v} = \dfrac{0.05\text{ g}}{100\text{ mL}} = \dfrac{1\text{ g}}{j}$

$j = 2{,}000$, thus 1:2000 w/v

$0.01\%\text{ w/v} = \dfrac{0.01\text{ g}}{100\text{ mL}} = \dfrac{1\text{ g}}{j}$

$j = 10{,}000$; 1:10,000 w/v

T. $0.25\text{ mL} \times \dfrac{1\text{ g}}{1000\text{ mL}} = 0.00025\text{ g} = 250\,\mu\text{g}$

CHAPTER 7 DILUTION AND CONCENTRATION

***Learning Objectives:** After completing this chapter the student should be able to:*

1. Calculate the amount of a preparation to be diluted to yield a preparation of lower strength.
2. Determine the quantity of a preparation to be combined with another preparation containing the same active ingredient to yield a preparation of intermediate strength.
3. Determine the concentration of a mixture prepared by combining preparations of different concentration.
4. Calculate the amount of active ingredient needed to prepare a concentrate that is to be diluted by the patient prior to use.
5. Solve problems involving concentration and dilution of pharmaceutical preparations through several different approaches.

Today's pharmacists are being prepared to be patient-focused clinical practitioners. For this reason, there has been an increase in prescriptions of individualized doses of various medications considering the unique characteristics of each specific patient. Individualized therapy frequently requires dilution or compounding of quality dosage forms by the pharmacist. Dilution of liquid medications for oral or parenteral use, reconstitution of dry powders, reduction of the strength of an ointment or cream, and re-encapsulation of drugs to allow a lower strength of a drug preparation, are all practices requiring knowledge and skills in calculations from today's pharmacists. The main reason for individualized therapy is the need of different dosage forms and/or different strengths from available manufactured products.

In previous chapters we have dealt with preparations in which the drug content is expressed as a *concentration*. It may be necessary to dilute concentrated systems prior to use. Sometimes the dilution is performed by the pharmacist and sometimes by the patient. On occasion, the strength of an active ingredient may have to be raised. Sometimes, components of varying strength must be blended to arrive at a product whose strength satisfies a required standard. All of these problems may be handled by essentially the same calculation technique.

Pharmaceutical Calculations, Fourth Edition, By Joel L. Zatz and Maria Glaucia Teixeira
ISBN 0-471-67623-3

---

1. An important equation for us in this chapter is

Quantity of solute = concentration × quantity of preparation

Calculate the amount of drug present in 300 mL of a 4.0% w/v solution of that drug.

Solution. $0.040\,\text{g/mL} \times 300\,\text{mL} = 12\,\text{g}$

---

2. Using the same equation, calculate the amount of alcohol present in 35 mL of a 60% solution of alcohol in water.

Solution. $0.60 \times 35\,\text{mL} = 21\,\text{mL}$

---

3. Calculate the amount of calamine in 13 g of a 10% w/w calamine ointment.

Solution. $(0.10)(13\,\text{g}) = 1.3\,\text{g}$

---

4. Calculate the number of milligrams of potassium permanganate in 90 mL of a 1:500 w/v potassium permanganate solution.

Solution. 180 mg

CALCULATIONS

$$90\text{ mL} \times \frac{1\text{ g}}{500\text{ mL}} = 0.18\text{ g} = 180\text{ mg}$$

---

5. Try these problems.

A. How many milliliters of solute are there in,

(1) 350 mL of 5% v/v solution?

(2) 4.20 L of a 1:2000 v/v solution?

B. How many grams of solute are there in

(1) 220 g of a 1:40 w/w solution?

(2) 170 mL of a 15.0% w/v solution?

C. How many milligrams of drug are there in

(1) 25.0 mL of a 3.25% w/v solution?

(2) 900 g of a 1:150 w/w mixture?

Solutions.

A. (1) 17.5 mL

(2) 2.1 mL

B. (1) 5.5 g

(2) 25.5 g

C. (1) 813 mg

(2) 6000 mg

PROBLEM SOLVING APPROACHES

6. You will be presented now with a simple and clear-cut way to deal with dilution and concentration of pharmaceutical preparations. In case your learning style does not correlate with algebraic methods, you may "connect" to one of the alternative techniques that we will provide later in this chapter. Some of these mathematical methods will require a few more steps during calculations but if you feel more comfortable and confident when using them, there is no problem. We will present the basics of each approach and follow with some examples solved by each one of the methods proposed.

ALGEBRAIC METHODS

Mass Balance Equation

7. It is sometimes necessary for the pharmacist to dilute preparations to make a product of lower strength. Here is an example:

℞ Ichthammol ointment 8% w/w 90.0 g

Ft. ung.

The pharmacist has some 20% w/w ichthammol ointment. He wishes to dilute it with petrolatum, an inert semisolid that contains no drug of any kind, to the proper strength. The quantity of the 20% ointment and of petrolatum to be mixed must be determined.

The key to solving problems of this type is the realization that the amount of an ingredient in the finished product must be equal to the contributions of the components of the formula. In terms of our example, we know that the ichthammol in our final product, the 8% ointment, must be the sum of the ichthammol contributed by the 20% ichthammol ointment and by the petrolatum. We may therefore write

g ichthammol from 20% ointment + g ichthammol from petrolatum = g ichthammol in 8% ointment

To conserve space, we are going to rewrite this equation, which is called a *mass balance equation*, using a bit of shorthand. The preparation that contains the ingredient will be written in parentheses and "ichthammol" will be abbreviated as "ich."

$$\text{g ich (20\% oint)} + \text{g ich (petrolatum)} = \text{g ich (8\% oint)}$$

Of course, there is no ichthammol in petrolatum, so our equation becomes

$$\text{g ich (20\% oint)} = \text{g ich (8\% oint)}$$

Recall that,

$$\text{g ich} = \text{(concentration)(quantity of preparation)}$$

Let j equal the quantity of 20% ointment to be used:

$$\text{g ich (20\% oint)} = 0.20j$$

$$\text{g ich (8\% oint)} = (0.08)(90.0\,\text{g}) = 7.2\,\text{g}$$

Substituting these quantities in our mass balance equation yields

$$(0.20)j = 7.20\,\text{g}$$

$$j = \frac{7.20\,\text{g}}{0.20} = 36\,\text{g}$$

Use 36 g of the 20% ichthammol ointment; the amount of petrolatum needed is 90 g − 36 g = 54 g.

---

8. In summary, to solve problems involving dilution or concentration:

(1) Write the mass balance equation.

(2) Substitute in the mass balance equation.

(3) Solve the resulting algebraic expression.

Now try this problem.

We wish to dilute an ointment containing 14% sulfur with petrolatum to make 60 g of an ointment containing 10% sulfur. How many grams of 14% sulfur ointment and how many grams of petrolatum will be necessary to make the dilution?

(1) Write a mass balance equation that shows where the sulfur in the finished product (the 10% ointment) comes from.

Solution. g S (14% oint) = g S (10% oint) (petrolatum contains no sulfur).

---

9. Calculate the amount of sulfur in the 10% ointment using the equation quantity of solute = concentration × quantity of preparation

---

Solution. g S (10% oint) = (0.100)(60 g) = 6 g

CALCULATIONS

Let j equal the quantity of 14% ointment to be used. Calculate the number of grams of sulfur in the 14% ointment.

g S (14% oint) = $0.14j$

(2) Substitute in the mass balance equation and **(3)** solve.

$0.14j = 6\,\text{g}$

$j = 42.9\,\text{g}$

Use 42.9 g of the 14% sulfur ointment. (The amount of petrolatum is 60.0 g − 42.9 g = 17.1 g)

---

10. It is necessary to prepare 180 mL of a 1 : 200 solution of potassium permanganate ($KMnO_4$). What quantity of a 5% stock solution of $KMnO_4$ should be diluted with water?

Fill in the blank in the mass balance equation:

____________________ = g $KMnO_4$ (1 : 200 sol)

Solution. g $KMnO_4$ (5% sol)

---

11. Let j equal the number of milliliters of 5% $KMnO_4$ solution to be diluted. Fill in the blanks:

A. g $KMnO_4$ (5% sol) = __________

B. g $KMnO_4$ (1 : 200 sol) = ____________

Solutions.

A. $(0.05\,\text{g/mL})j$

B. $\dfrac{1\,\text{g}}{200\,\text{mL}} \times 180\,\text{mL} = 0.9\,\text{g}$

---

12. Now, substitute in the mass balance equation and solve.

Solution. Use 18 mL of the 5% solution.

CALCULATIONS

$(0.05\,\text{g/mL})j = 0.9\,\text{g}$

$$j = \frac{0.9\text{ g}}{0.05\text{g/mL}} = 18\text{ mL}$$

---

13. How many milliliters of a 10% w/v Merthiolate solution should be diluted with water to make 440 mL of a 0.25% w/v Merthiolate solution?

Solution. 11 mL

CALCULATIONS

Let j equal the milliliters of 10% Merthiolate solution:

g merth (10%) = g merth (0.25%)

$(0.10\,\text{g/mL})j = (440\,\text{mL})(0.0025\,\text{g/mL}) = 1.1\,\text{g}$

$j = 11\,\text{mL}$

---

14. How many milliliters of water must be added to 180 mL of 36% w/v acetic acid solution in order to make up a solution of 10% w/v strength? (Assume that no shrinkage or expansion of volume occurs on mixing.)

Solution. 468 mL

CALCULATIONS

Let j equal the quantity of water to be added. The volume of the 10% solution will then be 180 mL + j.

g acetic acid (36% sol) = g acetic acid (10% sol)

$(0.36\,\text{g/mL})(180\,\text{mL}) = (0.1\,\text{g/mL})(180\,\text{mL} + j)$

$64.8\,\text{g} = 18.0\,\text{g} + (0.1\,\text{g/mL})j$

$$j = \frac{64.8\text{ g} - 18.0\text{ g}}{0.1\text{g/mL}} = 468\text{ mL}$$

---

15. How many milliliters of Zephiran concentrate (17% w/v of Zephiran) are required to prepare 2L of a 1:1500 solution of Zephiran?

Solution. 7.84 mL

CALCULATIONS

Let j equal the milliliters of Zephiran concentrate needed:

g Zephiran (17% sol) = g Zephiran (1:1500 sol)

$(0.17\,\text{g/mL})j = (1/1500\,\text{g/mL})(2000\,\text{mL})$

$j = 7.84\,\text{mL}$

---

16. If 12.5 g of a 10% zinc oxide ointment are diluted with 17.5 g of white petrolatum, what is the percentage strength of the resulting product?

Solution. 4.17%

CALCULATIONS

Let j equal the concentration of the diluted ointment:

g zinc oxide (10% oint) = g zinc oxide (dil oint)

$(0.1)(12.5\,\text{g}) = (j)(30.0\,\text{g})$

$$j = \frac{(0.1)(12.5\ \text{g})}{30.0\ \text{g}} = 0.0417 = 4.17\%$$

---

17. Hydrochloric acid USP is a 36% w/w solution of HCl in water with a specific gravity of 1.18. How many milliliters should be used to prepare 200 mL of a 5% w/v solution of HCl?

Solution. 23.6 mL

(If you ran into trouble, see the hint and solution presented below. If you were successful, continue after the solution.)

Hint. The mass balance equation is

g HCl (36% w/w sol) = g HCl (5% w/v sol)

We cannot set j equal to volume of 36% w/w HCl solution and multiply 0.36 by j to get the amount of HCl. j is a volume; 0.36 refers to concentration on a weight basis. The two quantities are just not compatible. However, it is possible to calculate the weight of the 36% solution and then convert that to volume using its density. Try the problem again. If you still have difficulty, consult the solution.

CALCULATIONS

Let j equal the weight of hydrochloric acid needed:

$$0.36j = (0.05\ \text{g/mL})(200\ \text{mL})$$

$$j = \frac{(0.05\ \text{g/mL})(200\ \text{mL})}{0.36} = 27.8\ \text{g}$$

$$27.8\ \text{g} \times \frac{1\ \text{mL}}{1.18\ \text{g}} = 23.6\ \text{mL}$$

Use 23.6 mL of hydrochloric acid USP.

---

18. A pharmacist wishes to prepare 150 mL of ammonia (NH_3) solution, 10% w/v. How many milliliters of strong ammonia solution (28.5% w/w) should be used? (Sp g of strong ammonia solution is 0.900.)

Solution. 58.4 mL

CALCULATIONS

Let j equal the weight of 28.5% solution to be used:

g NH_3 (28.5% w/w sol) = g NH_3 (10% w/v sol)

$$0.285j = (0.1\ \text{g/mL})(150\ \text{mL})$$

$$j = 52.6\ \text{g}$$

$$52.6\ \text{g} \times \frac{1\ \text{mL}}{0.900\ \text{g}} = 58.4\ \text{mL}$$

---

19. A crude drug is required to contain 0.28% w/w of an alkaloid, the active compound. How many kilograms of a batch of crude drug containing 0.30% w/w alkaloid must be combined with 500 g of crude drug containing 0.20% w/w alkaloid in order that the resulting mixture will meet the required standard? (*Hint*: Use the same procedure as for previous problems in this chapter; write the mass balance equation and then substitute in it.)

Solution. 2.0 kg

CALCULATIONS

The mass balance equation is

g alkaloid (0.20%) + g alkaloid (0.30%) = g alkaloid (0.28%)

Let j equal the quantity of 0.30% crude drug required. Can you go on from here? Try the problem again. If you still have trouble, go back to frame 4.

Let j equal the quantity of 0.30% crude drug necessary:

g alkaloid (0.20%) = (500 g)(0.0020)

g alkaloid (0.30%) = $0.0030j$

g alkaloid (0.28%) = (500 g + j)(0.0028)

Now complete the solution.

The solution is as follows:

g alkaloid (0.20%) + g alkaloid (0.30%) = g alkaloid (0.28%)

$(500\,\text{g})(0.0020) + 0.0030j = (500\,\text{g} + j)(0.0028)$

$1.0\,\text{g} + 0.0030j = 1.4\,\text{g} + 0.0028j$

$j = 2000\,\text{g} = 2.0\,\text{kg}$

---

20. How many milliliters of syrup containing 85.0% w/v sucrose should be mixed with 115 mL of syrup containing 60.0% w/v sucrose to prepare a syrup containing 76.0% w/v sucrose? (Assume that there is no expansion or shrinkage of volume when the two liquids are mixed.)

Solution. 204 mL

CALCULATIONS

Let j equal the volume of 85% sucrose:

g sucrose (85%) + g sucrose (60%) = g sucrose (76%)

$(0.85\,g/mL)(j) + (0.60\,g/mL)(115\,mL) = (0.76\,g/mL)(115\,mL + j)$

$0.85j + 69\,g = 87.4\,g + 0.76j$

$j = 204\,mL$

---

21. ℞ Belladonna tincture 20.0 mL (67% C_2H_5OH)

Elixir phenobarbital 70.0 mL (15% C_2H_5OH)

Alcohol USP qs

Syrup tolu qs ad 180.0 mL

How much alcohol USP (95% C_2H_5OH) should be used in this prescription in order that the concentration of ethanol (C_2H_5OH) in the finished product be 20% v/v?

Solution. 12.7 mL

CALCULATIONS

Let j equal the milliliters of alcohol USP needed:

mL eth (67%) + mL eth (15%) + mL eth (95%) = mL eth (product)

$(20.0\,mL)(0.67) + (70.0\,mL)(0.15) + 0.95j = (180)(0.20)$

$13.4\,mL + 10.5\,mL + 0.95j = 36.0\,mL$

$j = 12.7\,mL$

---

22. If 300 g of an ointment containing 4% (w/w) sulfur is combined with 220 g of a 10% (w/w) sulfur ointment, what is the strength of the mixture that results?

Solution. 6.54% w/w

CALCULATIONS

Let j equal the strength of finished product:

g S (4% oint) + g S (10% oint) = g S (mixture)

(300 g)(0.04) + (220 g)(0.1) = (520 g)(j)

$j = 0.0654 = 6.54\%$

---

23. A pharmacist combines 140 mL of a 0.90% sodium chloride solution with 250 mL of 3.4% sodium chloride solution. Assuming no expansion or contraction in volume, calculate the percentage strength of the mixture.

Solution. 2.5%

CALCULATIONS

Let j equal the concentration of the mixture:

g NaCl (0.90%) + g NaCl (3.4%) = g NaCl (mixture)

(140 mL)(0.0090 g/mL) + (250 mL)(0.034 g/mL) = (390 mL)(j)

$j = 0.025$ g/mL = 2.5%

---

24. Try these practice problems that review the material covered so far. Check your answers at the end.

A. If 4.0 mL of a 5.0% benzethonium chloride solution are diluted with water to 500 mL, what will the ratio strength of the resulting solution be?

B. How much 14% sodium solution should be diluted with water to make 350 mL of 6% sodium acetate solution?

C. How much lanolin (an inert ointment base) should be added to 300 g of cortisone ointment, 3%, to make an ointment containing cortisone, 1:250?

D. A pharmacist needs 60 mL of diluted acetic acid (a 6% w/v solution of $C_2H_4O_2$ in water) for a prescription. The pharmacist has only acetic acid USP on hand. This is a 36% w/w solution of $C_2H_4O_2$ in water with a specific gravity of 1.045. How many milliliters of acetic acid USP should the pharmacist use?

E. How much pure zinc oxide should be mixed with a 10.0% zinc oxide ointment to make 3.75 kg of a 12.0% zinc oxide ointment?

F. Three samples of a plant have the following potencies:

Sample 1 (220 g) contains 2.40% alkaloids

Sample 2 (50.0 g) contains 1.97% alkaloids

Sample 3 (450 g) contains 3.85% alkaloids

If the three samples are combined, what is the alkaloid content, expressed as a percent, of the mixture?

G. What is the percentage of benzalkonium chloride in a solution made by mixing 350 mL of a 1:100 benzalkonium chloride solution with 250 mL of a 1:300 benzalkonium chloride solution?

H. ℞

Phenobarbital elixir (15% C_2H_5OH)	30 mL
High-alcoholic elixir (78% C_2H_5OH)	qs
Elixir terpin hydrate (42% C_2H_5OH)	45 mL
Syrup qs ad 120 mL	

How many milliliters of high-alcoholic elixir should be used to make the C_2H_5OH content of the final solution 35%?

Solutions.

A. 1:2500
B. 150 mL
C. 1950 g
D. 9.6 mL
E. 83.3 g
F. 3.27%
G. 0.722%
H. 23.8 mL

Concentration × Quantity: The CQ Equation

25. The mass balance method gives us a general approach that can be used to solve any type of problem involving dilution, concentration and mixtures. The concentration × quantity (CQ) equation is a short cut that can be derived from the general mass balance approach and is useful in solving many dilution problems.

On every occasion that a pharmaceutical preparation is *diluted* with an inert diluent that contains no drug of any kind, the amount of the active ingredient (the drug) remains constant. Consequently, any increase in the quantity of preparation caused by addition of diluent only will bring a reduction in its concentration.

For example,

20% NaCl solution = 20 grams of NaCl per 100 mL of solution in water.

If 100 mL of water is added:

$$20\,\text{g}/100\,\text{mL} + 100\,\text{mL water} = 20\,\text{g}/200\,\text{mL} = 10\,\text{g}/100\,\text{mL} = 10\%$$

You can see that as the preparation's quantity is doubled, its concentration is reduced by half. This principle, in which the final concentration is inversely proportional to the quantity of solution prepared, is expressed in the following equation:

$$C_1\,Q_1 = C_f\,Q_f$$

where

C_1 = drug concentration in the original preparation
Q_1 = quantity of the original preparation
C_f = final concentration after dilution
Q_f = final quantity after dilution

From this equation, the strength of a solution or the amount of solution of a desired strength can be determined from an original solution. It applies to both the determination of concentrated and diluted solutions.

Now, let us apply this equation to one of the dilution problems we considered earlier (beginning of frame 7).

℞ Ichthammol ointment 8% w/w 90.0 g
Ft. ung.

The pharmacist has some 20% w/w ichthammol ointment. He wishes to dilute it with petro latum, an inert semisolid that contains no drug of any kind, to the proper strength. The quantity of the 20% ointment and of petrolatum to be mixed must be determined.

The use of the CQ equation considers that you have a volume of a known strength and you need to prepare another different strength.

In our example, we know that the ichthammol in our final product (8% ointment) will come from the 20% ichthammol ointment. We may therefore write,

$$C_1 Q_1 = C_f Q_f$$

$$(20\%)(Q_1) = (8\%)(90\,g)$$

$$Q_1 = 36\,g$$

Use 36 g of the 20% ichthammol ointment; the amount of petrolatum needed is 90 g − 36 g = 54 g.

---

26. On some occasions, the pharmacist needs to determine the amount of diluted solution of a desired strength that can be made from a certain volume of a concentrated solution. The following exemplifies this situation.

A pharmacist has 100 mL of a 0.25% (w/v) stock solution. How many milliliters of a 1:2000 (w/v) solution can be made from the stock?

$$1{:}2000 = 0.05\%$$

Using the CQ equation,

$$C_1 Q_1 = C_f Q_f$$

$$(0.25\%)(100\,mL) = (0.05\%)C_f$$

$$C_f = 500\,mL$$

Now try the following problems using the CQ equation. The first two problems were also solved previously by mass balance equation and you can compare both methods.

We wish to dilute an ointment containing 14% sulfur with petrolatum to make 60 g of an ointment containing 10% sulfur. How many grams of 14% sulfur ointment and how many grams of petrolatum will be necessary to make the dilution?

Solution. Use 42.9 g of the 14% sulfur ointment. The amount of petrolatum is 17.1 g.

CALCULATIONS

$(14\%)(Q_1) = (10\%)(60\,g)$

$Q_1 = 42.9\,g$ of 14%

Petrolatum: 60 g − 42.9 g = 17.1 g of petrolatum

27. How many milliliters of a 10% w/v Merthiolate solution should be diluted with water to make 440 mL of a 0.25% w/v Merthiolate solution?

Solution. 11 mL

CALCULATIONS

$(0.25\%)(440\,\text{mL}) = (10\%)Q_1$

$Q_1 = 11\,\text{mL}$ of 10%

Water = 440 − 11 = 429 mL

28. How many milliliters of a 1:2000 (w/v) solution of sodium chloride can be made from 120 mL of a 0.5% solution?

Solution. 1200 mL

CALCULATIONS

1:2000 = 0.05%

$(120\,\text{mL})(0.5\,\%) = Q_f(0.05\%)$

$Q_f = 1200\,\text{mL}$

29. If 250 mL of a syrup containing 50% (w/v) of sucrose are diluted to a liter with water, what will be the percentage strength? (Assume that there is no expansion or shrinkage of volume when the two liquids are mixed.)

Solution. 12.5%

CALCULATIONS

$(250\,\text{mL})(50\%) = (1000\,\text{mL})C_f$

$C_f = 12.5\%$

30. For more practice with the CQ equation, try the following problems.

A. A pharmacist diluted 200 mL of a 20% (w/v) aqueous solution to 1 liter. What will be the percentage strength of the resultant solution?

B. How many milliliters of water should be added to a 50 mL vial of 23.4% NaCl additive solution to prepare a 0.9% sodium chloride solution?

C. How many milliliters of sterile water for injection (SWFI) should be added to 500 mL of 50% (w/v) dextrose solution to reduce the concentration to 30% (w/v)?

D. ℞ Aluminum acetate solution (1:20 w/v) 50 mL

Sig. Apply to affected area 2x day.

If the pharmacist has on hand a 12.5% (w/v) aluminum acetate solution, how many milliliters should be used to fill the prescription?

E. How many milliliters of a 25% (w/v) stock solution of a drug are needed to prepare 50 mL of a solution containing 20 mg of drug per milliliter?

Solutions.

A. 4%

B. 1250 mL of water

C. 750 mL of water

D. 20 mL of 12.5%

E. 4 mL of 25%

CALCULATIONS

A. (20%)(200 mL) = j(1000 mL)

j = 4%

B. (23.4%)(50 mL) = (0.9%)j

j = 1300 mL of 0.9%

1300 mL − 50 mL = 1250 mL of water

C. (500 mL)(50%) = j(20%)

j = 1250 mL of 20%

Amount of SWFI = 1250 mL − 500 mL = 750 mL.

D. 1 : 20 = 5%

(5%)(50 mL) = (12.5%)j

j = 20 mL

E. 20 mg/mL = 2000 mg/100 mL = 2%

(50 mL)(2%) = j(25%)

j = 4 mL

---

31. Similarly to the mass balance equation, the CQ equation is in fact quite flexible and can be expanded to deal with the situation in which preparations containing different drug strengths are mixed together. In that case, the equation becomes,

$$C_1 Q_1 + C_2 Q_2 = C_f Q_f$$

where the numbers refer to individual components of the mixture and the dotted lines indicate that more CQ terms may be added if more than 2 preparations are combined. C_f and Q_f will reflect the volume of all components together and the resultant strength, respectively.

An application of this expanded formula can be seen in the following example.

If a pharmacist combines 20 mL azithromycin oral suspension 100 mg/5 mL and 30 mL of 200 mg/5 mL, what will be the concentration of azithromycin in the finished product?

$$C_1 Q_1 + C_2 Q_2 \ldots = C_f Q_f$$

If we consider

$$C_1 = 100\text{ mg/5 mL suspension}$$

$$Q_1 = 20\text{ mL}$$

$$C_2 = 200\text{ mg/5 mL suspension}$$

$$Q_2 = 30\text{ mL}$$

$$C_f = ?$$

$$Q_f = 20 + 30 = 50\text{ mL}$$

Substituting in the equation,

$$(100)(20) + (200)(30) = j(50)$$

$$j = 160\text{ mg/5 mL}$$

The final suspension will contain 160 mg/5 mL in 50 mL.

32. Another application of this expanded equation is the calculation of a volume of diluent required to prepare a diluted solution from a concentrated solution, such as in the following practice problem from previous section.

How many milliliters of water should be added to a 50 mL vial of 23.4% NaCl additive solution to prepare a 0.9% sodium chloride solution?

$$C_1 Q_1 + C_2 Q_2 \ldots = C_f Q_f$$

If we consider

$$C_1 = 23.4\% \text{ (concentrated solution)}$$
$$C_2 = \text{water (diluent)}$$
$$C_f = 0.9\% \text{ (diluted solution)}$$

Because water is a diluent and contains no drug, its contribution will be only to the final volume of diluted solution:

$$(23.4\%)(50\,\text{mL}) = (0.9\%)(50\,\text{mL} + j)$$
$$1170 = 45 + 0.9j$$
$$j = 1250\,\text{mL of water}$$

The final solution will be 1300 mL of 0.9% NaCl.

How many milliliters each of 15% (w/v) boric acid solution and 2.5% (w/v) boric acid solution are required to prepare 200 mL of a 5% (w/v) boric acid solution?

Because there is a need for a definite volume of the final dilution (5%), each individual component of the mixture will contribute with part of the volume. Let j be the amount needed of the stronger solution, then (200 mL – j) will be the weaker source of drug.

$$(15\%)(j) + (2.5\%)(200\,\text{mL} - j) = (5\%)(200\,\text{mL})$$
$$15j + 500 - 2.5j = 1000$$
$$j = 80\,\text{mL of } 15\%$$
$$200\,\text{mL} - j = 200 - 80 = 120\,\text{mL of } 2.5\%$$

Mix 80 mL of 15% (w/v) boric acid solution and 120 mL of 2.5% (w/v) boric acid solution to obtain 200 mL of 5% (w/v) boric acid solution.

---

33. Now try these.

A. A pharmacist mixed 40 g of a 5% (w/w) zinc oxide ointment with 60 g of a 7.5% (w/w) zinc oxide ointment. What is the concentration (w/w) of zinc oxide ointment in the finished product?

B. How many grams of zinc oxide should be added to the product prepared in previous example to obtain an ointment containing 12% (w/w) zinc oxide?

C. How many milliliters of a 25% (w/v) dextrose solution and how many milliliters of a 10% (w/v) dextrose solution are required to prepare 500 mL of a 12% (w/v) solution?

D. How many milliliters each of 1% (w/v) concentrate of benzalkonium chloride and 0.25% (w/v) concentrate of benzalkonium chloride should be used in preparing 1 liter of solution containing 0.55% (w/v) of benzalkonium chloride?

E. How many milliliters of water should be added to 200 mL of a topical solution containing 1 : 50 (v/v) methyl salicylate to make a solution with 1 : 400 (v/v) methyl salicylate?

Solutions.

A. 6.5% (w/w)

B. 6.25 g

C. 66.7 mL of 25% and 433.3 mL of 10%

D. 400 mL of 1% and 600 mL of 0.25%

E. 1400 mL

CALCULATIONS

A. $(40\,g)(5\%) + (60\,g)(7.5\%) = (100\,g)(j)$

$j = 6.5\%$

B. $(100\,\text{g})(6.5\%) + j(100\%) = (100 + j)(12\%)$

$550 = 88j$

$j = 6.25\,\text{g}$

C. $(25\%)j + (10\%)(500\,\text{mL} - j) = (500\,\text{mL})(12\%)$

$15j = 1000$

$j = 66.7\,\text{mL}$ of 25%

$500\,\text{mL} - 66.7\,\text{mL} = 433.3\,\text{mL}$ of 10%

D. $(1\%)j + (0.25\%)(1000\,\text{mL} - j) = (1000\,\text{mL})(0.55\%)$

$0.75j = 300$

$j = 400\,\text{mL}$ of 1%

$1000\,\text{mL} - 400\,\text{mL} = 600\,\text{mL}$ of 0.25%

E. $1:50 = 2\%$

$1:400 = 0.25\%$

$(200\,\text{mL})(2\%) = (200\,\text{mL} + j)(0.25\%)$

$350 = 0.25j$

$j = 1400\,\text{mL}$ of water

ALLIGATION ALTERNATE

34. This method is particularly useful when mixing two or more preparations of kno strengths to prepare a mixture of an intermediate desired strength. The final mixt will be an average of the individual strengths, which are calculated as proportio parts. Alligation alternate can also be used to solve any type of dilution or conc tration problem, including concentrations expressed in mg/mL, ratios, mixtures liquids of known specific gravities, etc. The strengths of all preparations being mi and the final mixture have to be expressed in a common denomination when sett up the alligation. When diluting a preparation the strength of the diluent is conside 0%. When increasing the strength of a mixture by adding more drug, the strength this component is 100%. A final proportion allows a correlation between the parts a any specific denomination needed.

We will use one of our previous examples to explain this method. How many millilit of water should be added to a 50 mL vial of 23.4% NaCl additive solution to prepar 0.9% sodium chloride solution?

$$\text{mL}\ (H_2O) + 50\,\text{mL}\ (23.4\%) = 0.9\%$$

$$\text{mL}\ (H_2O) = ?$$

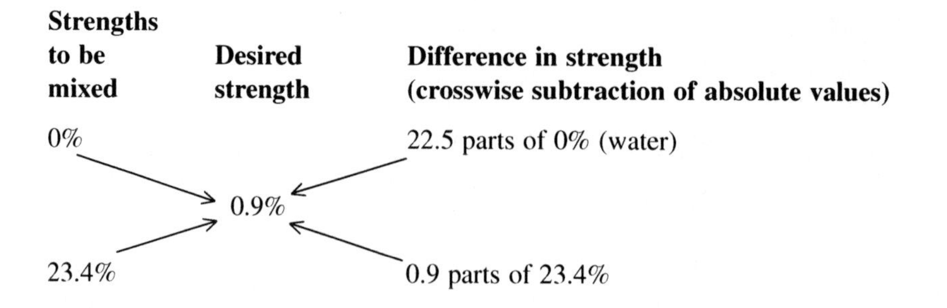

The proportional parts are obtained by subtracting the strengths to be mixed from the desired strength, in a crosswise manner, as directed by the arrows. Only *absolute values* are recorded. (The algebraic sign is ignored.)

$$23.4 - 0.9 = 22.5 \text{ parts}$$
$$0 - 0.9 = 0.9 \text{ parts}$$

To express the result, read across as follows:

22.5 parts of water need to be mixed with 0.9 parts of 23.4% solution to make a 0.9% solution.

You have on hand 50 mL of a 23.4% solution, which will correspond to 0.9 parts, then

$$\frac{50 \text{ mL}}{0.9 \text{ parts}} = \frac{j}{22.5 \text{ parts}}$$
$$j = 1250 \text{ mL of water}$$

---

35. Sometimes, a dilution may be achieved by combining a preparation of higher than the desired final strength with a preparation of lower strength. The final mixture, which results from the combination of relative amounts of each component, will lie somewhere between the strengths of each preparation. A more concentrated product may also be prescribed in which case the pharmacist will need to add more drug to an original strength.

The following example will make these ideas more clear.

℞ Zinc oxide
Starch aa 2 g
Petrolatum 60 g

After compounding the formula above, the pharmacist received a call from the prescriber to change the concentration of zinc oxide to 7.5% (w/w).

How many grams of zinc oxide should the pharmacist add to the original formula?

Initially we need to find the strength of zinc oxide in the formula:

$$\frac{2 \text{ g}}{64 \text{ g}} \times 100\% = 3.125\%$$

The 3.125% (w/w) zinc oxide (ZnOx) ointment will be mixed with pure zinc oxide (100% w/w) to increase the strength to 7.5% (w/w).

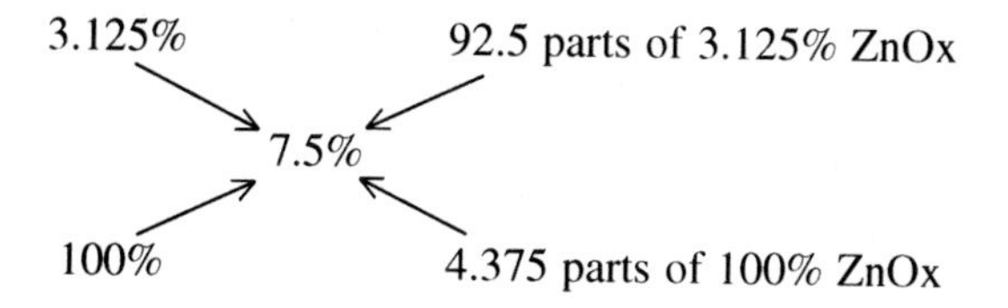

Since the pharmacist had already prepared the original formula, he has on hand 64 g of 3.125%, which will correspond to 92.5 parts. By setting a proportion, we can determine the amount of pure zinc oxide needed to raise the concentration of the product to 7.5%.

$$\frac{64 \text{ g}}{92.5 \text{ parts}} = \frac{j}{4.375 \text{ parts}}$$
$$j = 3.03 \text{ g of zinc oxide}$$

Check:

$$\frac{(2\text{ g}+3.03\text{ g})\text{ ZnOx}}{(64\text{ g}+3.03\text{ g})\text{ ointment}}\times 100\% = 7.5\%$$

or, by $C_1Q_1 + C_2Q_2 = C_fQ_f$

$$(3.125\%)(64\text{ g}) + (100)j = 7.5\%\ (64\text{ g} + j)$$
$$200 + 100j = 480 + 7.5j$$
$$280 = 92.5j$$
$$j = 3.03\text{ g of ZnOx}$$

Using the same *formula* as in the previous example, how many grams of a 25% (w/w zinc oxide ointment could be added to change the strength to 7.5% (w/w) of zinc oxide

We have:

64 g of 3.125% zinc oxide ointment

25% (w/w) zinc oxide ointment

We need to determine the proportional parts to combine these two ointments and have a final product of 7.5%.

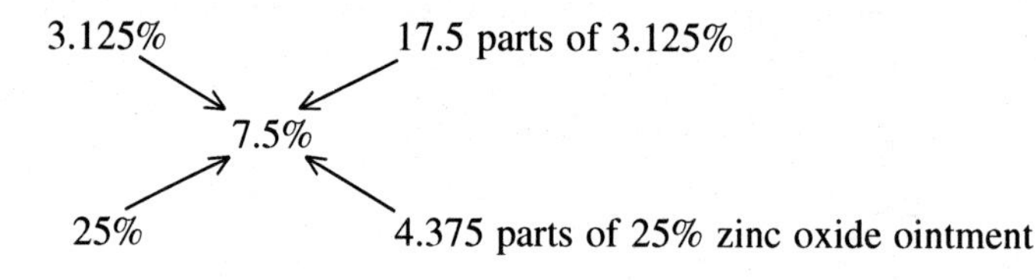

Then,

$$\frac{64\text{ g}}{17.5\text{ parts}} = \frac{j}{4.375\text{ parts}}$$
$$j = 16\text{ g of 25\% zinc oxide ointment}$$

Check:

$$64\text{ g of 3.125\% ZnOx ointment} = 2\text{ g ZnOx}$$
$$16\text{ g of 25\% ZnOx ointment} = 4\text{ g ZnOx}$$
$$\frac{6\text{ g ZnOx}}{80\text{ g ointment}}\times 100\% = 7.5\%$$

Solving by the CQ method:

$$(3.125\%)(64) + (25\%)(j) = (64\text{ g} + j)7.5\%$$
$$200 + 25j = 480 + 7.5j$$
$$280 = 17.5j$$

j = 16 g of 25% ZnOx ointment should be added to original formula to change the concentration to 7.5%.

SO WHICH METHOD SHOULD I USE?

36. Take your choice and use the method you are most comfortable with. You can apply one of the algebraic approaches, which are applicable in every situation. Or you can rely on alligation, a venerable technique developed over a century ago which has become part of the heritage of pharmacy, even though most pharmacy students today are familiar with the techniques of algebra.

Your result should be the same, regardless of the method chosen. The main thing is to understand what you are doing, lay out your work in an organized manner and check your answer.

---

37. Now, try this one.

You have on hand 400 g of a 2.5% zinc oxide ointment and 600 g of a 5% zinc oxide ointment.

A. If the two ointments are mixed, what is the concentration of zinc oxide in the finished product?

B. How many grams of zinc oxide should be added to the product to obtain an ointment containing 15% of zinc oxide?

Solutions.

A. 4%

B. 129.4 g of zinc oxide

CALCULATIONS

A. $400\text{ g} \times \frac{2.5\text{ g}}{100\text{ g}} = 10\text{ g}$

$600\text{ g} \times \frac{5\text{ g}}{100\text{ g}} = 30\text{ g}$

40 g ZnOx/1000 g ointment mixture × 100% = 4%

or

400 g × 2.5% = 1000

600 g × 5% = 3000

1000 4000

4000/1000 = 4%

B. 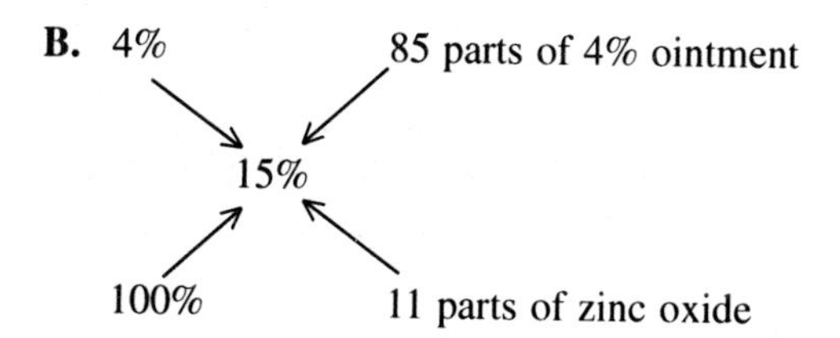

85 parts of 4% ZnOx ointment = 1000 g

$$\frac{85 \text{ parts}}{1000 \text{ g}} = \frac{11 \text{ parts}}{j}$$

$j = 129.4$ g of zinc oxide should be added.

or,

$0.15 = (40\,\text{g ZnOx} + j)/1000\,\text{g oint.} + j$

$150 + 0.15j = 40 + j$

$110 = 0.85j$

$j = 129.4$ g of zinc oxide should be added to 1000 g of 4% ointment to increase the concentration to 15% zinc oxide.

---

38. Consider the following problem and solve it by alligation alternate.

How many milliliters of a 2% (w/v) lidocaine hydrochloride (Lid.HCl) solution and how many milliliters of normal saline (NS) should be mixed to make 500 mL of a diluted solution containing 4 mg of the drug per milliliter?

We can approach this problem in two different ways:

(1) Convert 4 mg/mL into percentage

(2) Convert 2% into mg/mL

Solving by (1)

4 mg/mL = 400 mg/100 mL = 0.4 g/100 mL = 0.4%

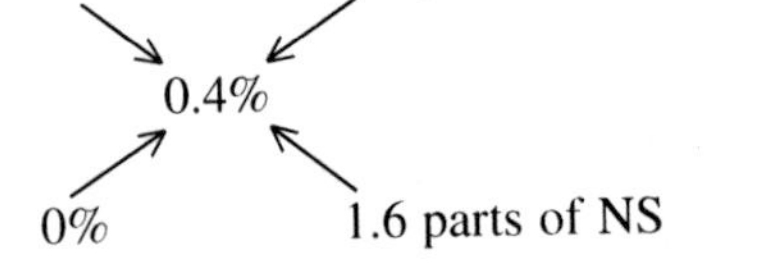

0.4 parts + 1.6 parts = 2 parts mixture to make 0.4%

We need 500 mL of 0.4% (4 mg/mL) Lid.HCl

$$\frac{2 \text{ parts}}{500 \text{ mL}} = \frac{0.4 \text{ parts}}{j}$$

$j = 100$ mL of 2% solution

2 parts/500 mL = 1.6 parts/j

$j = 400$ mL of NS

Solving by (2)

2% = 2 g/100 mL = 2000 mg/100 mL = 20 mg/mL

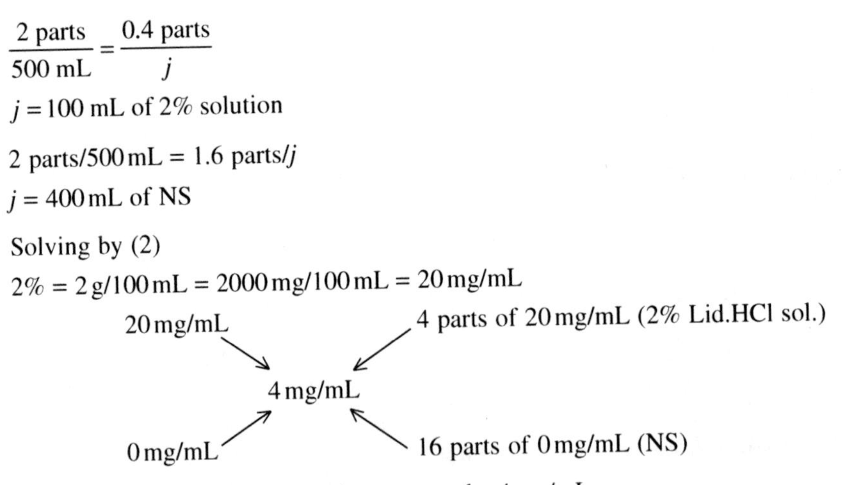

4 parts + 16 parts = 20 parts mixture to make 4 mg/mL

We need 500 mL of 4 mg/mL

$$\frac{20\ \text{parts}}{500\ \text{mL}} = \frac{4\ \text{parts}}{j}$$

$j = 100$ mL of 20 mg/mL solution

$$\frac{20\ \text{parts}}{500\ \text{mL}} = \frac{16\ \text{parts}}{j}$$

$j = 400$ mL of NS

---

39. Sometimes the pharmacist needs to mix liquids of known specific gravities to achieve a desired intermediate specific gravity. Alligation alternate will help the calculation of the amounts of each component.

How many grams of sorbitol solution having a specific gravity of 1.25 and how many grams (milliliters) of water should be mixed to prepare 200 mL of a sorbitol product having a specific gravity of 1.20?

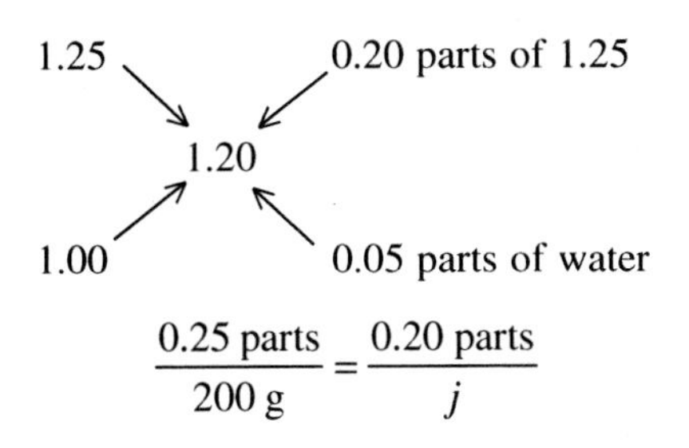

$$\frac{0.25\ \text{parts}}{200\ \text{g}} = \frac{0.20\ \text{parts}}{j}$$

$j = 160$ g of sorbitol solution with specific gravity of 1.25

$$\frac{0.25\ \text{parts}}{200\ \text{g}} = \frac{0.05\ \text{parts}}{k}$$

$$k = 40\ \text{g (mL) of water}$$

Check:

$$0.05 \times 1 = 0.05$$

$$0.20 \times 1.25 = 0.25$$

$$0.30/0.25 = 1.2 \text{ specific gravity needed.}$$

STOCK SOLUTIONS

40. Occasionally, the pharmacist is asked to dispense a stock solution that will be diluted by the patient prior to use. The calculations may be solved by any of the methods discussed.

A phenylmercuric nitrate concentrate has a concentration of 4% w/v. What is the final concentration, in terms of ratio strength, in a solution made by diluting 1 teaspoonful of the concentrate to 200 mL?

Let j equal the concentration of diluted solution:

$$(5\ \text{mL})(0.04\ \text{g/mL}) = (200\ \text{mL})(j)$$

$$j = 0.001\ \text{g/mL} = 1:1000$$

How many milliliters must 1 tablespoonful of the concentrate be diluted to in order t yield a 1:1500 solution of phenylmercuric nitrate?

Solution. 900 mL

CALCULATIONS

Let j equal the volume of diluted solution:

$$(0.04\,g/mL)(15\,mL) = \frac{1\,g}{1500\,mL}\,j$$

$$j = 900\,mL$$

---

41. ℞ Potassium permanganate solution 60.0
Make of such strength that 1 teaspoonful diluted to a liter will yield a 1:10,00 solution.

Calculate the number of grams of potassium permanganate ($KMnO_4$) needed to mak the solution.

Two approaches to this problem are possible. One way is to calculate the concentratio of the prescription solution and then find the amount of $KMnO_4$ needed to make the solutior The second method, which is somewhat easier, is to calculate the number of gram of $KMnO_4$ in the diluted solution. This quantity must also be contained in 1 teaspoonfu of the stock solution dispensed to the patient. If you can get this far, you can calculate th amount of $KMnO_4$ required for 60.0 mL of the stock solution.

Solution. 1.20 g

CALCULATIONS

$$g\ KMnO_4 \text{ in } 1\text{ tsp concentrate} = g\ KMnO_4 \text{ in diluted sol} = \frac{1\,g}{10^4\,mL} \times 1000\,mL = 0.1\,g$$

Therefore, each teaspoonful of the stock solution contains 0.1 g of $KMnO_4$. We mus prepare a total of 60.0 mL, or 12 teaspoonfuls.

$$\frac{0.100\,g}{\text{teaspoonful}} \times 12\text{ teaspoonfuls} = 1.20\,g$$

---

42. ℞ Silver nitrate qs
Ft sol, 30.0 mL
Each teaspoonful diluted to a pint yields a 1:5000 solution.

How many grams of silver nitrate ($AgNO_3$) should be used to make this solution? (Since the measuring devices in the average household are not as accurate as those available to the pharmacist, a pint may be approximated by 500 mL in the calculation.)

Solution. 0.6 g

CALCULATIONS

The amount of $AgNO_3$ in the diluted solution is

$$\frac{1\ \text{g}}{5000\ \text{mL}} \times 500\ \text{mL} = 0.1\ \text{g}$$

This is also the content of each teaspoonful of the concentrate:

$$\frac{0.1\ \text{g}}{\text{teaspoonful}} \times 6\ \text{teaspoonfuls} = 0.6\ \text{g}$$

---

43. If you would like additional practice, here are some more problems dealing with situations in which the patient must make the dilution.

A. ℞ Benzethonium chloride solution 60 mL
One teaspoonful diluted to 500 mL yields a 1:1500 solution.

How many grams of benzethonium chloride are needed?

B. A phenylmercuric acetate concentrate is to be prepared. One tablespoonful of the concentrate is diluted to 500 mL with water to yield a 1:2000 solution of phenylmercuric acetate. What must be the percentage concentration of phenylmercuric acetate in the concentrate?

C. ℞ Phenylmercuric nitrate solution 180 mL

One teaspoonful diluted to a pint yields a 1:2500 solution.

How many grams of phenylmercuric nitrate should be used for the solution? (Use 1 pt = 500 mL, since the patient will make the dilution.)

Solutions.

A. 4.00 g

B. 1.67%

C. 7.2 g

REVIEW PROBLEMS

Do all of the problems before verifying your answers. Use your preferred method for calculations.

A. ℞ Benzalkonium chloride qs

Aqua dest. qs ad 180.0 mL

One tablespoonful diluted to a liter yields a 1:750 solution.

How many grams of benzalkonium chloride should be used to make this solution?

B. How many milliliters of 95% alcohol should be diluted with water to make 65.0 mL of 40% alcohol?

C. A lotion is made by mixing 250 mL of witch hazel (14% C_2H_5OH), 1 L of diluted alcohol (49% C_2H_5OH), and sufficient water to make 2 L. What is the percentage strength of C_2H_5OH in the lotion?

D. How much 20% sulfathiazole ointment should be added to 200 g of 6% sulfathiazole ointment to make an ointment of 10% strength?

E. Belladonna tincture is required to contain 30 mg of alkaloids in each 100 mL of the tincture. If 250 mL of the tincture is prepared and the assay shows the alkaloids content to be 0.035%, how much solvent should be used to dilute the tincture so that it meets the standard requirement?

F. How much 4.5% sodium acetate solution can be made from 32 mL of a 12% sodium acetate solution?

G. If 142 g of petrolatum are mixed with 25 g of 0.20% mercury bichloride ointment, what will the final concentration of mercury bichloride be?

H. How many grams of petrolatum should be added to 15% sulfur ointment to make 120 g of a 6.5% sulfur ointment?

I. How many milliliters of a 15% solution of sodium chloride should be used to mak
1 L of a 0.9% solution of sodium chloride?

J. How many milliliters of 0.5% sodium sulfate solution should be mixed with 5.0%
sodium sulfate solution to make 1 L of a 2% solution?

K. How many milliliters of a 22% v/v solution of glycerin in water can be made b
diluting 90 mL of an 80% v/v glycerin solution with water?

L. ℞

Phenobarbital elixir	40.0 mL
High alcoholic elixir	40.0 mL
Syrup qs ad	120.0 mL

What is the ethanol percentage in the finished product? Phenobarbital elixir contain
15% C_2H_5OH; high alcoholic elixir contains 78% C_2H_5OH. Syrup contains n
ethanol.

M. How many grams of pure coal tar should be added to 36.0 g of 4.0% coal ta
ointment to make a 10% coal tar ointment?

N. We need 90 mL of phosphoric acid solution, 10% w/v. How many milliliters of phosphoric acid USP (85% w/w; sp g = 1.71) should be used?

O. ℞

Phenobarbital elixir (15% C_2H_5OH)	20.0 mL
Alcohol USP (95% C_2H_5OH)	
Syrup tolu (no C_2H_5OH)	qs ad 90.0 mL

How much alcohol USP should be used in this prescription to bring the C_2H_5OH content of the final preparation to 20%?

P. What is the specific gravity of a mixture containing 500 mL of water, 1000 mL of glycerin (sp. gr. = 1.45) and 1000 mL of alcohol (sp. gr. = 0.85)? (Assume no contraction occurs when the liquids are mixed)?

Q. How many grams of sorbitol solution having a specific gravity of 1.385 and how many grams of water (sp. gr. = 1.00) should be used to prepare 400 g of a sorbitol solution with specific gravity of 1.250?

R. How many grams of Neomycin sulfate should be added to 320 g of a 5% (w/w) neomycin sulfate ointment to prepare an ointment containing 20% (w/w) neomycin sulfate?

S. How many milliliters of phenobarbital elixir containing 25 mg/tsp and 35 mg/tsp should be used to prepare 200 mL of elixir containing 6.2 mg/mL?

T. How many grams of white petrolatum and how many grams of a 20% (w/w) benzocaine ointment should be used to obtain 60 g of 8% (w/w) benzocaine ointment?

U. How many grams of a 2.5% (w/w) hydrocortisone cream should be mixed with 250 g of a 0.75% (w/w) hydrocortisone cream to obtain a 1% (w/w) hydrocortisone cream?

V. In what proportion should 95% and 10% alcohol be mixed to make 70% alcohol? (Assume no contraction occurs when the liquids are mixed)

W. What is the percentage of alcohol in the following prescription? (Assume no contraction occurs when the liquids are mixed)

℞		
	Witch Hazel (15% alcohol, v/v)	20 mL
	Alcohol 95% v/v	10 mL
	Boric acid solution ad	50 mL

X. How many milliliters of a 5% (w/v) aluminum acetate solution should be used to prepare 300 mL of a 1 : 400 (w/v) aluminum acetate solution?

Y. A physician's order calls for 0.005% (w/v) isoproterenol hydrochloride IV injection in 50 mL normal saline. The pharmacist has on hand a 1 : 800 (w/v) isoproterenol hydrochloride aqueous solution. How many milliliters of normal saline should be added to the available solution to prepare the prescribed injection?

Z. ℞ Povidone iodine 50 mL

Make a solution such that 5 mL diluted to 250 mL equals a 1 : 2000 solution.

Sig. Vaginal douche

How many milliliters of a 10% solution of povidone iodine should be used in preparing the prescription?

Solutions to Review Problems

A. 16.0 g

B. 27.4 mL

C. 26.3%

D. 80 g

E. 42 mL

F. 85 mL

G. 0.030%

H. 68 g

I. 60 mL

J. 667 mL of 0.5% solution

K. 327 mL

L. 31% v/v

M. 2.4 g

N. 6.2 mL

O. 15.8 mL

P. 1.12

Q. 259.7 g of sorbitol sp. gr. 1.385 and 140.3 g of water

R. 60 g

S. 120 mL of 35 mg/tsp and 80 mL of 25 mg/tsp

T. 36 g petrolatum and 24 g of 20%

U. 41.7 g of 2.5% hydrocortisone cream

V. 12:5 (12 parts of 95% and 5 parts of 10%)

W. 25% v/v

X. 15 mL

Y. 48 mL NS

Z. 12.5 mL

Calculations for Review Problems

A. benzethonium chloride = 1/750 g/mL × 1000 mL = 1.33 g

Each tablespoonful of the stock solution contains 1.33 g:

1.33 g/tablespoonful × 12 tablespoonfuls = 16.0 g

B. Let j equal the volume of alcohol, 95%:

mL ethanol (95%) = mL ethanol (40%)

$0.95j = (65.0\text{ mL})(0.4)$

$j = 27.4\text{ mL}$

C. Let j equal the strength of the lotion:

mL eth (w. h.) + mL eth (dil alc) = mL eth (lotion)

$(250\text{ mL})(0.14) + (1000\text{ mL})(0.49) = (2000\text{ mL})j$

$$j = \frac{35\text{ mL} + 490\text{ mL}}{2000\text{ mL}} = 0.263 = 26.3\%$$

D. Let j equal the grams of 20% ointment:

g sulfa (20% oint) + g sulfa (6% oint) = g sulfa (10% oint)

$j(0.20) + (200\text{ g})(0.06) = (200\text{ g} + j)(0.10)$

$j = 80\text{ g}$

E. Let j equal the milliliters of solvent:

g alkaloids(0.035% tr) = g alkaloids(0.030% tr)

$(250\text{ mL})(0.00035\text{ g/mL}) = (250\text{ mL} + j)(0.00030\text{ g/mL})$

$0.0875\text{ g} = 0.075\text{ g} + (0.00030\text{ g/mL})(j)$

$j = 42\text{ mL}$

F. Let j equal the milliliters of 4.5% solution:

g sod acetate(4.5%) = g sod acetate(12%)

$(0.045\text{ g/mL})j = (0.12\text{ g/mL})(32\text{ mL})$

$j = 85\text{ mL}$

G. Let j equal the final concentration:

$g\ HgCl_2(0.20\%) = g\ HgCl_2(\text{final product})$

$(25\,g)(0.002) = (167\,g)j$

$j = 0.00030 = 0.030\%$

H. Let j equal the amount of petrolatum to add:

$(0.15)(120\,g - j) = (0.065)(120\,g)$

$18\,g - 0.15j = 7.8\,g$

$j = 18\,g - 7.8\,g/0.15 = 68\,g$

I. Let j equal the milliliters of 15% solution:

$g\ NaCl(15\%) = g\ NaCl(0.9\%)$

$(0.15\,g/mL)j = (0.009\,g/mL)(1000\,mL)$

$j = 60\,mL$

J. Let j equal the milliliters of 0.5% solution. Since the finished product measures 1000 mL,

$(1000\,mL - j) =$ volume of the 5% solution needed

g sod sulf(0.5%) + g sod sulf(5%) = g sod sulf(2%)

$(j)(0.005\,g/mL) + (1000\,mL - j)(0.05\,g/mL) = (1000\,mL)(0.02\,g/mL)$

$(j)(0.005\,g/mL) + 50\,g - (j)(0.05\,g/mL) = 20\,g$

$j = 667\,mL$

K. Let j equal the volume of diluted solution:

$(0.8)(90\,mL) = 0.22j$

$$j = \frac{72\ mL}{0.22} = 327\ mL$$

L. Let j equal the concentration of the finished product:

mL C_2H_5OH (pheno elix) + mL C_2H_5OH (h.a. elix)

= mL C_2H_5OH (product)

$(40.0\,mL)(0.15) + (40.0\,mL)(0.78) = 120\,mL\,(j)$

$j = 0.31 = 31\%$ v/v

M. Let j equal the grams of pure coal tar:

g coal tar + g coal tar (4% oint) = g coal tar (10% oint)

$j + (36.0\,g)(0.04) = (36.0 + j)(0.10)$

$j + 1.44\,g = 3.60\,g + 0.1j$

$0.9j = 2.16\,g$

$j = 2.4\,g$

N. Let j equal the grams of phosphoric acid USP to be used:

g H_3PO_4 (concentrate) = g H_3PO_4 (10% solution)

$0.85j = (90\,mL)(0.10\,g/mL)$

$j = 10.6\,g$

$$10.6\ g \times \frac{1\ mL}{1.71\ g} = 6.20\ mL$$

O. Let j equal the milliliters of alcohol USP to be added:

mL eth (alc USP) + mL eth (pheno elix) = mL eth (product)

$0.95j + (0.15)(20\,\text{mL}) = (0.20)(90\,\text{mL})$

$j = 15.8\,\text{mL}$

P. $500 \times 1.00 = 500$

$1000 \times 1.45 = 1450$

$1000 \times 0.85 = 850$

$\overline{2500}$ $\quad\quad \overline{2800}$ (totals)

$2800/2500 = 1.12$ is the specific gravity of the mixture

Q. 1.385 → 1.250 ← 0.250 parts of 1.385

1.000 → 1.250 ← 0.135 parts of water

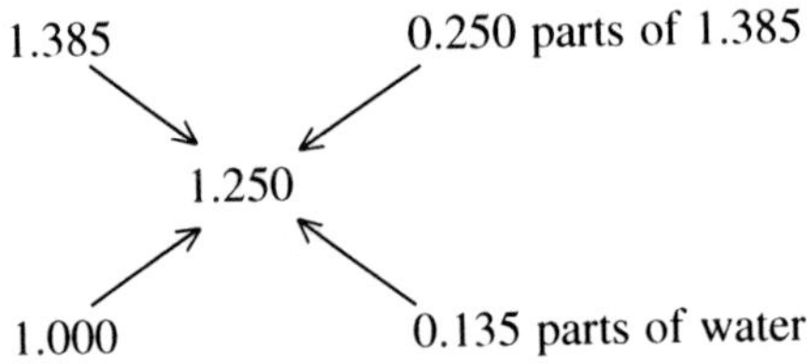

0.250 parts of 1.385 + 0.135 parts of water = 0.385 parts of 1.250

0.385 parts of 1.250 should correspond to 400 g

$$\frac{0.385\text{ parts}}{400\text{ g}} = \frac{0.250\text{ parts}}{j}$$

$j = 259.7$ g of sorbitol with specific gravity of 1.385

$$\frac{0.385\text{ parts}}{400\text{ g}} = \frac{0.135\text{ parts}}{k}$$

$k = 140.3$ g of water

R. $j(100\%) + (320\,\text{g})(5\%) = (320\,\text{g} + j)(20\%)$

$80j = 4800$

$j = 60\,\text{g}$ of neomycin sulfate

or 100 → 20 ← 15 parts of neomycin sulfate

5 → 20 ← 80 parts of 5%

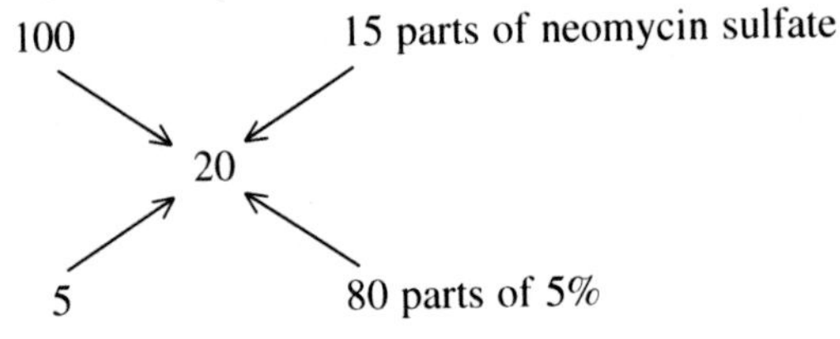

$$\frac{80\text{ p}}{320\text{ g}} = \frac{15\text{ p}}{j}$$

$j = 60$ g of neomycin sulfate

S. $25\,\text{mg/tsp} = 25\,\text{mg}/5\,\text{mL} = 5\,\text{mg/mL}$

$35\,\text{mg/tsp} = 35\,\text{mg}/5\,\text{mL} = 7\,\text{mg/mL}$

$(7)(j) + (5)(200 - j) = (200)(6.2)$

$2j = 240$

$j = 120\,\text{mL}$ of 7 mg/mL or 35 mg/tsp

$200 - j = 200 - 120 = 80\,\text{mL}$ of 5 mg/mL or 25 mg/tsp

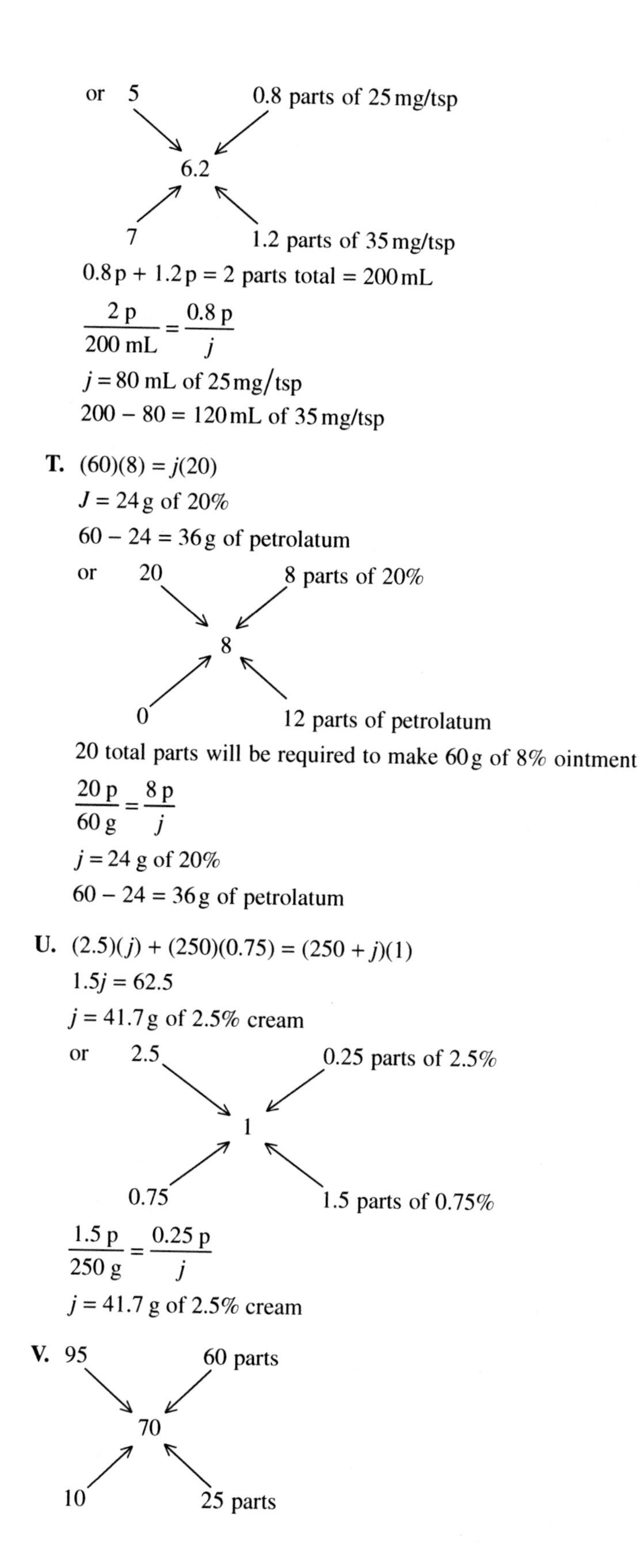

or 5 → 6.2 ← 0.8 parts of 25 mg/tsp

7 → 6.2 ← 1.2 parts of 35 mg/tsp

$0.8\,\text{p} + 1.2\,\text{p} = 2 \text{ parts total} = 200\,\text{mL}$

$$\frac{2\,\text{p}}{200\,\text{mL}} = \frac{0.8\,\text{p}}{j}$$

$j = 80$ mL of 25 mg/tsp

$200 - 80 = 120$ mL of 35 mg/tsp

T. $(60)(8) = j(20)$

$J = 24$ g of 20%

$60 - 24 = 36$ g of petrolatum

or 20 → 8 ← 8 parts of 20%

0 → 8 ← 12 parts of petrolatum

20 total parts will be required to make 60 g of 8% ointment

$$\frac{20\,\text{p}}{60\,\text{g}} = \frac{8\,\text{p}}{j}$$

$j = 24$ g of 20%

$60 - 24 = 36$ g of petrolatum

U. $(2.5)(j) + (250)(0.75) = (250 + j)(1)$

$1.5j = 62.5$

$j = 41.7$ g of 2.5% cream

or 2.5 → 1 ← 0.25 parts of 2.5%

0.75 → 1 ← 1.5 parts of 0.75%

$$\frac{1.5\,\text{p}}{250\,\text{g}} = \frac{0.25\,\text{p}}{j}$$

$j = 41.7$ g of 2.5% cream

V. 95 → 70 ← 60 parts

10 → 70 ← 25 parts

60:25 = 12:5, thus, 12 parts of 95% alcohol to 5 parts of 10% alcohol

or $95(x) + 10(y) = 70(x + y)$

$25x = 60y$

$x/y = 60/25 = 12/5$

W. 20 mL × 15 mL/100 mL = 3 mL

10 mL × 95 mL/100 mL = 9.5 mL

20 mL boric acid (no alcohol)

$$\frac{12.5 \text{ mL alcohol}}{50 \text{ mL solution}} \times 100\% = 25\%$$

X. 1:400 = 0.25%

$(300)(0.25) = j(5)$

$j = 15$ mL

Y. 1:800 = 0.125%

$(0.005)(50) = (0.125)j$

$j = 2$ mL of 0.125%

50 mL − 2 mL = 48 mL NS

OR: $(0.125)(50 - j) = (0.005)(50)$

$6 = 0.125j$

$j = 48$ mL of NS

Z. 1:2000 = 0.05%

$(0.05)(250) = j(5)$

$j = 2.5\%$, strength of 5 mL, also of 50 mL

$(2.5)(50) = (10)k$

$k = 12.5$ mL of 10% solution.

ISOTONICITY, pH, AND BUFFERS

***Learning Objectives:** After completing this chapter the student should be able to:*

1. Determine the sodium chloride equivalent of a substance from the volume of isotonic solution that can be made from 1 g of that substance.
2. Apply the sodium chloride equivalent method to determine the amount of NaCl needed to make a solution isotonic with body fluids.
3. Apply the sodium chloride equivalent method to determine the amount of a substance other than sodium chloride (tonicity agent) needed to make a solution isotonic with body fluids.
4. Calculate the volume of water to be added to a given amount of drug to make an isotonic solution (White–Vincent method).
5. Use the freezing point depression of 1% w/v solutions of drugs in calculations of isotonic solutions.
6. Perform pH calculations from hydrogen ion concentration and vice versa.
7. Calculate the pH value of a buffer system.
8. Prepare a buffer of desired pH by calculating the concentration (molar ratio) of buffer components.

This chapter will introduce calculations involved with *isotonicity* and review pH and buffers. Isotonicity is needed for minimization of tissue damage and drug efficacy, especially when drug solutions are mixed with blood and fluids of the eye, nose, or bowel. Buffer systems prevent a drift in solution pH, which could cause drug degradation or inactivity. Buffering capacity and pH are also extremely important in maintenance of the body's acid–base balance.

ISOTONICITY

1. To prevent irritation and cell destruction when placing a drug solution in contact with a body fluid, the properties of the solution should be matched in certain ways to that of the fluid. To minimize cell damage and maximize drug efficacy, the solution should be *isotonic* (equal tone) with the body fluid.

In general, pharmaceutical solutions that should be prepared as isotonic solutions include:

(i) Parenteral solutions to be given *intravenously*, *intramuscularly*, *intracutaneously*, or *intrathecally*. Some exceptions are known. Solutions given by subcutaneous injection are preferably given as isotonic solutions, but since injection is into fatty tissue,

Pharmaceutical Calculations, Fourth Edition, By Joel L. Zatz and Maria Glaucia Teixeira
ISBN 0-471-67623-3

this is not essential. Solutions injected into a central vein will get quick dilution by the extensive blood flow and do not necessarily need to be isotonic.

(ii) Aqueous solutions applied to the mucosal membrane of the *nose*, or to areas of *broken skin* may be irritating if not isotonic.

(iii) *Ophthalmic* solutions. Isotonicity is essential to prevent irritation or damage to the eye, whereas buffering is desirable to maintain optimum pH for drug stability and for therapeutic activity. The eye will tolerate solutions having pH values over a wide range provided they are administered in *small volumes*. A number of *buffer systems* have been formulated for use in ophthalmic solutions, providing a wide range of pH and compatibility with drugs commonly administered into the eye.

(iv) Some solutions for rectal use (*enemas*).

An important factor in achieving isotonicity is proper osmotic pressure. Osmotic pressure differences provide the driving force for *osmosis*, the movement of water from regions of low to high solute concentration. Body fluids owe their osmotic pressure to the salts, proteins, and other solutes that are normal components. If a solution that is injected intravenously has too low an osmotic pressure (hypotonic solution), water will flow into red blood cells, causing them to swell and even burst (hemolysis). If the osmotic pressure is too high (hypertonic solution), the cells will lose water and shrink (crenation). Movement of fluid into or out of cells is minimized when the drug solution that is introduced has the same osmotic pressure as that of the fluid. Such solutions are said to be isotonic with the particular fluid. Solutions that have the same osmotic pressure as that of a 0.9% sodium chloride solution are considered to be isotonic with blood, tears, and several body fluids.

Osmotic pressure is a major determinant of tonicity but it is not the only one. Solutions containing certain drugs can damage red blood cells despite the fact that they have the correct osmotic pressure. Nevertheless, monitoring osmotic pressure is a necessary step in assuring isotonicity.

The osmotic pressure of a solution depends on the number of ions and molecules that are dissolved. All *colligative properties* of a solution depend on the number of particles in solution and theoretically could be used to determine isotonicity. As they are all related to each other, freezing point depression has often been used as a means of estimating the effect of solutes on osmotic pressure and making solutions isotonic. Solutions having the same osmotic pressure as that of a 0.9% sodium chloride solution (normal saline solution) are considered to be isotonic.

Most often, simple solutions of drugs at therapeutic concentrations in water have an osmotic pressure that is too small. It is necessary to add another material (usually sodium chloride is chosen) to make the solution isotonic with body fluids. Why is sodium chloride a reasonable choice for adjusting osmotic pressure? Both sodium and chloride are predominant ions in biological fluids, so this salt does not cause irritation or otherwise disturb normal processes. There are several methods of calculating the amount of sodium chloride or other tonicity agent that should be used to make a solution isotonic.

Sodium Chloride Equivalents (E-values)

2. For a particular drug, there is a single concentration at which the drug solution will be isotonic with blood serum and tears. If we start with 1 g of drug, it will be possible to prepare a certain volume of isotonic solution. Table 8.1 is a modified USP table that lists the volume of isotonic solution that can be made from 1 g of several drugs.

TABLE 8.1 Volumes of water for isotonicity of selected drugs[1]

Drug (1 g)	Volume of Isotonic Solution (mL)
Atropine sulfate	14.3
Boric acid	55.7
Butacaine sulfate	22.3
Chloramphenicol sodium succinate	15.7
Chlorobutanol (hydrous)	26.7
Cocaine hydrochloride	17.7
Dibucaine hydrochloride	14.3
Ephedrine hydrochloride	33.3
Ephedrine sulfate	25.7
Epinephrine bitartrate	20.0
Eucatropine hydrochloride	20.0
Fluorescein sodium	34.3
Homatropine hydrobromide	19.0
Neomycin sulfate	12.3
Penicillin G potassium	20.0
Phenacaine hydrochloride	22.3
Phenylephrine hydrochloride	35.7
Phenylethyl alcohol	27.7
Physostigmine sulfate	14.3
Pilocarpine hydrochloride	26.7
Pilocarpine nitrate	25.7
Polymyxin B sulfate	10.0
Procaine hydrochloride	23.3
Proparacaine hydrochloride	16.7
Scopolamine hydrobromide	13.3
Silver nitrate	36.7
Sodium biphosphate	44.3
Sodium borate	46.7
Sodium phosphate (dibasic, heptahydrate)	32.3
Streptomycin sulfate	7.7
Sulfacetamide sodium	25.7
Sulfadiazine sodium	26.7
Tetracaine hydrochloride	20.0
Tetracycline hydrochloride	15.7
Zinc sulfate	16.7

[1]Adapted from the *United States Pharmacopeia*. XXII Revision. Easton, PA, Mack Publishing Company, 1990.

A more complete table can be found in the latest edition of the USP under the heading "Ophthalmic Solutions." Also, rather complete tables listing *sodium chloride equivalents* can be found in several reference sources, such as *Remington*: *The Science and Practice of Pharmacy* (Gennro). We provide in this text an abridged table (Table 8.2) of sodium chloride equivalents of some drugs regularly used in compounding isotonic formulas.

TABLE 8.2 Sodium chloride equivalents of some drug substances (E values)

Substance	E-value
Atropine sulfate, E-value = 0.13	0.12
Benzalkonium chloride	0.16
Boric acid	0.52
Chloramphenicol	0.10
Chlorobutanol	0.24
Cromolyn sodium	0.11
Dextrose, anhydrous	0.18
Dextrose, monohydrate	0.16
Ephedrine sulfate	0.23
Epinephrine bitartrate	0.18
Gentamicin sulfate	0.05
Glycerin	0.34
Homatropine hydrobromide	0.17
Lidocaine hydrochloride	0.22
Morphine sulfate, pentahydrate	0.11
Naphazoline hydrochloride	0.27
Oxymetazoline hydrochloride	0.20
Penicillin G potassium	0.18
Phenacaine hydrochloride	0.20
Phenylephrine hydrochloride	0.32
Physostigmine sulfate	0.13
Pilocarpine hydrochloride	0.24
Potassium biphosphate	0.43
Potassium nitrate	0.58
Procaine hydrochloride	0.21
Silver nitrate	0.33
Sodium borate	0.42
Sodium chloride	1.00
Sodium phosphate dibasic, anhydrous	0.53
Sodium phosphate dibasic, heptahydrate	0.29
Sodium phosphate monobasic, anhydrous	0.49
Sodium phosphate monobasic, monohydrate	0.42
Tetracaine hydrochloride	0.18
Tetracycline hydrochloride	0.12
Timolol maleate	0.14
Tobramycin	0.07
Zinc chloride	0.62
Zinc sulfate	0.15

The sodium chloride equivalent of a substance is described as the amount of sodium chloride with equivalent osmotic pressure as that of 1 g of the substance. In other words, the amount of sodium chloride represented by a substance can be determined by multiplying the number of grams of substance included in a prescription by its sodium chloride equivalent.

From Table 8.1, 14.3 mL of isotonic solution in water can be prepared from 1 g of atropine sulfate. Recalling that 0.9% aqueous sodium chloride solution is isotonic, the amount of sodium chloride in 14.3 mL would be,

$$14.3\text{ mL} \times \frac{0.009\text{ g}}{1\text{ mL}} = 0.13\text{ g NaCl}$$

Thus 14.3 mL of an isotonic solution of atropine sulfate contain 1 g of atropine sulfate while 14.3 mL of an isotonic solution of sodium chloride contain 0.13 g of NaCl. Therefore, 0.13 g of NaCl generates the same osmotic pressure as 1 g of atropine sulfate. In terms of osmotic pressure, each gram of atropine sulfate can be replaced by 0.13 g of NaCl. The number 0.13 is called the *sodium chloride equivalent* of atropine sulfate.

7.7 mL of a solution containing 1 g of streptomycin sulfate is isotonic with tears. Calculate the sodium chloride equivalent of streptomycin sulfate.

Solution. 0.069

CALCULATION

For isotonic sodium chloride solution,

$$7.7\text{ mL} \times \frac{0.9\text{ g}}{100\text{ mL}} = 0.069\text{ g NaCl}$$

Therefore, 0.069 g of NaCl is osmotically equivalent to 1 g of streptomycin sulfate; 0.069 is the sodium chloride equivalent of this drug.

---

3. If 20 mL of a solution containing 1 g penicillin G potassium is isotonic with tears, What is the sodium chloride equivalent of penicillin G potassium?

Solution. 0.18

CALCULATION

$$20\text{ mL} \times \frac{0.9\text{ g}}{100\text{ mL}} = 0.18\text{ g NaCl}$$

0.18 g NaCl is osmotically equivalent to 1 g penicillin G potassium.

0.18 is the sodium chloride equivalent of penicillin G potassium.

ISOTONICITY BY SODIUM CHLORIDE EQUIVALENT METHOD

4. Let us try a problem that shows how sodium chloride equivalents can be put to work. It is necessary to prepare 300 mL of a 1.0% solution of atropine sulfate. How many grams of sodium chloride should be dissolved in the solution to make it isotonic with tears?

If no atropine sulfate were present, the amount of sodium chloride needed for isotonicity would be

$$300 \text{ mL} \times 0.009 \text{ g/mL} = 2.70 \text{ g}$$

However, the solution does contain

$$300 \text{ mL} \times 0.01 \text{ g/mL} = 3.0 \text{ g atropine sulfate}$$

This amount of atropine sulfate exerts a certain osmotic pressure. Using the sodium chloride equivalent (0.13), we can determine the amount of sodium chloride that has the same osmotic pressure as 3.0 g of atropine sulfate:

$$3.0 \text{ g} \times 0.13 = 0.39 \text{ g NaCl}$$

The atropine sulfate in the solution exerts an osmotic pressure equal to that of 0.39 g of NaCl. Therefore, to find the amount of sodium chloride to add, it is necessary to subtract 0.39 g from the amount of NaCl that would have been needed had the salt been the only solute in the system

$$2.70 \text{ g} - 0.39 \text{ g} = 2.31 \text{ g}$$

To prepare the solution, dissolve 3 g of atropine sulfate and 2.31 g of sodium chloride in sufficient water to make a total volume of 300 mL.

How many milligrams of sodium chloride should be used to prepare the following prescription?

℞	Ephedrine sulfate		0.5 g
	NaCl		q.s.
	Purified water	ad	50 mL
	Make isotonic solution.		
	Sig. Eye drops		

Solution. 335 mg

CALCULATIONS

If no ephedrine sulfate were present, the amount of sodium chloride needed for isotonicity would be

$$50 \text{ mL} \times 0.009 \text{ g/mL} = 0.45 \text{ g}$$

The solution contains 0.5 g ephedrine sulfate, which has a sodium chloride equivalent (E value) of 0.23.

The amount of sodium chloride that has the same osmotic pressure as 0.5 g of ephedrine sulfate is:

$$0.5 \text{ g} \times 0.23 = 0.115 \text{ g NaCl}$$

The ephedrine sulfate in the solution exerts an osmotic pressure equal to that of 0.115 g of NaCl. Therefore, to find the amount of sodium chloride to add, it is necessary to subtract 0.115 g from the amount of NaCl that would have been needed had the salt been the only solute in the system.

$$0.45 \text{ g} - 0.115 \text{ g} = 0.335 \text{ g} = 335 \text{ mg}$$

To prepare the solution, dissolve 500 mg of ephedrine sulfate and 335 mg of sodium chloride in sufficient water to make a total volume of 50 mL.

---

5. To recapitulate, the sodium chloride equivalent method involves three steps:

(i) Calculate the number of grams of NaCl needed to make the desired volume isotonic.

(ii) Using values of the respective sodium chloride equivalents calculate the amount of NaCl osmotically equivalent to each of the other formula components and total these.

(iii) Subtract the result in (ii) from (i) to yield the amount of NaCl that must be used.

How many grams of sodium chloride should be used to make 90 mL of a 0.5% pilocarpine hydrochloride solution isotonic? The sodium chloride equivalent for pilocarpine hydrochloride is 0.24.

Solution. 0.702 g

$$90 \text{ mL} \times \frac{0.5 \text{ g}}{100 \text{ mL}} = 0.45 \text{ g pilocarphine hydrochloride}$$

Step 1: $90 \text{ mL} \times \frac{0.009 \text{ g}}{1 \text{ mL}} = 0.81 \text{ g NaCl}$

Step 2: $0.45 \text{ g} \times 0.24 = 0.108 \text{ g Nacl}$

Step 3: $0.81 \text{ g} - 0.108 \text{ g} = 0.702 \text{ g NaCl}$

---

6. ℞ Phenylephrine HCl 0.25%

Zinc sulfate 0.5%

Sodium chloride qs

Purified water, qs ad 60.0 mL

How many grams of sodium chloride should be used to make this solution isotonic? Ho
would you prepare this solution? (The sodium chloride equivalent (E) of phenylephrin
hydrochloride is 0.32; that of zinc sulfate is 0.15.)

Solution. 0.447 g

To prepare the solution, dissolve 0.15 g of phenylephrine hydrochloride, 0.3 g of zin
sulfate, and 0.447 g of sodium chloride in sufficient water to make 60 mL.

CALCULATIONS

If NaCl were the only solute, the amount needed would be

$$60.0 \text{ mL} \times \frac{0.009 \text{ g}}{1 \text{ mL}} = 0.54 \text{ g}$$

For phenylephrine HCl:

$$60 \text{ mL} \times 0.0025 \text{ g/mL} = 0.15 \text{ g}$$

$$0.15 \text{ g} \times 0.32 = 0.048 \text{ g NaCl}$$

For zinc sulfate:

$$60 \text{ mL} \times 0.005 \text{ g/mL} = 0.3 \text{ g}$$

$$0.3 \text{ g} \times 0.15 = 0.045 \text{ g NaCl}$$

The total amount of sodium chloride equivalent to the osmotic pressure exerted by bot
drugs is: 0.048 g + 0.045 g = 0.093 g NaCl

The amount of sodium chloride needed is: 0.54 g − 0.093 g = 0.447 g

---

7. How much sodium chloride should be included in 1 L of solution containing 1% gen
tamicin sulfate and 1:1000 benzalkonium chloride? Sodium chloride equivalents
gentamicin sulfate, 0.05; benzalkonium chloride, 0.16.

Solution. 8.34 g

CALCULATIONS

$$1000 \text{ mL} \times \frac{0.009 \text{ g}}{1 \text{ mL}} = 9 \text{ g NaCl}$$

$$\text{Gentamicin sulfate: } 1000 \text{ mL} \times \frac{1 \text{ g}}{100 \text{ mL}} \times 0.05 = 0.5 \text{ g}$$

Benzalkonium chloride: $1000 \text{ mL} \times \frac{1 \text{ g}}{1000 \text{ mL}} \times 0.16 = 0.16 \text{ g}$

$9 \text{ g} - (0.5 + 0.16) \text{ g} = 8.34 \text{ g}$ NacL needed

---

8. ℞ Procaine HCl, 2%

Ft. 30 mL

M. Ft. isotonic solution with sodium chloride.

Explain how the formula will be compounded.

Solution.

Weigh 0.6 g of procaine HCl

Weigh 0.144 g of NaCl

Dissolve in enough water to make 30 mL of isotonic solution

CALCULATIONS

$30 \text{ mL} \times \frac{0.009 \text{ g}}{1 \text{ mL}} = 0.27$ g of NaCl in 30 mL of an isotonic NaCl solution

$30 \text{ mL} \times \frac{2 \text{ g}}{100 \text{ mL}} = 0.6$ g of procaine HCl required

$0.6 \text{ g} \times 0.21 = 0.126$ g of NaCl represented by procaine HCl

$0.27 - 0.126 = 0.144$ g of NaCl required to make the solution isotonic

ISOTONIC BUFFERED SOLUTIONS

9. Sometimes, ophthalmic solutions need to be made isotonic and buffered to a certain pH. Isotonic buffered diluting solutions are added to the isotonic solution of a drug to produce the volume required. The method entails adding a sufficient volume of distilled water to a given amount of drug to make an isotonic solution of the drug. This isotonic solution is made to a final volume with an *isotonic vehicle*. Because both the drug solution and the vehicle are isotonic, the combination remains isotonic.

The USP describes two methods but we will focus on the White–Vincent method.

The *White–Vincent method* is based on the following equation:

$$V = W \times \text{NaCl equivalent} \times \frac{[100]}{[0.9]}$$

where, W is the weight of the drug (in grams) and V is the volume (in mL) of isotonic solution that can be prepared with W grams of drug.

As an example, we will prepare an isotonic solution and make it to a final volume with an isotonic vehicle.

℞ Epinephrine HCl (E = 0.29) 0.5%
Sterile preserved water, q.s. to make 15 mL
M. Ft. collyr. isotonic and buffered to pH 5.0

$$\text{weight of drug required} = \frac{[0.5]}{[100]} \times 15 = 0.075 \text{ g}$$

$$V = 0.075 \text{ g} \times 0.29 \times \frac{[100]}{[0.9]} = 2.4 \text{ mL}$$

Conclusion: Dissolve 0.075 g epinephrine HCl in sterile H_2O up to 2.4 mL and complete the volume to 15 mL with an isotonic vehicle buffered to pH 5.0.

Use the White–Vincent method in the following example.

---

10. ℞ Ingredient Z ($E = 0.18$) 2%
Sterile preserved water, q.s. 30 mL
Isotonic acetate buffer, q.s.
Sig. For the nose as decongestant.

Solution. Dissolve 0.6 g ingredient Z in water up to 12 mL and complete the volume to 30 mL with isotonic acetate buffer.

CALCULATIONS

$$W = \frac{2 \text{ g}}{100 \text{ mL}} \times 30 \text{ mL} = 0.6 \text{ g}$$

$$V = 0.6 \text{ g} \times 0.18 \times \frac{100 \text{ mL}}{0.9 \text{ g}} = 12 \text{ mL}$$

---

11. ℞ Homatropine hydrobromide (E = 0.17) 1%
M. Ft. collyr. isotonic 60 mL

The USP recommends ophthalmic solutions of salts of homatropine to be buffered at pH 6.8 and suggests a formula for a suitable isotonic phosphate buffer.

Solution. Dissolve 0.6 g homatropine HBr in sterile H_2O up to 11.3 mL and dilute to 60 mL with isotonic phosphate buffer pH 6.8.

CALCULATIONS

$W = 1/100 \times 60 = 0.6\ \text{g}$

$$V = 0.6\ \text{g} \times 0.17 \times \frac{100\ \text{mL}}{0.9\ \text{g}} = 11.3\ \text{mL}$$

Then, the prescription would be filled as follows:

Homatropine hydrobromide		600 mg
Sterile water,	to make	11.3 mL
Isotonic phosphate buffer, pH 6.8 q.s.	ad	60 mL

---

12. When an ophthalmic solution of a drug does not require a buffer, and drug stability is maintained at a pH of about 5, as is the case with many anti-infective agents, local anesthetics agents, and antihistamines, a *1.9% boric acid solution* (boric acid vehicle, USP) serves as a suitable vehicle. To the isotonic solution of the drug calculated through the White–Vincent method, sufficient 1.9% boric acid solution is added to bring the solution to volume.

    Calculate the volume of tetracaine solution to prepare for dilution in the following prescription:

℞	Tetracaine hydrochloride (E = 0.18)	0.5%
	M. Ft. collyr. isotonic	30 mL

Solution. Dissolve 0.15 g tetracaine HCl in sterile H_2O up to 3 mL and dilute to 30 mL with boric acid vehicle.

CALCULATIONS

$$W = \frac{0.5\ \text{g}}{100\ \text{mL}} \times 30\ \text{mL} = 0.15\ \text{g}$$

$$V = 0.15 \times 0.18 \times \frac{100\ \text{mL}}{0.9\ \text{g}} = 3\ \text{mL}$$

---

13. ℞ Oxytetracycline hydrochloride (E = 0.12) 0.05 g

    M. Ft. collyr. Isotonic 30 mL

Use boric acid vehicle to make to final volume.

Solution. Dissolve 50 mg oxytetracycline HCl in sterile H_2O up to 0.7 mL and dilute t 30 mL with boric acid vehicle.

CALCULATIONS

$W = 0.05\text{ g}$

$$V = 0.05 \times 12 \times \frac{100\text{ mL}}{0.9\text{ g}} = 0.7\text{ mL}$$

OTHER TONICITY AGENTS

14. Sometimes, an agent other than sodium chloride (boric acid, dextrose, sodium nitrate may be used to adjust osmotic pressure. Silver salts would precipitate in the presenc of chloride ion, so another salt such as sodium nitrate might be used. Also, for som ophthalmic solutions and injectables, sodium chloride cannot be used, or there is a indication to the contrary on the prescription. In these cases, the solution may be mad isotonic with *boric acid or dextrose.*

If a substance other than sodium chloride is to be employed to raise osmotic pressure, th *calculation* procedure requires *one extra step.*

To begin, follow the same procedure as though sodium chloride were to be used.

Extra step: After calculating the amount of sodium chloride needed, divide tha amount by the sodium chloride equivalent of the substance (tonicity agent) that will b used to raise the osmotic pressure of the solution. This produces the required result.

As an example, let us calculate the amount of sodium nitrate needed to make 200 mL o a 0.6% solution of silver nitrate isotonic. The sodium chloride equivalent of silver nitrat is 0.33. That of sodium nitrate is 0.68.

If only sodium chloride were present, the amount needed would be

$$200\text{ mL} \times \frac{0.009\text{ g}}{1\text{ mL}} = 1.8\text{ g NaCl}$$

The amount of silver nitrate needed is

$$200\text{ mL} \times \frac{0.6\text{ g}}{100\text{ mL}} = 1.2\text{ g}$$

This is osmotically equivalent to

$$1.2 \times 0.33 = 0.396\text{ g NaCl}$$

The amount of sodium chloride that would have to be added is

$$1.8\text{ g} - 0.396\text{ g} = 1.404\text{ g NaCl}$$

To find the amount of sodium nitrate that should be used to adjust osmotic pressure, w divide by the sodium chloride equivalent of that substance.

$$\frac{1.404}{0.68} = 2.06\text{ g } NaNO_3$$

Calculate the amount of boric acid that should be included in 300 mL of a 0.4% procaine hydrochloride solution to render the solution isotonic. Procaine hydrochloride, E = 0.21; boric acid, E-value = 0.52.

Solution. 4.7 g

CALCULATIONS

$$300 \text{ mL} \times \frac{0.009 \text{ g}}{1 \text{ mL}} = 2.7 \text{ g NaCl}$$

$$300 \text{ mL} \times \frac{0.4 \text{ g}}{100 \text{ mL}} = 1.2 \text{ g procaine hydrochloride}$$

$$1.2 \times 0.21 = 0.252 \text{ g NaCl}$$

$$2.7 \text{ g} - 0.252 \text{ g} = 2.45 \text{ g NaCl needed}$$

$$\frac{2.45 \text{ g}}{0.52} = 4.7 \text{ g boric acid}$$

---

15. ℞

Ephedrine sulfate (E = 0.23)		0.3 g
Chlorobutanol (E = 0.24)		0.15 g
Dextrose monohydrate (E = 0.16)		q.s.
Rose water	ad	30 mL

Make solution isotonic with nasal fluid.

Calculate the quantity of dextrose monohydrate to use

Solution. 1.03 g

CALCULATIONS

$$30 \text{ mL} \times \frac{0.009 \text{ g}}{1 \text{ mL}} = 0.27 \text{ g NaCl}$$

$$0.3 \text{ g} \times 0.23 = 0.069 \text{ g NaCl represented by ephedrine sulfate}$$

$$0.15 \text{ g} \times 0.24 = 0.036 \text{ g NaCl represented by chlorobutanol}$$

$$0.27 \text{ g} - (0.069 + 0.036) = 0.165 \text{ g NaCl}$$

$$\frac{0.165 \text{ g}}{0.16} = 1.03 \text{ g dextrose}$$

16. Sometimes, a prescription formula includes an isotonic solution as an ingredient. In this case, the prescription volume that needs to be made isotonic will be the total volume prescribed, less the volume of the isotonic ingredient. The prescription will be prepared by dissolving the drug(s) and tonicity agent in a sufficient volume of distilled water to make the solution isotonic. This isotonic solution is then made to the final volume in the prescription with the *isotonic solution* listed in the formula. Because both the drug solution and the isotonic ingredient are isotonic, the combination remains isotonic.

℞		
Tetracaine hydrochloride Sol. 2%		15 mL
Epinephrine bitartrate (E = 0.18)		0.1%
Boric acid (E = 0.52)		q.s.
Sterile water	ad	30 mL
M. Ft. sol. isot. with tears.		

The solution of tetracaine HCl 2% is already isotonic. How many milliliters of a 2.5% solution of boric acid should be used in compounding the prescription?

Solution. 10 mL of 2.5% boric acid solution

CALCULATIONS

30 mL (total volume prescribed) – 15 mL (already isotonic) = 15 mL (to be made isotonic)

$$15 \text{ mL} \times \frac{0.009 \text{ g}}{1 \text{ mL}} = 0.135 \text{ g NaCl (total needed)}$$

Because the amount of drug will be present in the total volume,

$$30 \text{ mL} \times \frac{0.1 \text{ g}}{100 \text{ mL}} = 0.03 \text{ g epinephrine bitartrate}$$

$0.03 \text{ g} \times 0.18 = 0.0054$ g NaCl represented by ephedrine

$0.135 - 0.0054 = 0.1296$ g NaCl (to be added)

$$\frac{0.1296 \text{ g}}{0.52} = 0.249 \text{ g boric acid}$$

$$0.249 \text{ g} \times \frac{100 \text{ mL}}{2.5 \text{ g}} = 9.96 = 10 \text{ mL boric acid solution } 2.5\%$$

Conclusion: Dissolve 0.03 g epinephrine bitartrate in 10 mL of 2.5% boric acid solution and make the volume to 15 mL with sterile water. Add 15 mL isotonic tetracaine hydrochloride sol. 2% to obtain 30 mL prescription.

Now it's your turn:

℞	Oxytetracycline HCl (E = 0.12)		2%
	Chlorobutanol sol. 0.01%		10 mL
	Sodium chloride		q.s.
	Sterile water	ad	30 mL
	Make solution isotonic with tears.		

The 0.01% solution of chlorobutanol is already isotonic. How many milliliters of a 0.9% solution of NaCl should be used to prepare the prescription? Explain how you would prepare the solution.

Solution. Use 12 mL of 0.9% NaCl solution. Dissolve 0.6 g oxytetracycline HCl in 12 mL NaCl 0.9% and make volume to 20 mL with sterile water. Add 10 mL chlorobutanol sol. 0.01% to obtain 30 mL prescription.

CALCULATIONS

$$30 \text{ mL} - 10 \text{ mL} = 20 \text{ mL}$$

$$20 \text{ mL} \times \frac{0.009 \text{ g}}{1 \text{ mL}} = 0.18 \text{ g NaCl}$$

$$30 \text{ mL} \times \frac{2 \text{ g}}{100 \text{ mL}} = 0.6 \text{ g oxytetracycline hydrochloride}$$

$$0.6 \text{ g} \times 0.12 = 0.072 \text{ g NaCl}$$

$$0.18 - 0.072 = 0.108 \text{ g NaCl}$$

$$0.108 \text{ g} \times \frac{100 \text{ mL}}{0.9 \text{ g}} = 12 \text{ mL NaCl } 0.9\%$$

---

17. ℞

Cromolyn sodium (E = 0.11)		1%
Chlorobutanol 2.5% solution		10 mL
Sodium chloride		q.s.
Sterile water	ad	50 mL

The 2.5% solution of chlorobutanol is already isotonic. How many milliliters of a 2% solution of NaCl should be used to prepare the prescription? Explain how you would prepare the solution.

Solution. 15.3 mL of 2% NaCl solution.

Dissolve 0.5 g cromolyn sodium in 15.3 mL NaCl 2% and make volume to 40 mL with sterile water. Add 10 mL chlorobutanol sol. 2.5% to obtain 50 mL prescription.

CALCULATIONS

$$50 \text{ mL} - 10 \text{ mL} = 40 \text{ mL}$$

$$40 \text{ mL} \times \frac{0.009 \text{ g}}{1 \text{ mL}} = 0.36 \text{ g NaCl}$$

$$50 \text{ mL} \times \frac{1 \text{ g}}{100 \text{ mL}} = 0.5 \text{ g cromolyn sodium}$$

$$0.5 \text{ g} \times 0.11 = 0.055 \text{ g NaCl}$$

$$0.36 - 0.055 = 0.305 \text{ g NaCl}$$

$$0.305 \text{ g} \times \frac{100 \text{ mL}}{2 \text{ g}} = 15.25 \text{ mL} = 15.3 \text{ mL NaCl } 2\%$$

ISOTONICITY BY FREEZING-POINT DEPRESSION METHOD

18. If USP tables with sodium chloride equivalents are not available, the freezing point depression method is another option to make a solution isotonic. This method is less used than the sodium chloride equivalent method because the concept of freezing point depression seems more complex.

The freezing-point depression method makes use of the fact that the depression of the freezing point of a solution is proportional to the osmotic pressure of the solution. Both of these properties are colligative properties.

The temperature at which plasma freezes is –0.52 °C and so the depression of freezing point is 0.52 °C. A hypotonic solution of any solute will have a higher freezing point than this value, that is, the freezing-point depression will be less than 0.52 °C. By the addition of more solute (or a second solute), the freezing point of the solution can be lowered to –0.52 °C and this will render the solution isotonic with plasma.

Tables of freezing-point depression (*D*) of 1% w/v solutions of drugs are included in many pharmaceutical texts. In making use of such tables, it is assumed that:

(i) In a mixture of solutes, each solute exerts an independent effect upon the freezing-point depression. The total freezing-point depression is thus the sum of the freezing-point depressions of each of the solutes.

(ii) The depression of freezing point is proportional to the concentration.

℞ Procaine HCl, 2%

Ft. 30 mL

M. Ft. isotonic solution with sodium chloride

The freezing-point depression of a 1% w/v solution of procaine hydrochloride is given as 0.122 °C. Thus, its freezing point is –0.122 °C and this can be written as:

$$\frac{-0.122\,^{\circ}\mathrm{C}}{1\%}$$

therefore, a 2% w/v solution has a freezing point of

$$\frac{-0.122\,^{\circ}\mathrm{C}}{1\%} \times 2\% = -0.244\,^{\circ}\mathrm{C}$$

But, the required freezing point is –0.52 °C.

Thus, a further lowering of freezing point is required and equals:

$$0.52\,^{\circ}\mathrm{C} - 0.244\,^{\circ}\mathrm{C} = 0.276\,^{\circ}\mathrm{C}$$

Since 1% w/v sodium chloride solution produces a freezing point depression of 0.576 °C, the concentration required to produce a depression of 0.276 °C is:

$$\frac{0.276\,^{\circ}\mathrm{C}}{0.576\,^{\circ}\mathrm{C}} \times 1\% = 0.48\%$$

The required quantities to make 30 mL solution will be:

$$\text{Procaine hydrochloride}\ \frac{2\ \mathrm{g}}{100\ \mathrm{mL}} \times 30\ \mathrm{mL} = 0.6\ \mathrm{g}$$

$$\text{Sodium chloride}\ \frac{0.48\ \mathrm{g}}{100\ \mathrm{mL}} \times 30\ \mathrm{mL} = 0.144\ \mathrm{g}$$

Now, practice isotonicity using the freezing point depression method.

How many milligrams each of naphazoline hydrochloride and sodium chloride are required to prepare 15 mL of a 2.5% solution of naphazoline hydrochloride isotonic with nasal mucosa? Freezing point depression of a 1% solution of naphazoline hydrochloride is 0.16°C and that of 1% NaCl solution is 0.576 °C.

Solution. 375 mg naphazoline HCl and 31.5 mg NaCl

CALCULATIONS

Required freezing point: −0.52°C

$$\frac{0.16\ ^\circ\text{C}}{1\%} \times 2.5\% = 0.4\ ^\circ\text{C}\ \text{(freezing point of a 2.5\% solution)}$$

$$0.52\ ^\circ\text{C} - 0.4\ ^\circ\text{C} = 0.12\ ^\circ\text{C}\ \text{(further lowering of freezing point required)}$$

$$\frac{0.12\ ^\circ\text{C}}{0.576\ ^\circ\text{C}} \times 1\% = 0.21\%\ \text{(NaCl solution needed to lower 0.12 °C)}$$

$$\frac{2.5\ \text{g}}{100\ \text{mL}} \times 15\ \text{mL} = 0.375\ \text{g} = 375\ \text{mg of naphazoline HCl}$$

$$\frac{0.21\ \text{g}}{100\ \text{mL}} \times 15\ \text{mL} = 0.0315\ \text{g} = 31.5\ \text{mg NaCl}$$

---

19. The freezing point depression of a 1% solution of ephedrine sulfate is 0.13 °C and that of 1% NaCl solution is 0.576 °C. How many grams of each of ephedrine sulfate and NaCl are required to prepare 30 mL of a 1% ephedrine sulfate solution isotonic with tears?

Solution. 0.3 g ephedrine sulfate and 0.204 g NaCl

CALCULATIONS

$$0.52\ ^\circ\text{C} - 0.13\ ^\circ\text{C} = 0.39\ ^\circ\text{C}$$

$$\frac{1\%}{0.576\ ^\circ\text{C}} = \frac{z}{0.39\ ^\circ\text{C}}$$

$z = 0.68\%$ (NaCl sol. needed)

30 mL × 1% = 0.3 g ephedrine sulfate

30 mL × 0.68% = 0.204 g NaCl

---

20. How many milligrams of sodium chloride should be used to make the following prescription isotonic? Use the freezing point depression method.

℞	Naphazoline HCl (fz. pt = −0.16 °C)		1%
	NaCl		q.s.
	Purified water	ad	15 mL

Solution. 94 mg NaCl

CALCULATIONS

$0.52\,°C - 0.16\,°C = 0.36\,°C$

Fz. pt of 1% NaCl solution = 0.576 °C

$$\frac{1\%}{0.576\,°C} = \frac{z}{0.36\,°C}$$

$z = 0.625\%$ (NaCl sol. needed)

15 mL × 1% = 0.15 g naphazoline HCl

15 mL × 0.625% = 0.094 g NaCl

Finally, if it occurs that a prescription has more than one medicinal/pharmaceutic ingredient, their freezing points are added and then subtracted from the required value of agent used to provide isotonicity.

pH AND BUFFERS

21. pH is a symbol for the negative logarithm of the hydrogen ion concentration (moles/liter).

$$\boxed{pH = -\log[H^+] = -\log[H_3O^+]} \qquad \boxed{[H^+] = 10^{-pH}}$$

Since H^+ (hydrogen ion) does not exist in aqueous solutions but is hydrated, H_3O^+ (hydronium ion) is often used.

A solution with pH 7 is neutral at 22 °C, one of a pH > 7 is alkaline, and one with a pH < 7 is acid.

pH versus Hydrogen Ion Concentration

22. The following examples will illustrate the conversion of $[H^+]$ to pH and vice-versa. If the hydrogen ion concentration of a solution is 0.025 M, what would be its pH?

$$pH = -\log(0.025)$$

$$pH = -(-1.6)$$

$$pH = 1.6$$

An isotonic vehicle has a pH of 5.0. What would be its hydrogen ion concentration?

$$[H^+] = 10^{-pH}$$

$$[H^+] = 10^{-5.0} = 0.00001\ M$$

Now practice with the following problems before checking your answers.

A. Calculate the pH of the following solutions. Consider their $[H^+]$:

(a) 3.5×10^{-3} M

(b) 1.45×10^{-5} M

(c) 0.0075 M

B. Calculate the hydrogen ion concentration of the following solutions. Consider their pH:

(a) 4.25

(b) 2.50

(c) 10.45

Solutions.

A. (a) pH = 2.46
(b) pH = 4.84
(c) pH = 2.12

B. (a) $[H^+] = 5.6 \times 10^{-5}$ M
(b) $[H^+] = 3.16 \times 10^{-3}$ M
(c) $[H^+] = 3.54 \times 10^{-11}$ M

CALCULATIONS

A. (a) $pH = -\log(3.5 \times 10^{-3}) = \log 10^3 - \log 3.5 = 3 - 0.54 = 2.46$
(b) $pH = -\log(1.45 \times 10^{-5}) = -(\log 1.45 + \log 10^{-5}) = -(0.16 - 5) = 4.84$
(c) $pH = -\log(0.0075) = -(-2.12) = 2.12$

B. (a) $[H^+] = 10^{-4.25} = 5.6 \times 10^{-5}$ M
(b) $[H^+] = 10^{-2.5} = 3.16 \times 10^{-3}$ M
(c) $[H^+] = 10^{-10.45} = 3.54 \times 10^{-11}$ M

BUFFERS

23. *Buffer, buffer solution*, and *buffered solution* are terms related to hydrogen ion concentration or pH. These terms refer to the ability of a system, particularly an aqueous solution, to resist change of pH on adding acid or alkali, or on diluting it with solvent.

If an acid or base is added to water, the pH of the water will change markedly, for water cannot resist change of pH—it is completely without *buffer action*. Even a very weak acid such as carbon dioxide changes the pH of water (from 7 to 5.7) when its small concentration present in air is equilibrated with pure water. Solutions of neutral salts such as sodium chloride similarly lack ability to resist change of pH on adding acid or base; such solutions are *unbuffered*. This susceptibility of distilled water and neutral salt solutions to change of pH on adding very small amounts of acid or base is often of great concern in pharmaceutical operations.

Buffered solutions undergo negligible changes in pH on addition of small quantities of acid or base. *Buffer action* involves a conjugate acid–base pair in the solution, such as either a weak acid and a salt of the weak acid or a weak base and a salt of the weak base. A buffer solution made of acetic acid and sodium acetate has acetate ion as the conjugate base of acetic acid. An ammonium hydroxide/ammonium chloride buffer solution has the ammonium ion as the conjugate acid of ammonia (= ammonium hydroxide). With both buffer systems mentioned above, the change of pH is slight if the amount of hydronium or hydroxyl ion added does not exceed the capacity of the buffer system to neutralize it.

Another pharmaceutically important buffer system is monobasic potassium phosphate (KH_2PO_4) and dibasic potassium phosphate (K_2HPO_4). This buffer is actually a weak-acid–conjugate-base buffer (similar to acetic acid/sodium acetate) in which an ion, $H_2PO_4^-$, serves as the weak acid, and HPO_4^{2-} is its conjugate base.

When hydroxyl ion is added to this buffer the following reaction takes place:

$$H_2PO_4^- + OH^- \rightarrow HPO_4^{2-} + H_2O$$

And, when hydronium ion is added, the following occurs:

$$HPO_4^{2-} + H_3O^+ \rightarrow + HPO_4^- + H_2O$$

Most drug molecules are either weak acids or weak bases. These molecules can exist in either the unionized or the ionized state, and the degree of ionization depends on the dissociation constant (K_a) of the drug and the pH of the environment. An equation for the ionization of a weak acid, HA, can be derived from the following equations:

$$HA + H_2O = H_3O^+ + A^-$$

$$A^- + H_2O = HA + OH^-$$

$$K_a = \frac{[H_3O^+][A^-]}{[HA]}$$

---

24. The resulting equation, obtained by taking the logarithm of both sides of K_a equation and multiplying by −1, is the Henderson–Hasselbalch equation, also known as the *buffer equation*.

For a weakly *acidic* drug:

$$\boxed{pH = pK_a + \log\frac{[A^-]}{[HA]}}$$

or

$$\boxed{pH = pK_a + \log\frac{[salt]}{[acid]}}$$

A^- = its conjugate base (the salt)

HA = the weak acid

pK_a = the logarithm of the reciprocal of the acid dissociation constant

$$pK_a = -\log K_a$$

Similarly, the equation for the ionization of a weak base, B, can be derived from:

$$B + H_3O^+ = BH^+ + H_2O$$

$$BH^+ + OH^- = H_2O + B$$

For a weakly *basic* drug:

$$pH = pK_a + \log\frac{[B]}{[BH^+]}$$

or

$$pH = pK_a + \log\frac{[base]}{[salt]}$$

$$pH = pK_w - pK_b + \log\frac{[base]}{[salt]}$$

B = the weak base

BH^+ = its conjugate acid (salt)

pK_w = autoprotolysis constant or ion product of water

pK_b = the logarithm of the reciprocal of the base dissociation constant

$$pK_w = pK_a + pK_b$$

For acidic drugs at pH values below their pK_a, the drug molecule would exist predominantly in the unionized form, the form best suited for passage of the drug across the biologic membranes. In alkaline pH, where the drug is largely in the ionized, lipid-insoluble form, the rate of membrane penetration would be relatively low, and access to the site of biologic activity would be reduced.

---

25. The Henderson–Hasselbalch equation also applies to all buffer systems formed from a single conjugate acid–base pair, no matter the nature of the salts.

As an example of the application of this equation, we will calculate the pH of a buffer solution containing acetic acid and sodium acetate, each in 0.1 M concentration. The K_a of acetic acid is 1.75×10^{-5} at 25 °C.

The pK_a of acetic acid is calculated by:

$$pK_a = -\log K_a = -\log 1.75 \times 10^{-5} = -\log 1.75 - \log 10^{-5} = -0.24 - (-5) = 4.76$$

Substituting this value into the Henderson–Hasselbalch equation:

$$pH = 4.76 + \log\frac{0.1}{0.1} = 4.76$$

The Henderson–Hasselbalch equation predicts that any solutions containing the same molar concentration of both acid and its conjugate base will have the same pH. Thus, a solution of 0.01 M concentration of each component will have the same pH as one of 0.1 M concentration of each component. Actually, there will be some difference in the pH of the solutions but for most practical purposes the approximate values of pH calculated by the Henderson–Hasselbalch are satisfactory. It should be pointed out, however, that *the buffer of higher concentration of each component will have a much greater capacity for neutralizing added acid or base.*

Now, practice the application of the Henderson–Hasselbalch equation to predict the pH of some buffer solutions.

What is the pH of a buffer solution containing 0.05 M boric acid and 0.5 M sodium borate in 1 liter of solution? The K_a value of boric acid is 6.4×10^{-10} at 25 °C.

Solution. pH = 10.2

CALCULATIONS

$$pK_a = -(\log K_a)$$

$$pK_a = -(\log 6.4 + \log 10^{-10}) = -(0.81 - 10) = -(-9.19) = 9.19$$

$$pH = pK_a + \log\frac{[salt]}{[acid]} = 9.19 + \log\frac{0.5}{0.05} = 9.19 + 1 = 10.19$$

---

26. A buffer solution was prepared with 0.02 M ammonium hydroxide (ammonia) and 0.2 M ammonium chloride. What would be the pH of this buffer solution, assuming that K_b value of ammonia at 25 °C is 1.8×10^{-5} and K_w value for water is 10^{-14}?

Solution. pH = 8.25

CALCULATIONS

$$pH = pK_w - pK_b + \log\frac{[base]}{[salt]}$$

$$pK_w = -(\log K_w) = -(\log 10^{-14}) = -(-14) = 14$$

$$pK_b = -(\log K_b) = -(\log 1.8 + \log 10^{-5}) = -(0.255 - 5) = 4.75$$

$$pH = 14 - 4.75 + \log\frac{0.02}{0.2} = 9.25 - 1 = 8.25$$

27. The Henderson–Hasselbalch equation is also useful for calculating the *ratio of molar concentrations* of a buffer system required to produce a solution of specific pH.

As an example, suppose that an acetic acid–sodium acetate buffer of pH 4.5 is to be prepared. What molar ratio of the buffer components should be used? The pK_a of acetic acid is 4.76 at 25 °C.

Rearranging the Henderson–Hasselbalch equation:

$$\log\frac{[\text{salt}]}{[\text{acid}]} = \text{pH} - \text{p}K_a = 4.5 - 4.76 = -0.26$$

$$\frac{[\text{salt}]}{[\text{acid}]} = \text{antilog}(-0.26)$$

$$\frac{[\text{salt}]}{[\text{acid}]} = 0.55$$

or

$$\frac{[\text{salt}]}{[\text{acid}]} = 10^{-0.26} = 0.55 \text{ or } 0.55{:}1$$

The interpretation of this result is that the proportion of sodium acetate (salt) to acetic acid should be 0.55 moles of sodium acetate to 1 mole of acetic acid to produce a pH of 4.5. A solution containing 0.055 mole of sodium acetate and 0.1 mole of acetic acid per liter would also meet this requirement. The actual concentration selected would depend mainly on the desired buffer capacity.

What is the molar ratio of salt/acid needed to prepare a sodium benzoate/benzoic acid buffer solution with a pH of 3.20? The pK_a of benzoic acid is 4.20 at 25 °C.

Solution. 0.1 : 1

CALCULATIONS

$$\text{pH} = \text{p}K_a + \log\frac{[\text{salt}]}{[\text{acid}]}$$

$$3.20 - 4.20 = -1$$

$$\log\frac{[\text{salt}]}{[\text{acid}]} = -1$$

$$\frac{[\text{salt}]}{[\text{acid}]} = 10^{-1} = 0.1$$

Molar ratio: 0.1 : 1 (0.1 moles of sodium benzoate to 1 mole of benzoic acid will produce a pH of 3.20)

28. What molar ratio of base/salt is required to prepare a buffer solution having a pH of 8.0? Assume the pK_b of the base is 6.6 at 25 °C.

Solution. 4 : 1

CALCULATIONS

$$pH = pK_a + \log\frac{[base]}{[salt]}$$

$$pH = pK_w - pK_b + \log\frac{[base]}{[salt]}$$

pK_w for water = -10^{-14}, $pK_w = 14$

$$8 = 14 - 6.6 + \log\frac{[base]}{[salt]}$$

$$8 - 7.4 = 0.6 = \log\frac{[base]}{[salt]}$$

$$\frac{[base]}{[salt]} = 10^{0.6} = 4$$

Molar ratio: 4:1 (4 moles of base to 1 mole of salt will produce a pH of 8)

29. Another application of the molar ratio of the components of a buffer system is the calculation of the *amounts* of each component of a known concentration of a buffer solution in a desired volume.

For example, the molar ratio of salt/acid in a buffer solution with a pH of 5.8 is 4:1. Assuming the final buffer concentration is 5.5×10^{-3} mol/liter, how many grams of salt (MW = 120) and how many grams of acid (MW = 80) should be used to prepare 1 liter of the buffer?

Molar ratio of salt: acid = 4 : 1

Mole fraction of salt = 4/5

Mole fraction of acid = 1/5

Final buffer conc. = 5.5×10^{-3} mol/liter

Conc. of salt = $4/5 \times (5.5 \times 10^{-3}$ mol/L) = 4.4×10^{-3} mol/L

Conc. of acid = $1/5 \times (5.5 \times 10^{-3}$ mol/L) = 1.1×10^{-3} mol/L

Grams of salt: $(4.4 \times 10^{-3}) \times 120 = 0.53$ g of salt for 1 liter

Grams of acid: $(1.1 \times 10^{-3}) \times 80 = 0.09$ g of acid for 1 liter

The concentration of a pH 6.2 buffer is 2.5×10^{-5} mols/liter. If the molar ratio of salt/acid is 6:2 how many milligrams of salt (MW = 100) and how many milligrams of acid (MW = 50) should be used to prepare 1 liter of the buffer?

Solution. 1.9 mg of salt and 0.32 mg of acid for 1 liter.

CALCULATIONS

Molar ratio of salt: acid = 6 : 2

Mole fraction of salt = 6/8

Mole fraction of acid = 2/8

Final buffer conc. = 2.5×10^{-5} mol/liter

Conc. of salt = $6/8 \times (2.5 \times 10^{-5}$ mols/L) = 1.88×10^{-5} mol/L

Conc. of acid = $2/8 \times (2.5 \times 10^{-5}$ mols/L) = 0.63×10^{-5} mol/L

Grams of salt: $(1.88 \times 10^{-5}) \times 100 = 0.0019$ g = 1.9 mg of salt for 1 liter

Grams of acid: $(0.63 \times 10^{-5}) \times 50 = 0.00032$ g = 0.32 mg of acid for 1 liter

---

30. *Buffer capacity* is the ability of a buffer solution to resist changes in pH upon addition of acid or alkali. In a general way, the concentration of acid in a weak acid–conjugate base buffer determines the capacity to neutralize added base, while the concentration of salt of the weak acid determines the capacity to neutralize added acid. This capacity to neutralize added acid or base applies similarly for a weak base–conjugate acid buffer. Finally, the degree of buffer action, and therefore the buffer capacity, is dependent on the kind and concentration of the buffer components, the ratio of the components, the pH region involved, and the kind of acid or alkali added.

The ability to resist change in pH on adding acid or alkali is possessed also by relatively concentrated solutions of strong acids and strong bases. In general, solutions of strong acids of pH 3.0 or less, and solutions of strong bases of pH 11 or more, exhibit this kind of buffer action by virtue of the relatively high concentration of hydronium or hydroxyl ions present. The USP includes among its "Standard Buffer Solutions" a series of hydrochloric acid buffers, covering the pH range 1.2 to 2.2. Other buffer systems used in pharmaceutical formulations include boric acid/sodium borate, acetic acid/sodium acetate and sodium acid phosphate/disodium phosphate.

The successful completion of many pharmaceutical preparations and pharmacopeial tests and assays requires adjustment to or maintenance of a specified pH by the addition of buffer solutions. The most common buffer systems are used:

(a) To establish hydrogen-ion activity for the calibration of pH meters;

(b) In analytical procedures;

(c) In the preparation of dosage forms that approach isotonicity;

(d) To prevent irritation of mucous membranes or small veins;

(e) To maintain stability of various pharmaceutical dosage forms.

Buffers used in physiological systems are carefully chosen so as not to interfere with pharmacological activity of the drug or normal functions of the organism. It is essential that buffers used in chemical analysis be compatible with the substance determined and the reagents used.

REVIEW PROBLEMS

Solve all these before checking the answers.

A. ℞ Cocaine hydrochloride 1.2 g

Sodium chloride qs

Purified water qs ad 30 mL

How much sodium chloride should be used to make this eye solution isotonic? The sodium chloride equivalent of cocaine hydrochloride is 0.16.

B. A pharmacist has to prepare 500 mL of a 1% solution of procaine hydrochloride for use in the eye. How many grams of sodium chloride should be used to make the solution isotonic with tears? The sodium chloride equivalent of procaine hydrochloride is 0.21.

C. A solution for use as an eye drop is to contain 0.4 g of ephedrine sulfate and 1 g of tetracycline hydrochloride per 100 mL. How many grams of sodium chloride must each 100 mL of the solution contain if it is to be isotonic with tears? Sodium chloride equivalent of ephedrine sulfate is 0.23; that of tetracycline hydrochloride is 0.14.

D. A pharmacist has to prepare 30 mL of 1% tetracaine hydrochloride solution. How much boric acid should be used to make this solution isotonic? The sodium chloride equivalent of tetracaine hydrochloride is 0.18; the sodium chloride equivalent of boric acid is 0.50.

E. This prescription contains the formula for an ophthalmic solution.

℞		
	Ingredient X	1.5%
	Zinc sulfate	0.2%
	Sodium chloride	q.s.
	Water, sufficient to make	100 mL

Calculate the amount of sodium chloride required to render 1.50 L of the solution isotonic. Sodium chloride equivalents: ingredient X: 0.22; zinc sulfate: 0.16.

F.

℞			
	Atropine sulfate		1.2%
	Boric acid		q.s.
	Purified water	qs ad	15 mL
	M. ft. opth. sol.		

The sodium chloride equivalents are atropine sulf.: 0.13; boric acid: 0.50. Calculate the amount of boric acid needed to make this solution isotonic.

G. How many milliliters of water must be added to seven tablets of NaCl, 250 mg each to prepare an isotonic saline solution?

H. How many milligrams of sodium chloride are needed to compound the following prescription?

℞			
	Phenylephrine HCl (E = 0.32)		0.5%
	Purified water	q.s.	30 mL
	Make isotonic with NaCl		

The pharmacist obtains the chemical by dilution of a 1% isotonic phenylephrine hydrochloride solution.

I. How many milligrams of NaCl are needed to compound the following prescription?

℞		
Procaine HCl (E = 0.21)		0.3 g
NaCl		q.s.
SWFI	q.s.	30 mL

J. How many milligrams of NaCl are required to prepare the following prescription? Epinephrine solution is isotonic.

℞		
Atropine sulfate (E = 0.14)		1%
Epinephrine solution 1:2000		8 mL
Purified water	q.s.	30 mL

K. ℞

Homatropine hydrobromide (E = 0.17)		1%
Chlorobutanol (E = 0.24)		0.2%
Purified water	ad	50 mL

Make isotonic solution and buffer to pH 6.8.

You have on hand an isotonic buffer solution pH 6.8. How many milliliters of purified water and how many milliliters of buffer should be used in compounding the prescription?

For Problems L–N, determine if the following drug products are *hypotonic*, *isotonic*, or *hypertonic*.

L. A parenteral bag containing 250 mL of dextrose 5% in water (E = 0.18)

M. Nasal drops containing ephedrine hydrochloride (E = 0.29) 10 mg/mL and chlorobutanol (E = 0.24) 0.5% in rose water.

N. A solution containing 10 g chloramphenicol sodium succinate (E = 0.10) and 0.5 g lidocaine hydrochloride (E = 0.22) in 100 mL of aqueous solution.

O. If the hydronium ion concentration $[H_3O^+]$ of a solution is 1.45×10^{-6} M, what will be the pH of the solution?

P. The hydrogen ion concentration $[H^+]$ of a solution is 0.5 mM. What is its pH? (1 M = 1000 mM)

Q. Phosphate buffer, USP, is frequently used to prepare ophthalmic solutions. If a phosphate buffer has a pH of 6.8, what would be its hydrogen ion concentration?

R. A drug solution has a pH of 4.8. What would be its hydrogen ion concentration?

S. What is the pH of a buffer solution prepared with 0.25 M sodium acid phosphate and 0.025 M disodium phosphate? At 25 °C, the pK_a value of sodium acid phosphate is 7.21.

T. The dissociation constant of acetic acid at 25 °C is 1.75×10^{-5}. What would be the pH of a buffer solution containing 0.5 M of acetic acid and 1 M of sodium acetate?

U. What is the pH of a solution containing 0.2 mole of a weakly basic drug and 0.02 mole of its salt per liter of solution? The pK_a of the drug is 9.36 at 25 °C.

V. What is the pH of a solution, with a hydronium ion concentration of 2×10^{4} M?

W. What molar ratio of salt/acid is required to adjust the pH of a solution to 5.6 using sodium acetate/acetic acid buffer solution? pK_a of acetic acid is 4.76 at 25 °C.

X. What is the molar ratio of salt/acid needed to prepare a disodium phosphate/sodium acid phosphate buffer solution with a pH of 8.4? The pK_a of sodium acid phosphate is 5.4.

Y. What molar ratio of base/salt is required to prepare a buffer solution having a pH o 9.2? Assume the pK_b of the base is 5.6 at 25 °C.

Z. The molar ratio of salt/acid in a buffer solution with a pH of 6.8 is 5:1. Assuming the final buffer concentration is 6.7×10^{-2} mol/liter, how many grams of salt (MW = 142) and how many grams of acid (MW = 120) should be used to prepare 1 liter of the buffer?

Solutions to Review Problems

A. 0.078 g NaCl

B. 3.45 g NaCl

C. 0.668 g NaCl

D. 0.415 g boric acid

E. 8.07 g NaCl

F. 0.224 g boric acid

G. 194.4 mL

H. 135 mg

I. 207 mg

J. 156 mg

K. 37.9 mL purified water and 12.1 mL isotonic buffer pH 6.8

L. Isotonic

M. Hypotonic

N. Hypertonic

O. pH = 5.84

P. pH = 3.3

Q. 1.58×10^{-7} M

R. $[H^+] = 1.58 \times 10^{-5}$ M

S. pH = 6.21

T. pH = 5.06

U. pH = 10.36

V. pH = 3.7

W. 6.9:1

X. 1000:1

Y. 6.3 : 1

Z. 7.92 g of salt and 1.34 g of acid for 1 liter

Calculations for Review Problems

A. $30 \text{ mL} \times \frac{0.009 \text{ g}}{1 \text{ mL}} = 0.27 \text{ NaCl}$

$1.2 \text{ g} \times 0.16 = 0.192 \text{ g}$

$0.27 - 0.192 \text{ g} = 0.078 \text{ g NaCl needed}$

B. $500 \text{ mL} \times \frac{0.009 \text{ g}}{1 \text{ mL}} = 4.5 \text{ g NaCl}$

$5 \text{ g procaine HCl} \times 0.21 = 1.05 \text{ g NaCl}$

$4.5 \text{ g NaCl} - 1.05 \text{ g NaCl} = 3.45 \text{ g NaCl needed}$

C. 100 mL requires 0.9 g NaCl for isotonicity

e.s.: $0.4 \text{ g} \times 0.23 = 0.092 \text{ g NaCl}$

t.h.: $1 \text{ g} \times 0.14 = 0.14 \text{ g NaCl}$

$0.9 \text{ g} - (0.092 + 0.14) \text{ g} = 0.668 \text{ g NaCl needed}$

D. $30 \text{ mL} \times \frac{0.009 \text{ g}}{1 \text{ mL}} = 0.27 \text{ g NaCl}$

$30 \text{ mL} \times \frac{1 \text{ g}}{100 \text{ mL}} = 0.3 \text{ g t.h.}$

$0.3 \text{ g} \times 0.18 = 0.054 \text{ g NaCl}$

$0.27 - 0.054 \text{ g} = 0.216 \text{ g NaCl needed}$

$\frac{0.216 \text{ g}}{0.52} = 0.415 \text{ g boric acid}$

E. NaCl alone: $1500 \text{ mL} \times \frac{0.009 \text{ g}}{1 \text{ mL}} = 13.5 \text{ g}$

Ingredient X: $1500 \text{ mL} \times \frac{1.5 \text{ g}}{100 \text{ mL}} = 22.5 \text{ g}$

$22.5 \text{ g} \times 0.22 = 4.95 \text{ g NaCl}$

Zinc sulfate: $1500 \times \frac{0.2 \text{ g}}{100 \text{ mL}} = 3 \text{ g}$

$3 \text{ g} \times 0.16 = 0.48 \text{ g NaCl}$

$13.5 \text{ g NaCl} - (4.95 \text{ g} + 0.48 \text{ g}) = 8.07 \text{ g NaCl}$

F. NaCl alone: $15 \text{ mL} \times \frac{0.009 \text{ g}}{1 \text{ mL}} = 0.135 \text{ g}$

Atropine sulfate: $15 \text{ mL} \times \frac{1.2 \text{ g}}{100 \text{ mL}} = 0.18 \text{ g}$

$0.18 \text{ g} \times 0.13 = 0.023 \text{ g NaCl}$

$0.135 \text{ g} - 0.023 = 0.112 \text{ g NaCl}$

$\frac{0.112 \text{ g}}{0.50} = 0.224 \text{ g boric acid}$

G. $250\,\text{mg} \times 7\,\text{tab} = 1750\,\text{mg}$

$$1.75\text{ g} \times \frac{100\text{ mL}}{0.9\text{ g}} = 194.4\text{ mL}$$

H. Since the drug will come from a solution that is already isotonic:

$$\frac{0.5\text{ g}}{100\text{ mL}} \times 30\text{ mL} = 0.15\text{ g phenylephrine}$$

$$0.15\text{ g} \times \frac{100\text{ mL}}{1\text{ g}} = 15\text{ mL of 1\% phenylephrine; this is already isotonic.}$$

Thus, only 15 mL needs to be made isotonic in the total 30 mL prescribed. $15\,\text{mL} \times 0.009\,\text{g/mL} = 0.135\,\text{g} = 135\,\text{mg}$

Mix 135 mg of NaCl with 15 mL of 1% isotonic phenylephrine solution and add purified water to make 30 mL.

I. $0.3\,\text{g} \times 0.21 = 0.063\,\text{g NaCl}$

$$30\text{ mL} \times \frac{0.9\text{ g}}{100\text{ mL}} = 0.27\text{ g NaCl}$$

$0.27 - 0.063 = 0.207\text{ g} = 207\text{ mg NaCl}$

J. $0.01 \times 30\,\text{mL} \times 0.14 = 0.042\,\text{g NaCl}$ (represented by atropine sulfate)
Epinephrine sol. is isotonic, thus:

$30\,\text{mL} - 8\,\text{mL} = 22\,\text{mL}$ needs to be made isotonic

$$22\text{ mL} \times \frac{0.9\text{ g}}{100\text{ mL}} = 0.198\text{ g}$$

$0.198 - 0.042 = 0.156\text{ g} = 156\text{ mg NaCl}$

K. $$50\text{ mL} \times \frac{0.9\text{ g}}{100\text{ mL}} = 0.45\text{ g NaCl}$$

Homatropine HBr: $50\text{ mL} \times \frac{1\text{ g}}{100\text{ mL}} = 0.5\text{ g}$

$0.5\text{ g} \times 0.17 = 0.085\text{ g NaCl}$

Chlorobutanol: $50\text{ mL} \times \frac{0.2\text{ g}}{100\text{ mL}} = 0.1\text{ g}$

$0.1\text{ g} \times 0.24 = 0.024\text{ g NaCl}$
$0.45\text{ g} - (0.085 + 0.024) = 0.341\text{ g NaCl}$

$$0.341\text{ g NaCl} \times \frac{100\text{ mL } H_2O}{0.9\text{ g NaCl}} = 37.9\text{ mL water}$$

$50\text{ mL} - 37.9\text{ mL} = 12.1\text{ mL}$ isotonic buffer pH 6.8

L. $$250\text{ mL} \times \frac{5\text{g}}{100\text{ mL}} \times 0.18 = 2.25\text{ g NaCl}$$

Compare to 0.9% NaCl: $250\text{ mL} \times \frac{0.9\text{ g}}{100\text{ mL}} = 2.25\text{ g}$

$2.25\text{ g} = 2.25\text{ g}$
The solution is *isotonic*.

M. $10\,mg/mL = 0.01\,g/mL \times 0.29 = 0.0029\,g/mL$

$0.5\,g/100\,mL = 0.005\,g/mL \times 0.24 = 0.0012\,g/mL$

$0.0029 + 0.0012 = 0.0041\,g/mL$ NaCl

Compare to 0.009 g/mL NaCl to be isotonic: $0.0041\,g < 0.009\,g$

The solution is *hypotonic*.

N. $10\,g \times 0.10 = 1\,g/100\,mL$

$0.5\,g \times 0.22 = 0.11\,g/100\,mL$

$1 + 0.11 = 1.11\,g/100\,mL$

Compare to 0.9 g/100 mL NaCl to be isotonic: $1.11\,g > 0.9\,g$

The solution is *hypertonic*.

O. $pH = -\log(1.45 \times 10^{-6}) = -(\log 1.45 + \log 10^{-6}) = -(0.16 - 6) = 5.84$

P. $pH = -\log(0.0005) = -(-3.3) = 3.3$

Q. $[H^+] = 10^{-pH} = 10^{-6.8} = 1.58 \times 10^{-7}\,M$

R. $pH = -\log\,[H^+] = 4.8$

$\log\,[H^+] = -4.8$

$[H^+] = \text{antilog}\,(-4.8) = 1.58 \times 10^{-5}\,M$

OR: $[H^+] = 10^{-pH}$

$[H^+] = 10^{-4.8} = 1.58 \times 10^{-5}\,M$

S. $pH = 7.21 + \log\dfrac{0.025}{0.25} = 7.21 + \log 0.1 = 7.21 + (-1) = 6.21$

T. $pK_a = -(\log K_a)$

$pK_a = -(\log 1.75 + \log 10^{-5}) = -(0.24 - 5) = -(-4.76) = 4.76$

$pH = pK_a + \log\dfrac{[\text{salt}]}{[\text{acid}]} = 4.76 + \log\dfrac{1}{0.5} = 4.76 + 0.3 = 5.06$

U. $pH = pK_a + \log\dfrac{[\text{base}]}{[\text{salt}]} = 9.36 + \log(0.2/0.02) = 9.36 + \log 10$

$pH = 9.36 + 1 = 10.36$

V. $pH = -\log(2 \times 10^4) = \log 10^4 - \log 2 = 4 - 0.3 = 3.69 = 3.7$

W. $pH = pK_a + \log\dfrac{[\text{salt}]}{[\text{acid}]}$

$\log\dfrac{[\text{salt}]}{[\text{acid}]} = pH - pK_a = 5.6 - 4.76 = 0.84$

Antilog of $0.84 = 6.9$

Or $\dfrac{[\text{salt}]}{[\text{acid}]} = 10^{0.84} = 6.9$

Molar ratio = 6.9 : 1 (6.9 moles of salt to 1 mole of acid will produce a pH of 5.6)

X. $\text{pH} = \text{p}K_\text{a} + \log \dfrac{[\text{salt}]}{[\text{acid}]}$

$$8.4 = 5.4 + \log \frac{[\text{salt}]}{[\text{acid}]}$$

$$3 = \log \frac{[\text{salt}]}{[\text{acid}]}$$

$$\frac{[\text{salt}]}{[\text{acid}]} = 10^3 = 1000$$

Molar ratio: 1000 : 1 (1000 moles of disodium phosphate to 1 mole of sodium acid phosphate will produce a pH of 8.4)

Y. $\text{pH} = \text{p}K_\text{a} + \log \dfrac{[\text{base}]}{[\text{salt}]}$

$$\text{pH} = \text{p}K_\text{w} - \text{p}K_\text{b} + \log \frac{[\text{base}]}{[\text{salt}]}$$

K_w for water = 10^{-14}, pK_w = 14

$$9.2 = 14 - 5.6 + \log \frac{[\text{base}]}{[\text{salt}]}$$

$$9.2 - 8.4 = 0.8 = \log \frac{[\text{base}]}{[\text{salt}]}$$

$$\frac{[\text{base}]}{[\text{salt}]} = 10^{0.8} = 6.3$$

Molar ratio: 6.3 : 1 (6.3 moles of base to 1 mole of salt will produce a pH of 9.2)

Z. Molar ratio of salt: acid = 5 : 1

Mole fraction of salt = 5/6

Mole fraction of acid = 1/6

Final buffer conc. = 6.7×10^{-2} mol/liter

Conc. of salt = $5/6 \times (6.7 \times 10^{-2}$ mol/L$) = 5.58 \times 10^{-2}$ mol/L

Conc. of acid = $1/6 \times (6.7 \times 10^{-2}$ mol/L$) = 1.12 \times 10^{-2}$ mol/L

Grams of salt: $(5.58 \times 10^{-2}) \times 142 = 7.92$ g of salt for 1 liter

Grams of acid: $(1.12 \times 10^{-2}) \times 120 = 1.34$ g of acid for 1 liter

CHAPTER 9

INJECTABLE MEDICATIONS AND INTRAVENOUS FLUIDS—PART I

Electrolyte Solutions: Millimols, Milliequivalents, and Milliosmols

LEARNING OBJECTIVES: *After completing this chapter the student should be able to:*

1. Express quantities in terms of millimols, milliequivalents, and milliosmols.
2. Determine the amount of an electrolyte solution that will supply a desired concentration or quantity of a component ion expressed as millimols or milliequivalents.
3. Calculate the number of milliosmols or milliequivalents contributed by a given quantity of an electrolyte or nonelectrolyte.
4. Calculate theoretical osmolarity or osmolality from the quantities of solute(s) used to prepare a solution.
5. Convert percentage strength of a solute to osmolar units.

Calculations related to parenteral medications involve specialized terminology, units, and calculations. Frequently, medication orders include drugs prescribed in millimols, milliequivalents, or milliosmols. These specific types of calculations are associated with electrolyte solutions, reconstitution of dry powders, administration of insulin, heparin and some antibiotics, intravenous admixtures, and rate of flow. Injectable medications and intravenous fluids will be covered in two chapters to facilitate learning.

MOL, MILLIMOL, MILLIEQUIVALENT

1. Electrolytes are of critical importance in maintaining normal body function. When an essential ion is lost through disease, it must be replaced. Solutions of salts are usually used as sources of electrolytes for this type of therapy.

Sometimes the necessary amount of the salt will be explicitly stated on the prescription. Alternatively, the physician will prescribe some quantity of the needed ion in chemical units, and the pharmacist will be asked to calculate how much salt will contain that quantity.

Although the concentration of electrolytes in solution may be expressed in several units of weight such as grams per 100 mL (%) or milligrams percent (mg%), chemical units, particularly *milliequivalents*, are more commonly used.

Pharmaceutical Calculations, Fourth Edition, By Joel L. Zatz and Maria Glaucia Teixeira
ISBN 0-471-67623-3

As we review mols and millimols, keep in mind the following:

- One mol (or gram molecular weight) of a substance is defined as the formula weight for that substance, expressed in grams.
- The number of mols of a reactant or product is indicated by the coefficient, which precedes that species in the chemical equation.
- A chemical equation relates quantities in terms of mols. By converting mols to grams, relationships in weight may be obtained.
- A millimol (mmol) is the molecular weight expressed in milligrams. It is a measure of quantity, not concentration.
- Whenever a chemical change occurs, 1 equivalent of a positively charged ion combines with 1 equivalent of a negatively charged ion. When a salt dissociates, the same number of equivalents of positive and negative ions are produced.
- A milliequivalent (mEq), or 1/1000 of an equivalent, is more widely used than the equivalent. The number of milliequivalents is found by multiplying the number of millimols by the absolute value of the valence.

MOLS, CHEMICAL EQUATION, AND WEIGHT RELATIONSHIP

2. A mol is the molecular weight of a substance in grams. The number of mols of a substance is calculated by dividing the number of grams of substance by the molecular weight in grams.

$$\text{mols} = \frac{\text{grams}}{\text{MW}}$$

Let us go through a couple of problems as a review.

The chemical equation for the dissociation of calcium nitrate ($Ca\ (NO_3)_2$) is:

$$Ca\ (NO_3)_2 = Ca^{2+} + 2NO_3^-$$

Thus, one mol of calcium nitrate yields 1 mol of calcium ion and 2 mols of nitrate ion. How many mols result from the dissociation of 1 mol of sodium sulfate? (Na_2SO_4)

$$Na_2SO_4 = 2Na^+ + SO_4^{2-}$$

Each mol of sodium sulfate yields 2 mols of sodium and 1 mol of sulfate.

Now it's your turn to practice, using elementary inorganic chemistry skills and data provided in Table 9.1 and Appendix 4 (names, symbols, atomic weight of some elements with pharmaceutical importance).

How many mols of magnesium chloride ($MgCl_2$) are necessary to yield 1 mol of magnesium ion?

Solution. $MgCl_2 = Mg^{2+} + 2Cl^-$

Each mol of magnesium chloride yields 1 mol of magnesium ion.

TABLE 9.1 Properties of some important ions

Ion	Formula	Atomic/ Formula Weight	Valence
Aluminum	Al^{3+}	27	3
Ammonium	NH_4^+	18	1
Acetate	$C_2H_3O_2^-$	59	1
Bicarbonate	HCO_3^-	61	1
Calcium	Ca^{2+}	40	2
Carbonate	CO_3^{2-}	60	2
Chloride	Cl^-	35.5	1
Citrate	$C_6H_5O_7^{3-}$	189	3
Ferrous	Fe^{2+}	56	2
Ferric	Fe^{3+}	56	3
Gluconate	$C_6H_{11}O_7^-$	195	1
Lactate	$C_3H_5O_3^-$	89	1
Lithium	Li^+	7	1
Magnesium	Mg^{2+}	24	2
Phosphate (monobasic)	$H_2PO_4^-$	97	1
Phosphate (dibasic)	HPO_4^{2-}	96	2
Potassium	K^+	39	1
Sodium	Na^+	23	1
Sulfate	SO_4^{2-}	96	2

---

3. How many mols of magnesium chloride are necessary to yield 1 mol of chloride ion?

Solution. One-half mol of $MgCl_2$

$MgCl_2 = Mg^{2+} + 2Cl^-$

Each mol of magnesium chloride yields 2 mols of chloride ion.

---

4. A chemical equation relates quantities in terms of mols. By converting mols to grams, relationships in weight may be obtained. In the case of anhydrous sodium carbonate, for example,

$$Na_2SO_3 = 2Na^+ + CO_3^{2-}$$

Each mol of sodium carbonate contains 2 mols of sodium. The molecular weight of sodium carbonate is 106. Therefore, 106 g of sodium carbonate contain 46.0 g of sodium.

How many grams of anhydrous sodium carbonate contain 350 mg of sodium?

Solution. 0.807 g

CALCULATIONS

$$\frac{106.0\text{ g}}{46.0\text{ g}} = \frac{j}{0.350\text{ g}}$$

$j = 0.807$ g

---

5. Calculate the percentage of calcium in calcium carbonate.

Solution. 40%

Calcium has an atomic weight of 40.0.

$CaCO_3$ has a molecular weight of 100.0.

$$\frac{40.0}{100.0} = 40.0\%$$

MILLIMOL (MMOL)

6. A millimol (mmol) is the molecular weight expressed in milligrams. The number of millimols of substance is calculated by dividing the number of milligrams of substance by the molecular weight of the substance.

$$\text{mmols} = \frac{\text{mg}}{\text{MW}}$$

If you feel you need a review, check these examples and try the next two problems. Otherwise, go to frame 8.

How many milligrams of sodium chloride represent 0.5 mmol?

Molecular wt of NaCl = 58.5

58.5 mg NaCl = 1 mmol

0.5 mmol = 29.3 mg

How many millimols are there in 3.2 g of calcium fluoride?

MW of CaF_2 = 78.0

78.0 mg CaF_2 = 1 mmol

3200 mg × 1 mmol/78.0 mg = 41 mmol

A prescription calls for 24 mmol of potassium chloride. How many grams of KCl are required?

Solution. 1.79 g

CALCULATIONS

MW of KCl = 74.5 and 1788 mg = 1.79 g

74.6 mg = 1 mmol

$$24 \text{ mmol} \times \frac{74.6 \text{ mg}}{1 \text{ mmol}} = 1790 \text{ mg} = 1.79 \text{ g}$$

---

7. How many grams of sodium chloride should be used to prepare this solution?

℞ NaCl solution 90.0 mL

Each 5 mL should contain 0.6 mmol of Na^+.

Solution. 0.632 g

CALCULATIONS

$$\frac{0.6 \text{ mmol}}{\text{teaspoonful}} \times 18 \text{ teaspoonfuls} = 10.8 \text{ mmol } Na^+$$

$NaCl = Na^+ + Cl^-$

From this equation we see that each mol of sodium chloride yields 1 mol of Na^+. Therefore, 10.8 mmol of Na^+ will be supplied by 10.8 mmol of NaCl. Thus,

MW of NaCl = 58.5

58.5 mg NaCl = 1 mmol

$$10.8 \text{ mmol} \times \frac{58.5 \text{ mg}}{\text{mmol}} = 632 \text{ mg} = 0.632 \text{ g}$$

EQUIVALENT AND MILLIEQUIVALENT (mEq)

8. The mol and millimol are convenient units because they are directly related to the chemical equation. Another unit often used in connection with electrolytes is the *equivalent*. The number of equivalents is found by multiplying the number of mols by the absolute value of the valence.

Since total positive and negative charges in a salt must balance, we can define the valence of a salt as the sum of either the positive or negative charges. For example, the valence of $CaCl_2$, which contains two positive charges and two negative charges, is 2.

Write the valence of each of the following compounds:

A. Potassium gluconate ($KC_6H_{11}O_7$):

B. Sodium acetate ($NaC_2H_3O_2$):

C. Sodium phosphate dibasic (Na_2HPO_4):

D. Calcium carbonate ($CaCO_3$):

Solutions.

A. valence = 1 [K^+, $(C_6H_{11}O_7)^-$]
B. valence = 1 [Na^+, $(C_2H_3O_2)^-$]
C. valence = 2 [2 Na^+, $(HPO_4)^{2-}$]
D. valence = 2 [Ca^{2+}, $(CO_3)^{2-}$]

---

9. How many equivalents of K^+ are produced by the dissociation of 1 mol of potassium carbonate (K_2CO_3)? How many equivalents of $(CO_3)^{2-}$?

Solution. 2 equivalents of each ion.

$K_2CO_3 = 2K^+ + CO_3^{2-}$
Thus 2 mol of K^+ are produced

mols × valence = equivalents
$2 \times 1 = 2$ equivalents of K^+
Only 1 mol of carbonate is produced.

mols × valence = equivalents
$1 \times 2 = 2$ equivalents of CO_3^{2-}
Notice that the equivalents of cation and anion are equal in number.

---

10. The equivalent is often too large a unit for calculation. The milliequivalent (mEq), 1/1000 of an equivalent, is widely used. A *milliequivalent* is a unit that measures the chemical combining activity of an electrolyte in solution. It considers the total number of ions in solution and the valence (charge) of the ions. For this reason, milliequivalents may be calculated using a number of problem-solving approaches. We will offer three approaches in this text.

(a) Equation 9.1 provided below is probably the easiest approach to calculate milliequivalents because it involves basic chemistry concepts. Also, the equation is simple enough to be remembered at all times. Multiply the number of millimols by the absolute value of the valence.

$$\boxed{\text{mEq} = \text{mmols} \times \text{valence}} \tag{9.1}$$

Since the number of millimols of a drug in solution may be calculated by dividing the amount of drug, in mg, by its molecular weight, then the following may also be used:

$$\text{mEq} = \frac{\text{mg}_{\text{substance}}}{\text{MW}} \times \text{valence}.$$

In addition, if we rearrange its terms, the number of milligrams of drug can be determined if a given number of milliequivalents of drug are known:

$$\text{mg} = \frac{\text{mEq} \times \text{MW}}{\text{valence}}$$

(b) Determine the number of equivalents (equivalent weight) of the compound by multiplying the number of mols by the absolute value of the valence. Calculate the amount of milligrams in 1 mEq of compound (1/1000 of equivalent weight). Set up a proportion correlating mEq with mg of drug.

(c) Treat the problem as a conversion (between mass and chemical units) and use dimensional analysis.

Now, check the following example solved by each approach.

How many mg/mL are present in a solution containing 2 mEq/mL of KCl? (MW = 74.5)

Solving by (a):

$$\text{mEq} = \text{mmols} \times \text{valence} = \frac{\text{mg}_{\text{substance}}}{\text{MW}_{(\text{mg})}} \times \text{valence}$$

$$\text{mg} = \frac{\text{mEq} \times \text{MW}}{\text{valence}}$$

$$2\,\text{mEq/mL} \times \frac{74.5\,\text{mg}}{1\,\text{mEq}} = 149\,\text{mg/mL}$$

or

$$\text{mEq/mL} = \frac{\text{mg/mL}_{\text{substance}}}{\text{MW}} \times \text{valence}$$

$$\text{mg/mL} = \frac{\text{mEq/mL} \times \text{MW}}{\text{valence}} = 2 \times \frac{74.5}{1} = 149\ \text{mg/mL}$$

Solving by (b):

MW of KCl = 74.5 g

Equiv. weight of KCl = 74.5/1 = 74.5 g (valence = 1)

1 mEq of KCl = 1/1000 × 74.5 g = 0.0745 g = 74.5 mg

2 mEq = 2 × 74.5 mg = 149 mg/mL

Solving by (c):

$$\frac{74.5\text{ mg}}{1\text{ mmol}} \times \frac{1\text{ mmol}}{1\text{ mEq}} \times \frac{2\text{ mEq}}{1\text{ mL}} = 149\text{ mg/mL}$$

---

11. Try the following.

How many mEq of $MgSO_4$ (MW = 120) are represented in 1 g of anhydrous magnesium sulfate?

Solution. 16.7 mEq

CALCULATIONS

$$\text{MW of } MgSO_4 = 120$$
$$\text{EqW} = 120/2 = 60\text{ g}$$
$$1\text{ mEq} = 60\text{ mg}$$
$$1\text{ g} = 1000\text{ mg}$$

$$1000\text{ mg} \times \frac{1\text{ mEq}}{60\text{ mg}} = 16.7\text{ mEq}$$

or

$$\text{MW of } MgSO_4 = 120$$
$$1\text{ mmol } MgSO_4 = 120\text{ mg}$$
$$\text{EqW} = 120\text{ g}/2 = 60\text{ g}$$
$$1\text{ mEq} = 1/1000 \times 60\text{ g} = 0.060\text{ g} = 60\text{ mg}$$
$$1\text{ mmol} = 2\text{ mEq} \quad (120 = 2 \times 60)$$

$$\frac{2\text{ mEq}}{\text{mmol}} \times \frac{1\text{ mmol}}{120\text{ mg}} \times \frac{1000\text{ mg}}{1\text{ g}} \times 1\text{ g} = 16.7\text{ mEq}$$

---

12. The normal magnesium (Mg^{2+}, AtW = 24) level in blood plasma is 2.5 mEq/Liter. How much would it be in mg%?

Solution. 3 mg%

CALCULATIONS

Atomic weight of magnesium = 24 g

$$mEq/L = \frac{mg/L_{substance}}{MW_{(mg)}} \times val$$

$$mg/L = \frac{mEq/L}{valence} \times MW = \frac{2.5}{2} \times 24 = 30\ mg/L = 3\ mg/100\ mL = 3\ mg\%$$

or

Atomic weight of magnesium = 24 g

Eq. weight of Mg^{2+} = 24/2 = 12 g

1 mEq = 12 mg

2.5 mEq/L = 2.5 × 12 = 30 mg/L = 3 mg/100 mL = 3 mg%

---

13. A patient weighing 132 lb is to receive 2 mEq of NaCl/ kg of body weight. How many milliliters of a 0.9% sterile solution of NaCl should be administered? (NaCl, MW = 58.5)

Solution. 780 mL

CALCULATIONS

MW of NaCl = 58.5 g

$$EqW = \frac{58.5\ g}{1} = 58.5\ g$$

1 mEq = 0.0585 g

2 mEq/kg dose = 2 × 0.0585 = 0.117 g/kg

132 lb = 60 kg (1 kg = 2.2 lb)

0.117 g × 60 = 7.02g of NaCl needed

0.9% NaCl available = 0.9 g/100 ml

$$7.02\ g \times \frac{100\ mL}{0.9\ g} = 780\ mL$$

or

$$132\ lb \times \frac{1\ kg}{2.2\ lb} = 60\ kg$$

60 kg × 2 mEq/kg = 120 mEq needed

$$mg = \frac{mEq \times MW_{(mg)}}{valence} = \frac{120}{1} \times 58.5 = 7020\ mg = 7.02\ g$$

$$7.02\ \text{g} \times \frac{100\ \text{mL}}{0.9\ \text{g}} = 780\ \text{mL}$$

or

$$\frac{2\ \text{mEq}}{\text{kg}} \times \frac{1\ \text{kg}}{2.2\ \text{lb}} \times 132\ \text{lb} \times \frac{0.0585\ \text{g}}{1\ \text{mEq}} \times \frac{100\ \text{mL}}{0.9\ \text{g}} = 780\ \text{mL}$$

---

14. How many milliequivalents of fluoride ion are contained in 0.35 g of potassium fluoride? (KF, MW = 58.1)

Solution. 6.02 mEq

CALCULATIONS

$KF = K^+ + F^-$

$1\ \text{mEq}\ F^- = 1\ \text{mmol}\ F^- = 1\ \text{mmol KF} = 58.1\ \text{mg}$

$350\ \text{mg KF} \times 1\ \text{mEq}\ F^-/58.1\ \text{mg KF} = 6.02\ \text{mEq}\ F^-$

or

$$\text{mEq} = \frac{350\ \text{mg}}{58.1\ \text{mg}} \times 1 = 6.02\ \text{mEq}\ F^-$$

---

15. How many milliequivalents of magnesium ion are there in each teaspoonful of a 2% solution of magnesium chloride ($MgCl_2$, MW = 95)?

Solution. 2.10 mEq

CALCULATIONS

$(5\ \text{mL} \times 0.02\ \text{g/mL}) = 0.1\ \text{g}\ MgCl_2 = 100\ \text{mg}$

$MgCl_2 = Mg^{2+} + 2Cl^-$

$1\ \text{mmol}\ MgCl_2 = 95.3\ \text{mg} = 1\ \text{mmol}\ Mg^{2+} = 2\ \text{mEq}\ Mg^{2+}$

$$100\ \text{mg}\ MgCl_2 \times \frac{2\ \text{mEq}\ Mg^{2+}}{95.3\ \text{mg}\ MgCl_2} = 2.10\ \text{mEq}\ Mg^{2+}$$

or

$$\frac{100 \text{ mg}}{95.3 \text{ mg}} \times 2 = 2.10 \text{ mEq } Mg^{2+}$$

---

16. How many grams of $CaCl_2.2H_2O$ (MW = 147) should be used to prepare 180 mL of a calcium chloride solution containing 2.5 mEq of calcium ion in each teaspoonful?

Solution. 6.62 g

CALCULATIONS

$2.5 \text{ mEq/teaspoonful} \times 36 \text{ teaspoonfuls} = 90 \text{ mEq needed}$

$CaCl_2 \cdot 2H_2O = Ca^{2+} + 2Cl^- + 2H_2O$

$1 \text{ mmol } Ca^{2+} = 2 \text{ mEq } Ca^{2+} = 1 \text{ mmol } CaCl_2 \cdot 2H_2O = 147.0 \text{ mg} CaCl_2 \cdot 2H_2O$

$$90 \text{ mEq } Ca^{2+} \times \frac{147 \text{ mg } (CaCl_2 \cdot 2H_2O)}{2 \text{ mEq } Ca^{2+}} = 6620 \text{ mg} = 6.62 \text{ g } CaCl_2 \cdot 2H_2O$$

---

17. If you would like more practice, try your hand at these. Check atomic weight, formulas and valences in Table 9.1 and Appendix 4.

A. If 2.86 g of magnesium chloride are used to prepare 120 mL of a solution, how many milliequivalents of magnesium ion will each teaspoon contain?

B. How many grams of potassium chloride should be used to prepare 90 mL of a solution containing 0.8 mEq of K^+/mL?

C. ℞ Solution calcium chloride 5% 180 mL

Sig: 1 teaspoonful tid

How many milliequivalents of chloride ion does the patient receive each day?

D. ℞ Potassium sulfate to yield 1.0 mEq of K^+

Aqua qs ad 5.0 mL

d.t.d. #24

Sig: 5.0 mL b.i.d.

How many grams of potassium sulfate should be used for this prescription?

E. If city water contains 2.5 ppm of NaF, calculate the number of milliequivalents of fluoride ingested by a person who drinks 1.5 L of water.

F. A solution is prepared by dissolving 8.42 g of sodium chloride in sufficient water to make 180 mL of solution. In how many milliliters will 8 mEq of sodium ion be contained?

Solutions.

A. 2.50 mEq

B. 5.36 g and F. 10 mL

C. 13.5 mEq

D. 2.09 g

E. 0.089 mEq

F. 10.1 mL

OSMOLS, MILLIOSMOLS, AND OSMOLARITY

18. We have already discussed the physiological importance of osmotic pressure in Chapter 8. Osmotic pressure depends only on the number of particles (molecules or ions) dissolved in unit volume of solvent. If an ideally behaving nonelectrolyte is dissolved in water, each molecule produces one particle in solution. For real substances, the same holds true provided that there is no dimerization or polymerization of molecules in solution.

The number of particles is expressed in terms of osmols. One osmol (Osm) is defined as the weight, in grams, of a solute osmotically equivalent to one gram-molecular weight (1 mol) of an ideally behaving nonelectrolyte. A milliosmol (mOsm) is 1/1000 of an osmol.

One mol (gram-molecular weight) of an *ideal nonelectrolyte* is equivalent to 1 osmol and 1 milliequivalent equals 1 milliosmol. Consequently, 1 mOsm of a nonelectrolyte is equivalent to the molecular weight expressed in milligrams. In terms of equations, for ideal nonelectrolytes,

$$1\,\text{Osm} = 1\,\text{mol}$$

$$1\,\text{mOsm} = 0.001\,\text{Osm} = 1\,\text{mmol}$$

Consider sucrose (sugar), whose molecular weight is 342. Then 342 g represent 1 mol of sucrose; theoretically, 342 g also represent 1 Osm of sucrose. Calculate the number of osmols and milliosmols equivalent to 1 g of sucrose.

Solution. 0.00292 Osm; 2.92 mOsm

CALCULATIONS

$1\,\text{Osm} = 342\,\text{g}$

$$1\,\text{g} \times \frac{1\,\text{Osm}}{342\,\text{g}} = 0.00292\,\text{Osm}$$

$$0.00292\,\text{Osm} \times \frac{1000\,\text{mOsm}}{1\,\text{Osm}} = 2.92\,\text{mOsm}$$

---

19. How many grams of dextrose, molecular weight 180, would be needed to provide 120 mOsm?

Solution. 21.6 g

CALCULATIONS

1 mOsm = 180 mg (molecular weight expressed in mg)

$$120\,\text{mOsm} \times \frac{180\,\text{mg}}{1\,\text{mOsm}} = 21{,}600\,\text{mg} = 21.6\,\text{g}$$

---

20. We can calculate the theoretical number of osmols or milliosmols of *electrolytes* by taking the number of ions formed into consideration. In this and all the other examples to follow, water is the only solvent considered because of the application to biological fluids. For purposes of calculation the number of osmols is taken to be the sum of the number of mols produced by *complete ionization*. We must recognize that the number of osmols resulting from putting an electrolyte into solution will be less than the theoretical value if dissociation is not complete. However, the assumption of complete dissociation is a good one for many salts of clinical importance and the theoretical value is close to the measured value. As an example, sodium chloride dissociates into two ions:

$$\text{NaCl} = \text{Na}^{+} + \text{Cl}^{-}$$

From the equation above, we can see that 1 mol of NaCl dissociates to yield 2 mols of ions. Each mol of ions represents particles that contribute to the osmotic pressure. Each mol of the ions produced adds another osmol. In other words,

$$1\,\text{mol NaCl} = 2\,\text{Osm}\ (1\,\text{Osm Na}^{+} + 1\,\text{Osm Cl}^{-})$$

Since the formula weight of NaCl is 58.5,

$$58.5\,\text{g of NaCl} = 2\,\text{Osm and } 58.5\,\text{mg} = 2\,\text{mOsm of NaCl}$$

---

21. Calculate the number of milliosmols corresponding to 0.386 g of NaCl.

Solution. 13.2 mOsm

CALCULATIONS

$1\ \text{mol} = 58.5\ \text{g NaCl} = 2\ \text{Osm}$

$1\ \text{mmol} = 58.5\ \text{mg NaCl} = 2\ \text{Osm}$

$$386\ \text{mg} \times \frac{2\ \text{mOsm}}{58.5\ \text{mg}} = 13.2\ \text{mOsm}$$

---

22. Recall that *water of hydration* present within a crystal becomes part of the solvent when the solid is dissolved in water. It affects the molecular weight of the solid, but not the number of osmols in solution. (The *water of hydration contributes no particles*.)

How many grams of $CaCl_2 .2H_2O$ would produce 100 mOsm?

Solution. 4.9 g. Note that the amount required is larger than for the anhydrous salt (see next example).

CALCULATIONS

$CaCl_2 \cdot 2H_2O = Ca^{2+} + 2Cl^- + 2H_2O$

147 mg $CaCl_2 \cdot 2H_2O$ = 3 mOsm (the water adds no particles)

$$100 \text{ mOsm} \times \frac{147 \text{ mg}}{3 \text{ mOsm}} = 4900 \text{ mg} = 4.9 \text{ g}$$

---

23. How many milliosmols of calcium ion are there in 8.75 g of anhydrous calcium chloride, MW = 111 and Solution. 78.8 mOsm?

Solution. 78.5 mOsm

CALCULATIONS

$CaCl_2 = Ca^{2+} + 2Cl^-$

The equation shows that each mol of $CaCl_2$ produces 1 mol of Ca^{2+}.

Consequently, 111 mg of $CaCl_2$ yield 1 mOsm of Ca^{2+}.

$$8750 \text{ mg} \times \frac{1 \text{ mOsm } Ca^{2+}}{111 \text{ mg}} = 78.5 \text{ mOsm } Ca^{2+}$$

---

24. It is the concentration of osmols in solution rather than the absolute number that determines osmotic pressure. So while it is important to be able to calculate the number of osmols (or milliosmols) contributed by a solute, it is also necessary to know how to determine osmol concentration. Two common expressions for *osmol concentration* are used; they are *osmolarity and osmolality*.

Osmolarity is analogous to molarity. In fact, for *nonelectrolytes*, they are numerically the same. A 1-osmolar solution contains 1 Osm per liter of solution.

$$\text{Osmolarity} = \frac{\text{number of Osm}}{\text{L of solution}}$$

If 2 Osm are present in a liter, we have a 2 osmolar solution. If 150 mOsm are dissolved in enough water to make a liter, the solution is 150 milliosmolar. More commonly, this concentration would simply be expressed as 150 mOsm/L.

Osmolarity is a weight/volume type of situation in which the amount of solute and volume of total solution are specified. Frequently a measured osmolarity will differ slightly from a calculated osmolarity. This may be explained by the fact that ionically bonded substances may not dissociate completely resulting in a measured value a little lower than the calculated value.

---

25. The second unit is *osmolality*. A 1-osmolal solution contains 1 osmol per kilogram of water.

$$\text{Osmolality} = \frac{\text{number of Osm}}{\text{kg of water}}$$

For practical purposes, the density of water can be taken as 1 g/mL, so that 1 kg = 1 L. If 300 mOsm are dissolved in a liter of water, the solution is 300 milliosmolal.

For each of the following examples, calculate the concentration (either osmolarity or osmolality as appropriate) and tell which it is.

A. 22 mOsm dissolved in 100 mL of water.

B. 15 mOsm dissolved in enough water to make a total volume of 100 mL.

C. A solution containing 0.25 mOsm/mL of solution.

D. A solution containing 0.20 mOsm/mL of water.

Solutions.

A. 220 milliosmolal or 0.22 osmolal

B. 150 milliosmolar or 0.15 osmolar

C. 250 milliosmolar or 0.25 osmolar

D. 200 milliosmolal or 0.2 osmolal

Why have two sets of units? Osmolarity is useful because solutions are conveniently administered by volume and we want to know what quantities of components are contained within any given volume. This fits right in with a w/v arrangement. On the other hand, osmotic pressure is a function of osmolal concentration and instrumental measurements are better related to osmolal values than the osmolar designation.

26. For dilute solutions, the difference between osmolar and osmolal concentrations is usually irrelevant. This is because the solute occupies so little space in the solution that the volume of the solvent is almost the same as the solution's total volume. In such cases only, it is possible to make the approximation that osmolality and osmolarity are numerically equal. As an example, the osmolality and osmolarity of a 0.9% solution of sodium chloride differ by less than 1%. However, equating osmolality and osmolarity is not valid when dealing with concentrated systems.

For which of the following solutions are osmolality and osmolarity most likely to have a similar value?

(a) 10% fructose solution

(b) 0.1% dextrose solution

Solution. **(b)** is correct. The solute concentration is small enough so that the contribution of solute to total volume is negligible.

---

27. Let's try an example to see how osmolarity is calculated using dextrose as a solute. If we make 400 mL of a solution containing 30.0 g of hydrous dextrose (MW = 198), its concentration can be calculated as follows:

For dextrose, 1 mol = 1 Osm = 198 g.

$$30.0 \text{ g} \times \frac{1 \text{ mol}}{198 \text{ g}} = 0.152 \text{ mol} = 0.152 \text{ Osm} = 152 \text{ mOsm}$$

$$\frac{152 \text{ mOsm}}{400 \text{ mL}} = \frac{j}{1000 \text{ mL}}$$

$$j = 380 \text{ mOsm/L}$$

Calculate the concentration, in mOsm/L, of a 0.9% solution of sodium chloride in water.

Solution. 308 mOsm/L

CALCULATIONS

$$NaCl = Na^+ + Cl^-$$

$$1 \text{ mmol NaCl} = 58.5 \text{ mg NaCl} = 2 \text{ mOsm}$$

$$0.9\% = 0.9 \text{ g}/100 \text{ mL} = 9 \text{ g/L} = 9000 \text{ mg/L}$$

$$\frac{9000 \text{ mg}}{\text{L}} \times \frac{2 \text{ mOsm}}{58.5 \text{ mg}} = 308 \text{ mOsm/L}$$

28. Calculate the number of grams of potassium chloride needed to make 200 mL of a solution to contain 250 mOsm/L.

Solution. 1.86 g

CALCULATIONS

$KCl = K^+ + Cl^-$

1 mmol KCl (74.5 mg) is equivalent to 2 mOsm.

$250\ \text{mOsm}/\text{L} \times 0.2\ \text{L} = 50\ \text{mOsm}$ needed.

$$50\ \text{mOsm} \times \frac{74.6\ \text{mg}}{2\ \text{mOsm}} = 1870\ \text{mg} = 1.87\ \text{g}$$

---

29. How many milliosmols of sodium ion are there in each milliliter of a 1% solution of sodium sulfate (Na_2SO_4)? How many theoretical milliosmols total?

Solution. 0.141 mOsm Na^+/mL; a total of 0.211 mOsm/mL

CALCULATIONS

$Na_2SO_4(142\ \text{g}/\text{mol}) = 2Na^+ + SO_4^{2-}$

$1\ \text{mmol}\ (142\ \text{mg}) = 2\ \text{mOsm}\ Na^+ + 1\ \text{mOsm}\ SO_4^{2+}$

A 1% solution contains 1 g/100 mL = 0.01 g/mL = 10 mg/mL.

For Na^+:

$$10\ \text{mg salt}/\text{mL} \times \frac{2\ \text{mOsm}\ Na^+}{142\ \text{mg}} = 0.141\ \text{mOsm}/\text{mL}$$

Total:

$$10\ \text{mg salt}/\text{mL} \times \frac{3\ \text{mOsm}\ Na^+}{142\ \text{mg}} = 0.211\ \text{mOsm}/\text{mL}$$

---

30. In some instances, the quantity of solute contributing to the *osmolarity* of a solution is expressed *in millimols or milliequivalents* instead of a mass unit. For *nonelec-*

trolytes, each millimol theoretically contributes 1 milliosmol to a solution. For *electrolytes*, the number of milliosmols per millimol can easily be determined by inspection of the dissociation equation.

As an example, let us calculate the number of milliosmols per 100 mL of a solution containing 14 mmol of dextrose and 2.0 mmol of potassium chloride per 100 mL.

Dextrose: 14 mmol = 14 mOsm

KCl: $KCl = K^+ + Cl^-$

2 mOsm = 1 mmol

2 mmol = 4 mOsm

Total: 14 mOsm + 4 mOsm = 18 mOsm/100 mL

Calculate the number of mOsm/mL in a solution containing 130 mmol/L sodium chloride (NaCl) and 30 mmol/L magnesium chloride ($MgCl_2$).

Solution. 0.35 mOsm/mL

CALCULATIONS

$NaCl = Na^+ + Cl^-$; 1 mmol = 2 mOsm

$MgCl_2 = Mg^{2+} + 2Cl^-$; 1 mmol = 3 mOsm

$$\text{NaCl: } 130 \text{ mmol/L} \times \frac{2 \text{ mOsm}}{1 \text{ mmol}} = 260 \text{ mOsm/L}$$

$$\text{MgCl}_2\text{: } 30 \text{ mmol/L} \times \frac{3 \text{ mOsm}}{1 \text{ mmol}} = 90 \text{ mOsm/L}$$

$$260 + 90 = 350 \text{ mOsm/L} = 0.35 \text{ mOsm/mL}$$

---

31. How many milliosmols/L are there in 1 L of a NaCl solution containing 1 mEq of sodium ion in each 20 mL?

Solution. 100 mOsm

CALCULATIONS

$NaCl = Na^+ + Cl^-$

From the equation, each millimol of NaCl contributes 1 mEq of sodium ion and a total of 2 mOsm to the solution. This solution contains 2 mOsm for each mEq of Na^+. Consequently, 20 mL contain 2 mOsm, 100 mL contain 10 mOsm and a liter contains 100 mOsm.

---

32. We saw in a previous example (Frame 27) that normal saline solution, 0.9 g NaCl per 100 mL, had an osmolarity of 308 mOsm/L. If a solution is injected intravenously, it should have close to the same osmolarity (within about a 10% tolerance) to avoid cell destruction and tissue damage. If a small volume is injected slowly, minimal disruption occurs because it is diluted by plasma. However, injection of large volumes with incorrect osmolarity can cause problems. We can use calculations of osmolarity to get an idea of whether an osmotic pressure imbalance is likely.

As an example, let us calculate the osmolarity of Ringer's solution, a salt solution that may be injected for fluid replacement or as a means of drug administration. Ringer's solution contains 0.86% sodium chloride, 0.03% potassium chloride and 0.033% calcium chloride. The respective formula weights are 58.5, 74.6 and 111.

Verify by writing the dissociation equations that each millimol is equivalent to 2 mOsm in the case of NaCl and KCl and that for $CaCl_2$, each millimol is equivalent to 3 mOsm.

$$\text{NaCl: } 0.86\,\text{g}/100\,\text{mL} = 860\,\text{mg}/100\,\text{mL} = 8600\,\text{mg}/\text{L}$$

$$8600\ \text{mg}/\text{L} \times \frac{1\ \text{mmoL}}{58.5\ \text{mg}} \times \frac{2\ \text{mOsm}}{1\ \text{mmol}} = 294\ \text{mOsm}/\text{L}$$

$$\text{KCl: } 300\ \text{mg}/\text{L} \times \frac{1\ \text{mmoL}}{74.5\ \text{mg}} \times \frac{2\ \text{mOsm}}{1\ \text{mmol}} = 8.05\ \text{mOsm}/\text{L}$$

$$\text{CaCl}_2\text{: } 330\ \text{mg}/\text{L} \times \frac{1\ \text{mmoL}}{111\ \text{mg}} \times \frac{3\ \text{mOsm}}{1\ \text{mmol}} = 8.92\ \text{mOsm}/\text{L}$$

The total is 311 mOsm/L, which is quite close to the value of 308 mOsm/L for normal saline solution. Therefore, Ringer's solution is osmotically equivalent to normal saline solution.

Calculate the theoretical osmolarity of a solution containing 2.5% dextrose (MW = 198) and 0.45% sodium chloride and compare it to normal saline solution.

Solution. 280 mOsm/L. This is osmotically equivalent to normal saline solution.

CALCULATIONS

$$\text{Dextrose: } 25{,}000\ \text{mg}/\text{L} \times \frac{1\ \text{mOsm}}{198\ \text{mg}} = 126\ \text{mOsm}/\text{L}$$

$$\text{NaCl: } 4500\ \text{mg}/\text{L} \times \frac{2\ \text{mOsm}}{58.5\ \text{mg}} = 154\ \text{mOsM}/\text{L}$$

The total is 280 mOsm/L, which is within 10% of normal saline solution and so is osmotically equivalent.

---

33. The USP states that the ideal osmolar concentration (or osmolar strength = mOsmols in a liter of solution) may be calculated according to the equation:

$$\text{mOsmol/Liter} = \frac{\text{wt. of substance (g/L)}}{\text{MW(g)}} \times \text{no. particles} \times 1000$$

$$\text{mOsmols} = \text{mmols} \times \text{no. particles}$$

For an electrolyte drug the osmolar concentration depends upon the degree of dissociation of the substance. Assuming *complete dissociation*:

$$1 \text{ mmol of NaCl} = 2 \text{ milliosmols of total particles } (Na^+, Cl^-)$$

$$1 \text{ mmol of } CaCl_2 = 3 \text{ milliosmols of total particles } (Ca^{2+}, Cl^-, Cl^-)$$

Find 1 mmol of the substance (in *mg*); find the number of mOsmols of total particles (number of particles) in 1 mmol; correlate number of milliosmols with *mg* of drug through a proportion.

For example:

How many milliosmols of sodium are represented in 1 liter of 3% hypertonic NaCl solution?

$$\frac{30 \text{ g/L}}{58.5 \text{ g}} \times 1 \text{ particle } (Na^+) \times 1000 = 512.8 \text{ mOsmols of } Na^+$$

A solution of NaCl contains 77 mEq/liter. Calculate its osmolar strength (=osmolarity) in terms of mOsmols/Liter. Assume complete dissociation.

Solution. 154 mOsmols/Liter

CALCULATIONS

$1 \text{ mEq} = 0.0585\text{g}$

$77 \text{ mEq} = 77 \times 0.0585 = 4.5 \text{ g/Liter}$

$$\frac{4.5 \text{ g/L}}{58.5 \text{ g}} \times 2 \text{ particles} \times 1000 = 154 \text{ mOsmols/Liter}$$

---

34. Now, try these problems before verifying the answers. Choose the approach with which you feel most comfortable.

A. How many (a) millimols, (b) milliequivalents and (c) milliosmols of calcium chloride ($CaCl_2 \cdot 2H_2O$, MW = 147) are represented in 147 mL of a 10% (w/v) calcium chloride solution?

B. Calculate the number of milliosmols in 350 mL of normal saline solution.

C. Calculate the number mOsm/L in a solution containing 5% dextrose (MW = 198) and 0.2% sodium chloride.

D. A 10 mL pre-filled syringe contains a 4.2% (w/v) solution of sodium bicarbonate ($NaHCO_3$, MW = 84). What is the osmolarity (mOsmols/Liter) of this solution?

E. How many milliosmols are represented in a liter of a 5% (w/v) solution of anhydrous dextrose ($C_6H_{12}O_6$, MW = 180) in water.

Solutions.

A. **(a)** 100 mmols
(b) 200 mEq
(c) 300 mOsmols

B. 108 mOsm

C. 321 mOsm/L

D. 1000 mOsm/L

E. 277.7 mOsm

CALCULATIONS

A.

(a) $147\text{ mL} \times \dfrac{10\text{ g}}{100\text{ mL}} = 14.7\text{ g of calcium chloride} = 14,700\text{ mg}$

$1\text{ mmol} = 147\text{ mg}$

$\dfrac{1\text{ mmol}}{147\text{ mg}} = \dfrac{14,700\text{ mg}}{j}$

$j = 100\text{ mmols}$

(b) Eq. W. = 147/2 = 73.5 g (valence of calcium = 2)

$1\text{ mEq} = 1/1000 \times 73.5\text{ g} = 0.0735\text{ g}$

$\dfrac{1\text{ mEq}}{0.0735\text{ g}} = \dfrac{14.4\text{ g}}{j}$

$j = 200\text{ mEq}$

or

$\text{mEq} = \text{mmols} \times \text{valence}$

$\text{mEq} = 100\text{ mmols} \times 2 = 200\text{ mEq}$

(c) $\text{mOsmols} = \text{mmols} \times \text{particles}$

$\text{mOsmols} = 100\text{ mmols} \times 3 = 300\text{ mOsmols}$

or

$1\text{ mmol} = 147\text{ mg}$

147 mL of a 10% (w/v) calcium chloride solution = 14,700 mg

$CaCl_2$ dissociates into 3 total particles = 3 mOsmols of particles

$\dfrac{147\text{ mg}}{3\text{ mOsmols}} = \dfrac{14,700\text{ mg}}{j}$

$j = 300\text{ mOsmols}$

or

$\dfrac{14.7\text{ g}}{147\text{ g}} \times 3\text{ particles} \times 1000 = 300\text{ mOsmols in 147 mL of 10\% sol.}$

or

$1000\text{ mL} \times \dfrac{14.7\text{ g}}{147\text{ mL}} = 100\text{ g/L}$

$\dfrac{100\text{ g/L}}{147} \times 3 \times 1000 = 2048\text{ mOsmols/L} = 300\text{ mOsmols in 147 mL}$

B. $350\text{ mL} \times \dfrac{900\text{ mg}}{100\text{ mL}} \times \dfrac{2\text{ mOsm}}{58.5\text{ mg}} = 108\text{ mOsm}$

C. Dextrose: $50{,}000 \text{ mg/L} \times \frac{1 \text{ mOsm}}{198 \text{ mg}} = 253 \text{ mOsm/L}$

NaCl: $2000 \text{ mg/L} \times \frac{2 \text{ mOsm}}{58.5 \text{ mg}} = 68 \text{ mOsm/L}$

Total $= 321 \text{ mOsm/L}$

D. $4.2 \text{ g}/100 \text{ mL} = 42{,}000 \text{ mg/L}$

$$\frac{42{,}000 \text{ mg}}{84 \text{ mg}} \times 2 \text{ particles} = 1000 \text{ mOsm/L}$$

E. $1000 \text{ mL} \times \frac{5 \text{ g}}{100 \text{ mL}} \times \frac{1000 \text{ mg}}{\text{g}} \times \frac{1 \text{ mmol}}{180 \text{ mg}} \times \frac{1 \text{ mOsm}}{1 \text{ mmol}} = 277.7 \text{ mOsm}$

REVIEW PROBLEMS

A. How many grams of sodium chloride are equivalent to 1 mol?

B. How many mols are equivalent to 45.0 g of potassium carbonate?

C. How many grams of sodium carbonate decahydrate ($Na_2CO_3 \cdot 10H_2O$) contain 350 mg of sodium?

D. How many millimols are there in 1.50 g of Epsom salts ($MgSO_4.7H_2O$)?

E. How many grams of anhydrous sodium sulfate should be used to prepare the following prescription?

℞ Sodium sulfate solution 60.0 mL

Each milliliter should contain 0.1 mmol of Na^+.

F. How many mEq/Liter are present in a solution containing 10 mg% of Ca^{2+} ions?

G. If 10 mEq of Cl^- are desired, how many milligrams of $CaCl_2$ should be used?

H. How many milliliters of a 10% potassium chloride solution must a patient take to obtain 5.0 mEq of K^+?

I. How many grams of magnesium chloride should be used to prepare 60 mL of a solution intended to contain 1.5 mEq of magnesium ion per milliliter?

J. How many grams of potassium carbonate will contain the same quantity of potassium as 3.50 g of potassium chloride?

K. How many millimols of chloride ion are present in 1 teaspoonful of a 10% solution of magnesium chloride?

L. If the fluoride content of a dentifrice is to be 0.25%, how many grams of sodium fluoride should be used per kilogram of toothpaste?

M. How many milliliters of a 6.0% solution of magnesium sulfate contain 5.0 mmol of magnesium ion?

N. How many milliequivalents of chloride ion does each tablespoonful of a 3.5% potassium chloride solution contain?

O. How many milliliters of a 5% solution of $MgCl_2$ contain 20 mEq of Mg^{2+}?

P. How many grams of K_2SO_4 should be used to prepare 240 mL of a potassium sulfate solution to contain 1 mEq K^+/mL?

Q. How many milliequivalents of chloride ion are contained in each milliliter of a 0.9% sodium chloride solution?

R. ℞ Calcium chloride solution, 7.5 mEq Ca^{2+}/teaspoonful

Ft 120 mL

How many grams of calcium chloride should be used to prepare this prescription?

S. A patient receives 2 liters of an electrolyte fluid containing 246 mg of sodium acetate ($C_2H_3NaO_2$, MW = 82), 25 mg of magnesium chloride ($MgCl_2$, MW = 95) and 89 mg of potassium gluconate ($C_6H_{11}KO_7$, MW = 234) per 100 mL of solution. How many milliequivalents each of sodium, magnesium and potassium does the patient receive?

T. How many grams of calcium chloride ($CaCl_2$) would produce 100 mOsm?

U. Calculate the concentration, in mOsm/L, of a 5% w/v solution prepared by dissolving hydrous dextrose in water.

V. If the osmotic pressure of a potassium chloride solution is expressed as 220 mOsm/L, what is its percentage strength?

W. How many grams of $CaCl_2 \cdot 2H_2O$ should be dissolved in water to make 150 mL of a solution that contains 298 mOsm/L?

X. A patient weighing 154 lb is to receive 0.2 mEq/kg of body weight of calcium gluconate ($C_{12}H_{22}CaO_{14}$, MW = 430).

(a) How many milliliters of a 25% solution of calcium gluconate should be added to the patient's infusion solution to provide the prescribed amount?

(b) How many milliosmols of calcium are represented in the volume received by the patient?

Y. A pharmacist added 25 mL of a 7.5% solution of magnesium acetate ($C_4H_6MgO_4$, MW = 142) to a patient's infusion solution. How many millimols, milliequivalents and milliosmols did this patient receive?

Z. A 10 mL ampul contains a 10% solution of calcium gluconate ($C_{12}H_{22}CaO_{14}$, MW = 430).

(a) How many milliequivalents of calcium are present in each milliliter of the solution?

(b) What is the osmolarity (mOsmols/liter) of this solution?

AA. How many milliosmols are represented in a liter of a solution containing 156 mg of K^+ ions per 100 mL? (At Wt Potassium = 39)

AB. A solution of calcium gluconate ($C_{12}H_{22}CaO_{14}$, MW = 430) contains 25 milliequivalents per liter. What is the osmolar strength of this solution in terms of milliosmols per liter, assuming complete dissociation?

Solutions to Review Problems

A. 58.5 g

B. 0.326 mol

C. 2.18 g

D. 6.1 mmol

E. 0.426 g

F. 5 mEq/L

G. 555 mg

H. 3.73 mL

I. 4.28 g

J. 3.24 g

K. 10.5 mmol

L. 5.53 g

M. 10 mL

N. 7.05 mEq

O. 19 mL

P. 20.9 g

Q. 0.154 mEq

R. 9.99 g

S. 60 mEq sodium, 10.52 mEq magnesium, 7.6 mEq potassium

T. 3.7 g

U. 253 mOsm/L

V. 0.82%

W. 2.19 g

X. **(a)** 12 mL

(b) 6.98 mOsm of calcium

Y. 13.2 mmols; 26.4 mEq; 39.6 mOsm

Z. **(a)** 0.465 mEq/mL

(b) 697.67 mOsm/L

AA. 40 mOsm

AB. 37.5 mOsm/L

Calculations for the Review Problems

A. Na: 23.0

Cl: *35.5*

Total: 58.5

1 mol = 58.5 g

B. K: 78

C: 12.0

3 O: *48.0*

Total: 138

1 mol of potassium carbonate = 138 g

$$45.0\text{ g} \times \frac{1\text{ mol}}{138.2\text{ g}} = 0.326\text{ mol}$$

C. $Na_2CO_3 \cdot 10H_2O = 2Na^+ + CO_3^{2-} + 10H_2O$

The molecular weight of the salt is 286 [(2 × 23) + 12 + (3 × 16) + (10 × 18)]

$$\frac{286\text{ g}}{46.0\text{ g}} = \frac{j}{0.350\text{ g}}$$

$j = 2.18$ g

D. MW of $MgSO_4 \cdot 7H_2O = 246$

246 mg = 1 mmol

$$1500\text{ mg} \times \frac{1\text{ mmol}}{246\text{ mg}} = 6.1\text{ mmol}$$

E. 0.1 mmol /mL × 60 mL = 6 mmol Na^+

$$Na_2SO_4 = 2Na^+ + SO_4^{2-}$$

Each mol of sodium sulfate yields 2 mols of Na^+.

$$\frac{1\text{ mol }Na_2SO_4}{2\text{ mol }Na^+} = \frac{j}{6\text{ mmol }Na^+}$$

$j = 3$ mmol Na_2SO_4 needed

MW of $Na_2SO_4 = 142$

$$3\text{ mmol} \times \frac{142\text{ mg}}{\text{mmol}} = 426\text{ mg} = 0.426\text{ g}$$

F. At. weight of calcium = 40 g

$$\text{Eq. weight of } Ca^{2+} = 40/2 = 20 \text{ g}$$
$$1 \text{ mEq} = 1/1000 \times 20 = 0.02 \text{ g} = 20 \text{ mg}$$
$$10 \text{ mg\%} = 10 \text{ mg in } 100 \text{ mL or } 100 \text{ mg/Liter}$$

$$\frac{100 \text{ mg}}{\text{Liter}} \times \frac{1 \text{ mEq}}{20 \text{ mg}} = 5 \text{ mEq/Liter} \quad \text{or} \quad \frac{100 \text{ mg}}{20} = 5 \text{ mEq/Liter}$$

G. $CaCl_2 = Ca^{2+} + 2\ Cl^-$

$$1 \text{ mmol } CaCl_2 = 2 \text{ mmol } Cl^- = 2 \text{ mEq } Cl^- = 111 \text{ mg } CaCl_2$$

$$10 \text{ mEq } Cl^- \times \frac{111 \text{ mg } CaCl_2}{2 \text{ mEq Cl}} = 555 \text{ mg } CaCl_2$$

H. $KCl = K^+ + Cl^-$

$$1 \text{ mEq } K^+ = 1 \text{ mmol } K^+ = 1 \text{ mmol KCl} = 74.5 \text{ mg}$$

$$5.0 \text{ mEq } K^+ \times \frac{74.5 \text{ mg KCl}}{1 \text{ mEq } K^+} = 373 \text{ mg KCl}$$

$$\frac{0.373 \text{ g}}{0.10 \text{ g/mL}} = 3.73 \text{ mL}$$

I. $60 \text{ mL} \times 1.5 \text{ mEq/mL} = 90 \text{ mEq}$

$$MgCl_2 = Mg^{2+} + 2Cl^-$$
$$1 \text{ mmol } Mg^{2+} = 2 \text{ mEq } Mg^{2+} = 1 \text{ mmol } MgCl_2 = 95 \text{ mg}$$

$$90 \text{ mEq } K^+ \times \frac{95 \text{ mg}}{2 \text{ mEq}} = 4275 \text{ mg} = 4.28 \text{ g}$$

J. 3.50 g of potassium chloride contain

$$3.50 \text{ g KCl} \times \frac{39 \text{ g K}}{74.5 \text{ g KCl}} = 1.83 \text{ g of potassium}$$

The formula weight of potassium carbonate is 138 (138 g of potassium carbonate contain 78 g of potassium)

$$\frac{78}{138} = \frac{1.83}{j}$$
$$j = 3.24 \text{ g}$$

K. $5 \text{ mL} \times 0.1 \text{ g/mL} = 0.5 \text{ g } MgCl_2$

$$MgCl_2 = Mg^{2+} + 2Cl^-$$

Thus 95 mg (1 mmol) of $MgCl_2$ yield 2 mmol of Cl^-

$$\frac{95 \text{ mg}}{2 \text{ mmol}} = \frac{500 \text{ mg}}{j}$$
$$j = 10.5 \text{ mmol}$$

L. $1000\,g \times 0.0025 = 2.5\,g$ F

Atomic wt F = 19.0; molecular wt NaF = 42.0

$$\frac{19.0}{42.0} = \frac{2.5\text{ g}}{j}$$
$$j = 5.53\text{ g}$$

M. 1 mmol $MgSO_4$ = 120 mg = 1 mmol Mg^{2+}

$$5.0\text{ mmol} \times 120\text{ mg/mmol} = 600\text{ mg} = 0.6\text{ g}$$

$$\frac{0.6\text{ g}}{0.060\text{ g/mL}} = 10\text{ mL}$$

N. $15\,mL \times 0.035\,g/mL = 0.525\,g$

1 mmol Cl = 1 mEq Cl = 1 mmol KCl = 74.5 mg

$$525\text{ mg} \times \frac{1\text{ mEq}}{74.5\text{ mg}} = 7.05\text{ mEq}$$

O. $MgCl_2 = Mg^{2+} + 2\,Cl^-$

95 mg (1 mmol) of $MgCl_2$ yield 2 mEq of Mg^{2+}

$$20\text{ mEq} \times \frac{95\text{ mg}}{2\text{ mEq}} = 950\text{ mg}$$

$$\frac{5\text{ g}}{100\text{ mL}} = \frac{0.950\text{ g}}{j}$$
$$j = 19\text{ mL}$$

P. $1\,mEq/mL \times 240\,mL = 240\,mEq$ K^+

$$K_2SO_4 = 2K^+ + SO_4^{2-}$$

1 mmol (174 mg) of K_2SO_4 yields 2 mEq of K^+

$$240\text{ mEq} \times \frac{174\text{ mg}}{2\text{ mEq}} = 20{,}880\text{ mg} = 20.9\text{ g}$$

Q. Each milliliter contains $1\,mL \times 0/00900\,g/mL = 9.00\,mg$

1 mmol Cl = 1 mEq Cl = 1 mmol NaCl = 58.5 mg

$$9.00\text{ mg} \times \frac{1\text{ mEq}}{58.5\text{ mg}} = 0.154\text{ mEq}$$

R. $7.5\,mEq/5\,mL \times 120\,mL = 180\,mEq$

1 mmol Ca^{2+} = 2 mEq Ca^{2+} = 1 mmol $CaCl_2$ = 111 mg

$$180\text{ mEq} \times \frac{111\text{ mg}}{2\text{ mEq}} = 9990\text{ mg} = 9.99\text{ g}$$

S. $$2000\text{ mL} \times \frac{246\text{ mg}}{100\text{ mL}} \times \frac{1\text{ mEq}}{82\text{ mg}} = 60\text{ mEq Na}^+$$

$$2000\text{ mL} \times \frac{25\text{ mg}}{100\text{ mL}} \times \frac{1\text{ mEq}}{47.5\text{ mg}} = 10.52\text{ mEq Ng}^{2+}$$

$$2000\text{ mL} \times \frac{89\text{ mg}}{100\text{ mL}} \times \frac{1\text{ mEq}}{234\text{ mg}} = 7.6\text{ mEq K}^+$$

T. $CaCl_2 = Ca^{2+} + 2\ Cl^-$

From the equation, each mol of $CaCl_2$ yields 3 Osm. The formula weight of $CaCl_2$ is 111. Therefore, 111 g $CaCl_2$ = 3 Osm; 111 mg $CaCl_2$ = 3 mOsm

$$100\ \text{mOsm} \times \frac{111\ \text{mg}}{3\ \text{mOsm}} = 3700\ \text{mg} = 3.7\ \text{g}$$

U. 5% = 5 g/100 mL = 50 g/L

$$50\ \text{g/L} \times \frac{1\ \text{Osm}}{198\ \text{g}} = 0.253\ \text{Osm/L} = 253\ \text{mOsm/L}$$

The concentration is 253 mOsm/L. Alternately, recognizing that a 1 Osm solution contains 198 g/L,

$$\frac{198\ \text{g/L}}{1\ \text{Osm/L}} = \frac{50\ \text{g/L}}{j}$$

$$j = 0.253\ \text{Osm/L}\ \text{or}\ 253\ \text{mOsm/L}$$

V. $$220\ \text{mOsm} \times \frac{74.5\ \text{mg}}{2\ \text{mOsm}} = 8195\ \text{mg/L} = 0.819\ \text{g/100 mL or } 0.82\%$$

W. $CaCl_2 \cdot 2H_2O = Ca^{2+} + 2Cl^- + 2H_2O$

$$147\ \text{mg}\ CaCl_2 \cdot 2H_2O = 3\ \text{mOsm}$$

$$0.150\ \text{L} \times 298\ \text{mOsm/L} \times \frac{0.147\ \text{g}}{3\ \text{mOsm}} = 2.19\ \text{g}$$

X. **(a)** $$\frac{0.2\ \text{mEq}}{\text{kg}} \times \frac{1\ \text{kg}}{2.2\ \text{lb}} \times 154\ \text{lb} \times \frac{100\ \text{mL}}{25\ \text{g}} \times \frac{1\ \text{g}}{1000\ \text{mg}} \times \frac{215\ \text{mg}}{1\ \text{mEq}} = 12.04\ \text{mL} = 12\ \text{mL}$$

(b) $$12\ \text{mL} \times \frac{25{,}000\ \text{mg}}{100\ \text{mL}} \times \frac{1\ \text{mmol}}{430\ \text{mg}} \times \frac{1\ \text{mOsm}(Ca^{2+})}{\text{mmol}} = 6.98\ \text{mOsm of calcium}$$

Y. $$25\ \text{mL} \times \frac{7.5\ \text{g}}{100\ \text{mL}} = 1.875\ \text{g} = 1875\ \text{mg}$$

$$\text{mmols} = \frac{1875\ \text{mg}}{142\ \text{mg}} = 13.2\ \text{mmols}$$

$$\text{mEq} = \text{mmols} \times \text{valence} = 13.2 \times 2 = 26.4\ \text{mEq}$$

$$\text{mOsm} = \text{mmols} \times \text{particles} = 13.2 \times 3 = 39.6\ \text{mOsm}$$

Z. **(a)** $10\ \text{g/100 mL} = 0.1\ \text{g/mL}$

$$\frac{100\ \text{mg}}{430\ \text{mg}} \times 2 = 0.465\ \text{mEq/mL}$$

(b) $10\ \text{g/100 mL} = 100\ \text{g/Liter} = 100{,}000\ \text{mg/L}$

$$\frac{100{,}000\ \text{mg}}{430\ \text{mg}} \times 3 = 697.6\ \text{mOsm/L}$$

AA. 1 mmol = 39 mg = 1 mOsm

$$\frac{156 \text{ mg}}{100 \text{ mL}} = 1560 \text{ mg/Liter}$$

$$\frac{1560 \text{ mg}}{\text{Liter}} \times \frac{1 \text{ mOsm}}{39 \text{ mg}} = 40 \text{ mOsm}$$

AB. 1 mEq = 430/2 = 215 mg

$$25 \text{ mEq/L} = 25 \times 215 = 5375 \text{ mg/L} = 5.375 \text{ g/L}$$

$$\frac{5.375 \text{ g/L}}{430} \times 3 \text{ particles} \times 1000 = 37.5 \text{ mOsm/L}$$

or $$\frac{3 \text{ mOsm}}{\text{mmol}} \times \frac{1 \text{ mmol}}{430 \text{ mg}} \times \frac{215 \text{ mg}}{1 \text{ mEq}} \times 25 \text{ mEq} = 37.5 \text{ mOsm/L}$$

CHAPTER 10

INJECTABLE MEDICATIONS AND INTRAVENOUS FLUIDS—PART II

Reconstitution of Dry Powders, Units of Potency, IV Admixtures, and Rate of Flow

LEARNING OBJECTIVES: *After completing this chapter the student should be able to:*

1. Determine the concentration of a solution after reconstitution (or constitution).
2. Calculate the amount of diluent needed to produce a desired concentration when the powder does *not* contribute to the final volume of constituted solution.
3. Calculate the amount of diluent needed to produce a desired concentration when the powder *contributes* to the final volume of constituted solution.
4. Determine doses of drugs with potencies designated in *units*, e.g. insulin and heparin.
5. Calculate the equivalency of an antibiotic based on μg of activity per milligram.
6. Perform calculations related to special dosing for cancer chemotherapy.
7. Evaluate the amount of additive(s) needed to produce an admixture of a specified content.
8. Calculate the rate of flow needed to deliver a large volume parenteral or a specified amount of drug over a specified period of time.
9. Convert a desired duration of time of administration into a flow rate of drops per minute.

RECONSTITUTION OF DRY POWDERS

1. Many drugs (antibiotics, steroids, and biologics) that are not stable in solution are prepared as dry-filled solids or lyophilized powders. Prior to use, these dry powders must be reconstituted as a solution with a suitable diluent in the proper volume to give the specified concentration (usually provided in the package insert). Occasionally, the physician may prescribe a final concentration different from the one provided by the manufacturer. Also, in some cases, the pharmacist will need to determine if the powdered drug contributes to the final volume of the reconstituted solution before modifying the label instructions.

Let us go over some examples.

A pharmacist receives a medication order for an IM (intramuscular) injection of 250 mg Ancef® (cefazolin sodium) every 8 hours. What volume of reconstituted solution would the pharmacist use to obtain the prescribed dose from 500 mg and 1 g vials?

Pharmaceutical Calculations, Fourth Edition, By Joel L. Zatz and Maria Glaucia Teixeira
ISBN 0-471-67623-3

Reconstitution of the dry powder with compatible diluent is based on the amounts indicated by the manufacturer:

Vial size	Volume of diluent	Final volume
500 mg	2 mL	2.2 mL
1 g	2.5 mL	3 mL

From the 500 mg vial:

250 mg × 2.2 mL/500 mg = 1.1 mL of reconstituted solution

From the 1 g vial:

250 mg × 3 mL/1,000 mg = 0.75 mL of reconstituted solution

---

2. A medication order calls for 400,000 units of penicillin G sodium to be added to 1 liter of D5W. Available in the pharmacy is a vial containing 5,000,000 units. The manufacturer directions are to add 8 mL of diluent to get a concentration of 500,000 units/mL. How many milliliters of the reconstituted solution will be needed to prepare the order?

Solution. 0.8 mL

CALCULATIONS

400,000 units × mL/500,000 units = 0.8 mL

---

3. A pharmacist needs to prepare a medication order that calls for 7,500 units of polymixin B sulfate added to 100 mL normal saline. The source of polymixin sulfate is a vial containing 50,000 units of the dry powder. The directions on the vial are to add 9.4 mL of diluent to obtain a concentration of 5,000 units per milliliter. How many mL of the reconstituted solution should be withdrawn and added to the NS bag?

Solution. 1.5 mL

CALCULATIONS

7,500 units × mL/5,000 units = 1.5 mL

---

4. Penicillin G sodium 7,500 units/mL

Sterile water for injection (SWFI) ad 20 mL

The source of penicillin G sodium is a vial containing 5,000,000 units of the dry powder. The directions on the vial are to add 18 mL of diluent to obtain a concentration of 250,000 units per milliliter. How many mL of the reconstituted solution should be withdrawn to prepare the prescription?

Solution. 0.6 mL

CALCULATIONS

7,500 units/mL × 20 mL = 150,000 units needed

$$\frac{250{,}000 \text{ units}}{\text{mL}} = \frac{150{,}000 \text{ units}}{j}$$

$j = 0.6$ mL

Take 0.6 mL of the reconstituted solution and add SWFI q.s. to 20 mL.

Changing Final Concentration of Reconstituted Solution when Dry Powder does *not* Account for Volume

5. Most injectable drugs available as dry powders are not very bulky and their contribution to the final volume by the solid material is considered negligible. In these cases, calculations of volume of solvent needed to produce a desired concentration (other than the one provided by the manufacturer) are relatively simple and direct.

A pharmacist has on hand a vial containing 200,000 units of penicillin G potassium. Assuming that the volume occupied by the penicillin G potassium is negligible, how much diluent must he add to obtain a solution containing 50,000 units/mL?

$$200{,}000 \text{ units} \times \frac{\text{mL}}{50{,}000 \text{ units}} = 4 \text{ mL}$$

Using a vial containing 1,000 mg of methicillin sodium and sterile water for injection (SWFI) as the diluent, explain how a pharmacist would prepare the following order

℞			
℞	Methicillin sodium		40 mg/mL
	SWFI	ad	10 mL
	Sig. I.M. injection		

$$10 \text{ mL} \times 40 \text{ mg/mL} = 400 \text{ mg needed}$$

$$400 \text{ mg needed}/1{,}000 \text{ mg available} = 2/5$$

Dissolve 1,000 mg vial in 5 mL SWFI, use 2 mL of reconstituted solution (=400 mg) and q.s. to 10 mL.

By dimensional analysis:

$10 \text{ mL} \times \frac{40 \text{ mg}}{1 \text{ mL}} \times \frac{5 \text{ mL}}{1{,}000 \text{ mg}} = 2$ mL of reconstituted solution and q.s. to 10 mL with SWFI

In practice, any multiple of 2 and 5 will provide the correct reconstitution, assuming that the ratio 2:5 is maintained and remembering that the powder needs to be completely dissolved. Another limitation for reconstitution of injections is the use of syringes and needles, which limits the volumes measured.

---

6. A physician prescribes 20 g of an ointment containing 15,000 units of penicillin G potassium per gram of ointment. The pharmacist has vials containing 500,000 units of crystalline penicillin G potassium and 5-mL vials of SWFI. Assuming there will be no change in volume of reconstituted solution caused by the powdered drug, how would the pharmacist compound the prescription?

Solution. Reconstitute 500,000 units vial with 5 mL SWFI, take 3 mL and incorporate in sufficient ointment base to make a total of 20 g. (The ointment base would have to be capable of absorbing this volume of aqueous liquid.)

CALCULATIONS

15,000 units/g × 20 g = 300,000 units needed

300,000 units needed/500,000 units available = 3/5

Thus, 3/5 of a vial is needed. If one vial is dissolved in 5 mL of diluent, 3 mL will contain 300,000 units needed for the prescription.

One could also reconstitute the powder with 10 mL and use 6 mL to obtain 300,000 units. However, the minimum volume possible should be utilized to facilitate compounding and maintain the physical characteristics of the ointment.

---

7. A compounding pharmacist is asked to prepare a nose drops according to the formula provided below. Penicillin G potassium is available as powder for reconstitution in vials, each containing 200,000 units of crystalline penicillin G potassium. How would he obtain the drug needed in compounding the prescription?

℞	Penicillin G potassium		2,500 units/mL
	0.9% NaCl	q.s	30 mL

Solution. Reconstitute the powder with 20 mL diluent and use 7.5 mL to prepare the prescription.

CALCULATIONS

2,500 units/mL × 30 mL = 75,000 units needed

75,000 units/200,000 units = 7.5/20

The pharmacist will reconstitute a 200,000 units vial with 20 mL 0.9% NaCl, take 7.5 mL of this dilution and q.s. to 30 mL to get 2,500 units/mL required in prescription.

Changing Final Concentration of Reconstituted Solution when Dry Powder *Contributes* to Final Volume

8. Sometimes the dry powder accounts for a considerable volume of the final constituted solution, such as in oral powders for reconstitution and some bulky injectables. The calculations of volume of solvent needed to produce a desired concentration will then need to consider the powder bulk and the volume it will occupy.

The package insert of a vial containing 2 g of Rocephin® (ceftriaxone sodium) specifies that when 7.2 mL of normal saline are added to the dry powder, the final concentration is 250 mg/mL. How many milliliters of normal saline should be used to prepare the following solution?

℞ Rocephin® 2 g
Normal Saline q.s.
Sig. 50 mg in 1 mL by I.M. injection

$$\frac{2{,}000\text{ mg}}{250\text{ mg/mL}} = 8\text{ mL}$$

8 mL − 7.2 mL (diluent) = 0.8 mL (vol. occupied by powder)

$$2{,}000\text{ mg} \times \frac{1\text{ mL}}{50\text{ mg}} = 40\text{ mL (total volume needed for Rx)}$$

40 mL − 8 mL = 39.2 mL normal saline (to prepare the prescription)

---

9. Label instructions for a 2.5 g ampicillin product state that when 78 mL of purified water are added to the powder, 100 mL of constituted suspension containing 125 mg of ampicillin per 5 mL results. How many milliliters of purified water should the pharmacist add to the dry powder to prepare a prescription that calls for 100 mg/tsp?

Solution. 103 mL

CALCULATIONS

100 mL − 78 mL = 22 mL (volume occupied by the powder)

$$2{,}500\text{ mg} \times \frac{5\text{ mL}}{100\text{ mg}} = 125\text{ mL of purified water}$$

But, dry powder occupies 22 mL of volume, then,

125 mL − 22 mL = 103 mL of purified water should be added to get 100 mg/tsp.

---

10. Penicillin G sodium is available in the hospital pharmacy in vials containing 5,000,000 units of drug. The directions on the vial are to reconstitute the powder with 23 mL of SWFI to obtain a concentration of 200,000 units/mL. The pharmacist needs to prepare a medication order that calls for 125,000 units per milliliters for IM injection. How many milliliters of SWFI should the pharmacist add to the dry powder to prepare the desired strength?

Solution. 38 mL of SWFI

CALCULATIONS

5,000,000 units × mL/200,000 units = 25 mL (volume of reconstituted solution)

25 mL − 23 mL = 2 mL (volume occupied by the powder)

5,000,000 units × mL/125,000 units = 40 mL of SWFI

Dry powder occupies 2 mL of volume.

40 mL − 2 mL = 38 mL of SWFI should be added to get 125,000 units/mL.

Insulin, Heparin and Other Calculations Involving Units of Activity

11. Insulin, heparin, and some antibiotics, vitamins and biologics have their activity expressed as *units of activity* such as USP *units, International Units* (I.U), *μg of activity/g or mg,* etc. It is important to understand that one drug's unit of potency has no relationship with the units of another drug. For example, insulin units are different from heparin units, which are different from penicillin units, and so on. Basically, each drug uses a unit system that has its own conversion factor and is based on some biologic assay. Some examples are shown in Table 10.1.

Doses of drugs designated in units are *prescribed in units and measured in units.* As mentioned earlier in this text, it is important to always spell out the word "units" following the number, since a poorly written "U" may be mistaken for a zero.

---

12. *Insulin,* a hormone produced by the pancreas and needed for treatment of diabetes mellitus, is commercially available in several types (based on duration of action, time of onset of action or peak action) and from several sources (human, beef or pork). All insulin preparations available in the USA are standardized to include *100 or 500 insulin units per milliliter* of solution or suspension. Insulin products are designated as U-100 and U-500, meaning the strengths of the solutions or suspensions (U-500 is a concentrated solution available for hospital use only). Units of insulin are measured in insulin syringes, calibrated in units according to the strength of insulin to be used and requiring no calculation or conversion. In the absence of insulin syringes, a required dosage may be converted and measured in milliliters, using a 1 cc syringe.

TABLE 10.1 Some drugs expressed in units of activity and their strength equivalents (USP 26/2003, Official Monographs)

Drug	Units of activity
Amoxicillin	NLT 900 μg and NMT 1050 μg Amoxicillin per mg
Amphotericin B	NLT 750 μg Amphotericin B per mg
Antihemophilic factor	NLT 100 Antihemophillic factor units per gram of protein
Bacitracin	NLT 40 Bacitracin units per mg
Cefepime	NLT 825 μg and NMT 911 μg Cefepime per mg
Cefuroxime	NLT 745 μg and NMT 875 μg Cefuroxime per mg
Chymotrypsin	NLT 1000 USP Chymotrypsin units per mg
Chorionic gonadotropin	NLT 1500 USP Chorionic gonadotropin units per mg
Clindamycin hydrochloride	NLT 800 μg of Clindamycin per mg
Dactinomycin	NLT 950 μg and NMT 1030 μg Dactinomycin per mg
Digitalis	NLT 1 USP Digitalis unit per 100 mg dried leaf of *Digitalis purpurea*
Gentamicin sulfate	NLT 590 μg of Gentamicin per mg
Heparin sodium	NLT 140 USP Heparin units (not equivalent to IU) per mg
Hyaluronidase	NMT 0.25 μg of Tyrosine per USP Hyaluronidase unit
Insulin	NLT 26.5 USP Insulin units* per mg
Insulin human	NLT 27.5 USP Insulin Human units# per mg
Lincomycin hydrochloride	NLT 790 μg of Lincomycin per mg
Mumps skin test antigen	NLT 20 C_fU (complement-fixing units) per mL
Nystatin	NLT 4400 USP Nystatin units per mg
Pancreatine	NLT 25 USP units of Amylase activity, NLT 2 USP units of lipase activity and NLT 25 USP units of protease activity per mg
Penicillin G potassium	NLT 1440 Penicillin G units and NMT 1680 Penicillin G units per mg
Rubella virus vaccine live	NLT 1000 $TCID_{50}$(tissue culture infectious dose) per 0.5 mL
Tetracycline	NLT 975 μg of Tetracycline hydrochloride per mg
Typhoid vaccine	8 units per mL
Vancomycin hydrochloride	NLT 925 μg of Vancomycin per mg
Vasopressin	NLT 300 USP Vasopressin units per mg

* 1 USP Insulin unit = 0.0342 mg of pure Insulin derived from beef or 0.0345 mg of pure Insulin derived from pork.
# 1 USP Insulin Human unit = 0.0347 mg of pure Insulin human.

Heparin represents a group of mucopolysaccharides that prevent or slow the formation of blood clots. Heparin is administered by intravenous (bolus or heparin drip) or deep subcutaneous route for treatment or prophylaxis of venous thrombosis. Salt forms of heparin (heparin sodium or calcium) are measured in units and commercial preparations are standardized to contain 140 USP heparin units per milligram. Dosages of heparin solutions are adjusted based on the patient's blood coagulation tests (e.g. PTT = partial thromboplastin time) and the specific needs of the patient. For example, low doses (5,000 units, sub-Q) are used to provide effective prophylaxis for a variety of pre-surgical situations. Medium doses are used for patients with phlebitis, pulmonary emboli and during hip replacement surgery, while high doses are indicated for patients with massive pulmonary embolism. Whereas heparin is prescribed many times in a "per-day" basis or "as needed", a weight-based heparin protocol has been the most used way of prescribing heparin. Pediatric doses are always calculated in a weight-based protocol.

Penicillins and several other antibiotics are also among the most common drugs with potencies expressed in units.

Occasionally, some *biologics* (diagnostic antigens, immune sera, bacterial vaccines) have strengths expressed in units of antigen per milliliter. We will discuss this subject in more detail in Chapter 12 (Immunizations).

---

13. You will notice through this section that most calculations involving units are extremely simple and may be solved by dimensional analysis or simple proportion.

A parenteral nutrition (PN) formulation calls for the addition of 20 units of insulin for each 100 mL of solution. A patient is to receive 1 liter of PN per day. You have in the pharmacy, 10 mL vials of Humulin® R, U-500 (500 units/mL). How many milliliters of this solution should be added to the patient's PN to provide the required units in a daily dose?

1,000 mL × 20 units/100 mL = 200 units for daily dose

200 units × 1 mL/500 units = 0.4 mL

0.4 mL of U-500 should be added per day.

---

14. A patient undergoing hip replacement surgery was prescribed a heparin dose of 250 units/kg of body weight. How many milliliters of a solution of heparin sodium 10,000 units/mL should be administered, if the patient weighs 220 lb?

Solution. 2.5 mL

CALCULATIONS

220 lb × 1 kg/2.2 lb × 250 units/kg = 25,000 units

25,000 units × 1 mL/10,000 units = 2.5 mL

---

15. A pharmacist received a medication order to add 40,000 units of heparin sodium to 1 liter of Dextrose 5% in water for a 180 lb patient. The rate of the infusion was prescribed as 2,000 units per hour.

(a) What was the concentration of heparin sodium in the infusion, in units/mL?

(b) How long will the infusion run, in hours?

(c) What is the dose of heparin sodium received by the patient, on a unit/kg/minute basis?

Solution.

(a) 40 units/mL

(b) 20 hours

(c) 0.41 units/kg/min

CALCULATIONS

(a) 40,000 units/1,000 mL = 40 units/mL

(b) 40,000 units × hour/2,000 units = 20 hours

(c) 180 lb = 81.8 kg

20 hr = 1,200 min

40,000 units/1,200 min = 33.33 units/min

33.33 units/min/81.8 kg = 0.41 units/kg/min

DOSING IN CHEMOTHERAPY

16. The primary goal of chemotherapy is to destroy cancer cells or control their growth while minimizing the effects on normal cells. Anticancer drugs (chemotherapeutic agents) are indicated in some localized cancers as well as in such widespread cancers as leukemia and cancers that have metastasized to other organs. Most anticancer drugs are *cytotoxic* agents that preferentially kill fast-growing cells. Immunosuppressants are also categorized as anticancer drugs and warrant the same special procedures to minimize the potential for accidental exposure.

Basics on chemotherapy and dosing:

- These agents are generally administered by the oral and/or intravenous (injection or infusion) routes although other routes may be occasionally used.
- Combination chemotherapy, in which more than a single anticancer drug is included in a patient's therapy regimen, is often used as a way to maximize effectiveness.
- Treatment may be cyclic with drugs administered at the same time or alternately on the same or different days.
- Prescriptions have instructions written in a special format: D 1–5 = days 1 to 5; D 1, 5 = days 1 and 5.
- Drugs are frequently abbreviated and some drugs may have different acronyms depending on the protocol: CDDP or PT = cisplatin; DOX or A or H = doxorubicin; MTX or M = methotrexate; VLB or Ve = vinblastine.
- Combination chemotherapy follows drug/dosage regimen identified by abbreviations, e.g. for ovarian cancer, CTX-PT = Cyclophosphamide (CTX) + Diamminedichloroplatinum or Cisplatin (CDDP or PT) or for colorectal cancer FU/LU = Fluorouracil + Leucovorin.
- Dosing is generally based on body weight or surface area. Standard regimens are prescribed and must be adjusted depending on the patient's kidney and/or liver functions.
- It is the pharmacist's responsibility to interpret the medication order and follow the individualized dosage regimen, calculating doses appropriately and dispensing the dosage forms and quantities required.

Table 10.2 lists some common treatment agents and combinations.

We will now practice some cancer treatments to get used to the typical interpretation and abbreviations/acronyms. Calculations, in general, are quite easy and direct.

TABLE 10.2 Some representative agents for cancer treatment and cancer chemotherapy protocols

Agents and Classification	Abbreviations
Antimetabolites	
Fluorouracil	5-FU; FU
Mercaptopurine	6-MP
Thioguanine	6-TG
Methotrexate	MTX; M
Microtubule-Targeting	
Vincristine	V
Vinblastine	VLB; Ve, V
Topisomerase-Targeting	
Doxorubicin	DOX; A; H
Alkylating	
Cyclophosphamide	CTX
Carmustine	BCNU
Miscellaneous	
Cisplatin	CDDP; PT; C
Acronyms for Combination Regimens (Protocols)	**Types of Cancer**
AC (Doxorubicin + Cyclophosphamide)	Breast
CMF (Cyclophos. + Methothex. + Fluorour.)	Breast
CAF or FAC (Cyclophos. + Dox. + Fluorour.)	Breast
5-FU (Fluorouracil)	Colorectal
FU-LVR (Fluorouracil + Leucovorin)	Colorectal
CE (Cisplatin + Etoposide)	Lung (non-small cell)
CAV (Cyclophos. + Dox. + Vincristine)	Lung (small cell)
MOPP (Mechlorethamine + Vinc. + Procarb + Prednisone)	Lymphoma (Hodgkin's disease)
DTIC (Dacarbazine)	Melanoma
FAM (Fluorour. + Doxorubicin + Mitomycin)	Pancreatic
DES (Diethylstilbesterol)	Prostate
DOX (Doxorubicin)	Thyroid

* Sources: Pharmacotherapy Handbook, Wells et al., 5th Edition, 2003 and Cancer Chemotherapy Protocols, Ignoffo & Forni, Cetus Corp., 1989.

Regimen: FU/LU, repeat 2–6 cycles

Cycle: 28 days

Fluorouracil, 500 mg/m^2, IV, D 1, 8, 15, 22

Leucovorin, 200 mg/m^2, IV, D 1, 22

The interpretation of this medication order would be:

The patient will receive a combination regimen of fluorouracil and leucovorin, repeated for 2 to 6 cycles (doctor decides as therapy progresses). The cycle is defined as 28 days and will include fluorouracil at a dose of 500 mg/m^2, by intravenous infusion, during days

1, 8, 15 and 22 of therapy and leucovorin at a dose of 200 mg/m^2, by intravenous infusion, during days 1 and 22 of therapy.

Based on the above medication order, calculate for each drug the total intravenous dose per cycle for a patient having a BSA (body surface area) of 1.5 m^2.

Solution. 3,000 mg of fluorouracil and 600 mg of leucovorin

CALCULATIONS

Fluorouracil: 500 mg/m^2 × 1.5 m^2 (BSA) × 4 (days of treat.) = 3,000 mg
Leucovorin: 200 mg/m^2 × 1.5 m^2 × 2 = 600 mg

---

17. For a patient having a BSA of 2.00 m^2 calculate the total cycle dose for each drug in the regimen prescribed below.

Regimen: MVAC
Cycle: 28 days
Methotrexate: 40 mg/m^2/day, p.o., D 1–14

Vinblastine: 5 mg/m^2, IV, D 2, 8
Doxorubicin: 50 mg/m^2, IV, D 1, 8
Cisplatin: 100 mg/m^2, IV, D 1

Solution. 1,120 mg MTX, 20 mg VLB, 200 mg DOX (A) and 200 mg CDDP (C)

CALCULATIONS

40 mg × 2.00 (BSA) × 14 (days) = 1,120 mg

5 mg × 2.00 × 2 = 20 mg

50 × 2.00 × 2 = 200 mg

100 mg × 2.00 × 1 = 200 mg

For each of cisplatin and etoposide, calculate the total intravenous dose per cycle for a patient having a BSA of 2.2 m^2.

Regimen: CE

Cycle: 3 days; repeat every 3–4 weeks

Cisplatin: 100 mg/m^2, IV, day 1

Etoposide: 120 mg/m^2, IV, days 1–3

Solution. 220 mg cisplatin and 792 mg etoposide

CALCULATIONS

Cisplatin: $100\,mg/m^2 \times 2.2\,m^2 = 220\,mg$

Etoposide: $120\,mg/m^2 \times 2.2\,m^2 \times 3$ days $= 792\,mg$

INTRAVENOUS ADMIXTURES

18. The compounding of intravenous admixtures comprises the aseptic addition of one or more drugs to a sterile intravenous basic fluid (large-volume solutions). Commonly, the drug(s) is (are) referred to as an *additive* and includes small volume parenterals such as electrolytes, antibiotics or powders requiring constitution with an adequate solvent before transfer, vitamins, heparin and other drugs. The compounding of IV admixtures is a professional function and requires special training. Pharmacy admixture programs centralize the responsibility for compounding, dispensing and controlling all parenteral admixtures. This responsibility includes checking all components and calculations against the medication order.

Let us practice some of these calculations.

A parenteral admixture solution is to contain sodium, 60 mEq/L and calcium, 9 mEq/L. A sodium chloride solution with a concentration of 4 mEq/mL and a calcium gluconate solution containing 0.45 mEq/mL of calcium ion are available. How many milliliters of this solution are needed to contribute the sodium ion in preparing 2 L of the admixture?

Solution. 30 mL of NaCl sol. and 40 mL of calcium gluconate solution.

CALCULATIONS

$j \times 4\,mEq/mL = 2\,L \times 60\,mEq/L$

$j = 30\,mL$

$k \times 0.45\,mEq/mL = 2\,L \times 9\,mEq/L$

$k = 40\,mL$

---

19. Calculate the quantity of KCl solution, 2 mEq/mL, needed to supply potassium ion in 1.75 L of a solution to contain 40 mEq of potassium ion/liter.

Solution. 35 mL

CALCULATIONS

$j \times 2\,mEq/mL = 1.75\,L \times 40\,mEq/L$

$j = 35\,mL$

20. A commercial electrolyte concentrate supplies the following:

Sodium	0.8 mEq/mL
Potassium	0.4 mEq/mL
Calcium	0.096 mEq/mL
Magnesium	0.16 mEq/mL
Gluconate	0.096 mEq/mL
Chloride	1.2 mEq/mL

If 70 mL of the concentrate were used to prepare 1 L of a parenteral admixture, how many milliequivalents of sodium, potassium, calcium and chloride ions would be present?

Solution. 56 mEq sodium, 28 mEq potassium, 6.72 mEq calcium and 84 mEq chloride

CALCULATIONS

mEq Na^+ in concentrate = mEq Na^+ in admixture

Sodium: 70 mL × 0.8 mEq/mL = 56 mEq

Potassium: 70 mL × 0.4 mEq/mL = 28 mEq

Calcium: 70 mL × 0.096 mEq/mL = 6.72 mEq

Chloride: 70 mL × 1.2 mEq/mL = 84 mEq

---

21. A hospital pharmacy received the medication order below. Available in the pharmacy are 70% dextrose solution, 23.4% concentrated NaCl solution, cimetidine HCl 150 mg/mL and 500 mL containers of SWFI. How should the following fix be prepared?

℞ 300 mg Cimetidine HCl in 500 mL 3.5% Dextrose and 0.7% NaCl

Hint: Calculate the amounts of dextrose and concentrated NaCl solution needed. Withdraw their combined volume from a SWFI container and add the dextrose and concentrated NaCl solution to that container. Add the required amount of cimetidine solution and mix.

Solution. See below.

CALCULATIONS

$500\,\text{mL} \times 3.5\% = j \times 70\%$

$j = 25\,\text{mL}$ of 70% dextrose solution

$500\,\text{mL} \times 0.7\% = k \times 23.4\%$

$k = 14.96 = 15\,\text{mL}$ of 23.4% NaCl solution

$\text{SWFI} = 500 - (25 + 15) = 460\,\text{mL}$

Cimetidine: $300\,\text{mg} \times \text{mL}/150\,\text{mg} = 2\,\text{mL}$

Preparation of the medication order:

The pharmacist will withdraw and discard 40 mL from the 500 mL SWFI container. This volume will be replaced by 25 mL of dextrose 70% solution and 15 mL of 23.4% NaCl solution. This will allow 500 mL of 3.5% dextrose and 0.7% NaCl required in medication order. To this solution, the pharmacist will add 2 mL of cimetidine HCl 150 mg/mL solution.

---

22. A pharmacist received the following computerized medication order:

2/20/03	Mary Peterson	Dr. John Dayton
ID# 1234	Rm. 301 W	
6 mEq KCl 3 mEq Ca gluconate Insulin regular 50 units		
In 1000 cc D5W/NS × 8 h		

Available in the pharmacy:

KCl, 10 mL vial containing 2 mEq/mL

Ca gluconate, 10 mL vial containing 4.6 mEq

Humulin R (human recombinant DNA insulin), U-100, 10 mL vial

D5W/0.9%NaCl, 1000 mL bag

How would the pharmacist prepare the order?

Solution. Add 3 mL KCl, 6.5 mL Ca gluconate and 0.5 mL of Humulin-R to the 1000 mL bag of D5W/0.9% NaCl.

CALCULATIONS

KCl: 6 mEq × mL/2 mEq = 3 mL

Ca gluconate: 3 mEq × 10 mL/4.6 mEq = 6.5 mL

Insulin: 50 units × mL/100 units = 0.5 mL

---

23. A patient is to receive 250,000 units of penicillin G potassium in 500 mL D5W. The instructions for reconstitution of a 1,000,000 unit package are to add 1.6 mL of SWFI and result in 2.2 mL of solution. How many milliliters of the reconstituted solution should the pharmacist add to the D5W bag?

Solution. 0.55 mL

CALCULATIONS

1,000,000 units/2.2 mL = 250,000 units/j

j = 0.55 mL

---

24. A medication order for a hospitalized patient weighing 132 lb calls for 5,000 units/kg of body weight of Polymixin B sulfate in 500 mL D5W by IV drip. Polymixin B sulfate is available in 10-mL vials containing 500,000 units and may be reconstituted with 0.9% NaCl injection. How would the pharmacist reconstitute the drug and how many milliliters of the reconstitutions should he add to the bag of D5W to prepare the infusion?

Solution. Dissolve powder in 5 mL sodium chloride 0.9% and add 3 mL of this solution to the bag of D5W.

CALCULATIONS.

132 lb × kg/2.2 lb × 5,000 units/kg = 300,000 units needed

300,000 units needed/500,000 units available = 3/5

If a vial containing 500,000 units is reconstituted with 5 mL, 3 mL of this solution will contain 300,000 units needed to prepare the infusion.

RATE OF FLOW

25. In the hospital, patients often receive solutions intravenously (directly into a vein). It may be necessary to administer a large volume of solution continuously. The process is called *intravenous infusion*. This is done slowly, over a period of time, using an infusion set that can be adjusted to deliver a chosen number of drops per minute or an infusion pump delivering the solution in milliliters per minute or per hour. The *rate of flow* is specified by the physician and depends on the volume to be infused and desired duration of time of administration. Rate of flow of intravenous fluid may be requested in mL/min, drops/min, amount of drug per unit of time (mg per hour), or, as the approximate time required to administer the total volume of the infusion. If an intravenous infusion is to drip slowly by gravity flow, the pharmacist may need to convert the desired total interval into drops/min. Occasionally, the pharmacist determines the rate of flow of a parenteral fluid using a nomogram or an infusion rate table that accompanies some commercial products.

Here is a typical problem: A patient is to receive 900 mL of a solution over 12 hours (h). If the administration set delivers 20 drops/mL, at how many drops/min should the medication be infused?

From the information given, we can calculate the number of mL/min that must be administered. Then, by converting milliliters to drops, it will be possible to determine the number of drops/min.

$$\frac{900\text{ mL}}{12\text{ h}} = 75\text{ mL/h}$$

$$75\text{ mL/h} \times \frac{1\text{ h}}{60\text{ min}} = 1.25\text{ mL/min}$$

$$1.25\text{ mL/min} \times 20\text{ drops/mL} = 25\text{ drops/min}$$

---

26. A solution is to be administered by intravenous infusion at a rate of 45 mL/h. How many drops/min should be infused if the administration set delivers 40 drops/mL?

Solution. 30 drops/min

CALCULATIONS

$$45\text{ mL/h} \times \frac{1\text{ h}}{60\text{ min}} = 0.75\text{ mL/min}$$

$$0.75\text{ mL/min} \times 40\text{ drops/mL} = 30\text{ drops/min}$$

---

27. A patient is to receive 2000 mL of a solution by intravenous infusion over a period of 24 h. What rate of infusion (drops/min) should be utilized if the infusion set delivers 20 drops/mL?

Solution. 28 drops/min

CALCULATIONS

$$\frac{2000\text{ mL}}{24\text{ h}} = 83.3\text{ mL/h}$$

$$83.3\text{ mL/h} \times \frac{1\text{ h}}{60\text{ min}} \times \frac{20\text{ drops}}{1\text{ mL}} = 27.8\text{ drops/min}$$

This has to be rounded off to 28 drops/min.

---

28. One hundred micrograms of a drug, dissolved in 240 mL of solution, is to be infused at a rate of 25 μg/h. If a infusion set delivers 15 drops/mL, what should the rate of administration be (in drops/min)?

Solution. 15 drops/min

If you had trouble, read the following hint then see if you can complete the solution.

Hint: Remember that our calculations deal with flow rates. We know that the drug has to be infused at a rate of 25 μg/h. We have to first determine the rate of infusion, in mL/h, for the solution that contains the drug. Knowing that each 240 mL holds 100 μg of drug, find the volume that contains 25 μg. This is the volume that must be administered each hour.

CALCULATIONS

The volume of solution to be infused each hour can be found by proportion.

$$\frac{100\ \mu\text{g}}{240\text{ mL}} = \frac{25\ \mu\text{g}}{j}$$

$$j = 60\text{ mL}$$

$$60\text{ mL/h} \times \frac{1\text{ h}}{60\text{ min}} \times \frac{15\text{ drops}}{1\text{ mL}} = 15\text{ drops/min}$$

---

29. A solution for intravenous infusion contains 0.025 μg/mL of a drug. Calculate the flow rate, in drops/min, needed to administer the drug at a rate of 2 μg/h. (Infusion set delivers 25 drops/mL).

Solution. 33 drops/min

CALCULATIONS

$$\frac{0.025\ \mu g}{1\ mL} = \frac{2\ \mu g}{j}$$

$j = 80\ mL$

$$80\ mL/h \times \frac{1\ h}{60\ min} \times \frac{25\ drops}{mL} = 33\ drops/min$$

---

30. If 50 mg of a drug are added to a 500 mL LVP (large volume parenteral) fluid, what will be the rate of flow, in milliliters per hour, to deliver 2 mg of drug per hour?

Solution. 20 mL/h

CALCULATIONS

$50\ mg/500\ mL = 2\ mg \times j$

$j = 20\ mL$ per hour

---

31. A 12-kg infant is to receive a continuous infusion of a drug at 1.2 mL/hr to deliver 4 μg/kg per minute. How many milligrams of the drug should be added to a 100 mL infusion solution?

Solution. 240 mg

CALCULATIONS

$12\ kg \times 4\ \mu g/kg/min \times mg/1000\ \mu g \times hr/1.2\ mL \times 60\ min/hr \times 100\ mL = 240\ mg$

REVIEW PROBLEMS

A. The label of a 750 mg dry powder for constitution into pediatric drops states that when 12 mL of purified water are added to the powder, 15 mL of a pediatric suspension results, containing 50 mg/mL. How many milliliters of water should be added to have the dose in each 10 drops, if the dropper delivers 20 drops/mL, the infant weighs 33 lb (1 kg = 2.2 lb), and the dose of the drug is 2.5 mg/kg?

B. A pharmacist received the following prescription for an ophthalmic ointment:

Cyclosporine	2%	
SWFI	qs	
Hydrophilic ointment	ad	15 g

The only available source of cyclosporine is a vial containing 600 mg dry powder and the package insert states that the powder accounts for 0.5 mL of the volume of constituted solution. Using SWFI to constitute the powder how would the pharmacist reconstitute and compound the prescription?

C. The label of a 7.5 g bottle of ampicillin for oral suspension states that when 111 mL of purified water are added to the powder, 150 mL of a suspension results, containing 250 mg of ampicillin per teaspoon. How many *milliliters of purified water* should be used to prepare, in each teaspoon, the correct dose of ampicillin for a 55-lb child based on the dose of 6 mg/kg?

D. A patient is to receive 400,000 units of penicillin G potassium in 100 mL D5W. Available is a vial of penicillin G potassium 1,000,000 units and the manufacturer states that when 4.6 mL of a suitable diluent is added, a 200,000 units/mL solution will result. How many milliliters of reconstituted solution must be withdrawn and added to the bag of D5W?

E. If a physician orders 10 units of insulin to be added to a liter of D5W to be administered over 8 hours,

(a) What will be the infusion rate, in drops/min, if an IV set delivering 15 drops/mL is used?

(b) How many units of insulin will be infused in each half hour?

F. ℞ Regimen: CAF

Cycle: repeat every 28 days

Cyclophosphamide, 100 mg/m^2, PO, D 1–14

Doxorubicin, 30 mg/m^2, IV bolus, D 1, 8

Fluorouracil, 500 mg/m^2, IV bolus, D 1, 8

(a) What would be the interpretation of this medication order?

(b) Calculate, for each drug, the total dose per cycle for a patient having a BSA of 1.5 m^2.

G. What would be the interpretation of the medication order? How many milligrams of each drug would be needed for a complete combination chemotherapy treatment (total of 6 cycles)?

John Doe   Diagn.: SCLC (small cell lung cancer)

Age: 55

Wt. = 200 lb

Ht. = 6 ft

℞ Regimen: CAV

Cycle: repeat every 3 wk × 6

Cyclophosphamide, 100 mg/m^2, IV, D 1

Doxorubicin, 45 mg/m^2, IV, D 1

Vincristine, 1 mg/m^2, IV, D 1

H. A medication order calls for 2 g cefotaxime sodium to be added to 500 mL NS. Using a constituted injection that contains 180 mg/mL, how many milliliters should be added to prepare the order?

I. A medication order for a patient weighing 121 lb calls for Amikin® (amikacin sulfate) 0.25 mg/kg of body weight to be added to 200 mL D5W. Amikacin sulfate is available in a concentration of 50 mg/mL. How many milliliters should be added to the D5W solution?

J. An intravenous admixture calls for the addition of 20 mEq of sodium bicarbonate. How many milliliters of a 8.4% (w/v) sodium bicarbonate solution should be added?

K. A 55-lb child is to receive heparin sodium by intermittent IV infusion in a dose of 50 units/kg of body weight every four hours. How many milliliters of heparin sodium injection containing 5,000 units/mL will be administered daily?

L. A physician orders 50 μg/kg of digoxin for a 10-lb newborn baby to be diluted five-fold with D5W. How many milliliters of a 0.25 mg/mL digoxin injection should be used?

M. A patient receives a solution by intravenous infusion at a rate of 36 drops/min. How much solution is infused in 3 h if the infusion set delivers 30 drops/mL?

N. A drug is not to be administered in elderly patients at a rate greater than 50 mg/min to avoid circulatory failure. What should be the maximum infusion rate in mL/min, for a solution containing 20 mg/mL of drug in 100 mL normal saline?

O. Fifteen hundred milliliters of a solution is to be administered to a patient by intravenous infusion over a period of 24 h. At what rate, in drops/min, should the solution be given if the administration set delivers 25 drops/mL?

P. A solution contains 1.25 mg of a drug per milliliter. At what rate should the solution be infused (drops/min) if the drug is to be administered at a rate of 80 mg/h? (set delivers 30 drops/mL).

Q. A 50 mL vial containing 1 mg/mL of Alteplase is *added to* 100 mL of D5W and administered intravenously with an infusion set that delivers 15 drops/mL. How many drops per minute should be given to administer 25 mg of the drug per hour?

R. How many milliliters of a constituted injection containing 1 gram of drug in 4 mL should be used in filling a medication order requiring 275 mg of the drug to be added to 500 mL of 0.9% sodium chloride solution?

S. If the solution prepared in previous problem is administered at the rate of 1.6 mL per minute, how many milligrams of the drug will the patient receive in 1 hour?

T. If a physician orders 1,500 units of heparin to be added to a liter of Ringer's injection to be administered over 8 hours,

(a) How many drops per minute should be administered using an IV set that delivers 30 drops per milliliter?

(b) How many units of heparin would be administered in each 30-minute period?

U. A medication order calls for lidocaine hydrochloride at 1.5 mg/min. The pharmacy stocks premixed lidocaine HCl containing 0.4 % lidocaine in 250 mL dextrose 5% in water. How many milliliters per hour should the infusion pump be programmed to deliver the required dose?

V. A patient is to receive 500 mL dopamine hydrochloride drip at 20 μg/kg/min. The patient weighs 154 lb. The pharmacy has dopamine HCl premixed in concentration of 3.2 mg/mL in D5W. How many milliliters per minute should be the rate to program the infusion pump?

W. A patient is to receive daily 1000 mL parenteral nutrition (PN) containing 8 units of insulin per 100 mL of PN. If the parenteral nutrition is administered over 8 hours,

(a) How many milliliters of solution will be administered per minute?

(b) How many units of insulin would be administered in each 30-minute period?

X. How many milliliters of heparin sodium 10,000 units/mL should be administered to a 143-lb patient, if the recommended dose is 100 units/kg?

Y. Vitamin A (Aquasol A Parenteral®) is available in vials containing 50,000 I.U. (international units) per mL. If 1 I.U. is equivalent to the biological activity of 600 ng of β-carotene, how many milliliters would provide a dose containing 15,000 μg of β-carotene?

Z. A medication order calls for 300 mg of cefazolin sodium to be administered by intramuscular injection every 8 hours. The pharmacy has vials containing 250 mg, 500 mg and 1 g. Explain how the prescribed amount of drug could be obtained from each vial. Which vial would you choose?

Manufacturer's instructions for reconstitution:

250 mg	2 mL diluent	2 mL solution
500 mg	2 mL diluent	2.2 mL solution
1 g	2.5 mL diluent	3 mL solution

Solutions to Review Problems

A. Add 7 mL of water

B. Reconstitute the powder with 1.5 mL and use 1 mL for compounding the prescription.

C. Reconstitute powder with 211 mL purified water to get a dose/tsp.

D. 2 mL

E. **(a)** 31 drops/min

(b) 0.6 units

F. **(a)** see solution

(b) 2100 mg of C, 90 mg of A, 1500 mg of F

G. 1278 mg of C; 575.1 mg of A; 12.78 mg of V

H. 11 mL

I. 0.28 mL or 0.3 mL

J. 20 mL

K. 0.25 mL

L. 0.9 mL

M. 216 mL

N. 2.5 mL/min

O. 26 drops/min

P. 32 drops/min

Q. 19 drops/min

R. 1.1 mL

S. 52.8 mg/hr

T. **(a)** 63 drops/min

(b) 94 units

U. 22.5 mL/hr

V. 0.44 mL/min

W. **(a)** 2.08 mL/min

(b) 5 units

X. 0.65 mL

Y. 0.5 mL

Z. 2.4 mL of 250 mg-vial, 1.3 mL of 500 mg-vial and 0.9 mL of 1 g-vial. Best choice is 1 g-vial given that the smallest volume will be less painful for IM injection.

Calculations the Review Problems

A. 15 mL − 12 mL = 3 mL (volume occupied by the powder)

$$2.5\ \text{mg/kg} \times \frac{1\ \text{kg}}{2.2\ \text{lb}} \times 33\ \text{lb} = 37.5\ \text{mg (dose for infant)}$$

20 drops/mL × 10 drops = 0.5 mL (volume of product to contain the dose)

750 mg × 0.5 mL/37.5 mg = 10 mL (amount of product containing desired dose)

10 mL − 3 mL = 7 mL of water to be added.

B. 2 g/100 g × 15 g = 0.3 g = 300 mg

300 mg needed/600 mg available = 1/2

Then, reconstitute 600 mg dry powder with 2 mL − 0.5 mL (vol. of powder) = 1.5 mL and take 1 mL to be incorporated with hydrophilic ointment.

You may understand that mathematically any other volume used to reconstitute the powder would work, if half of the dilution were used. Pharmaceutically, one has to remember the limitations of hydrophilic ointment in absorbing aqueous solutions.

C. $150\,mL - 111\,mL = 39\,mL$ (volume occupied by powder)

$6\,mg/kg \times \frac{1\,kg}{2.2\,lb} \times 55\,lb = 150\,mg$ (dose for child)

$7500\,mg \times 5\,mL/150\,mg = 250\,mL$ (amount of product containing dose in 1 tsp)

$250\,mL - 39\,mL = 211\,mL$ (purified water used for reconstitution)

D. $400{,}000$ units $\times$ mL/200,000 units = 2 mL

E. **(a)** drops/min = 15 drops/mL $\times$ 1000 mL/8 hr $\times$ hr/60 min = 31.25 = 31 drops/min

(b) 10 units/8hr $\times$ 60 min/hr $\times$ 30 min = 0.625 = 0.6 units

F. **(a)** The patient will receive a combination regimen of Cyclophosphamide (C), Doxorubicin (A) and Fluorouracil (F). The cycle, to be repeated every 28 days, includes Cyclophosphamide at a dose of $100\,mg/m^2$ by oral route (per os) during days 1 through 14, Doxorubicin at a dose of $30\,mg/m^2$, by intravenous bolus injection, during days 1 and 8 and Fluorouracil at a dose of $500\,mg/m^2$, by intravenous bolus injection, during days 1 and 8 of therapy.

(b) $C = 100\,mg/m^2 \times 1.5\,m^2 \times 14$ days $= 2100\,mg$

$A = 30\,mg/m^2 \times 1.5\,m^2 \times 2 = 90\,mg$

$F = 500\,mg/m^2 \times 1.5\,m^2 \times 2 = 1500\,mg$

G. BSA (from Nomogram in Ch. 4) = $2.13\,m^2$

Cyclophosphamide: $100\,mg/m^2 \times 2.13\,m^2 \times 1$ day $\times$ 6 cycles = 1278 mg

Doxorubicin: $45\,mg/m^2 \times 2.13\,m^2 \times 1$ day $\times$ 6 cycles = 575.1 mg

Vincristine: $1\,mg/m^2 \times 2.13\,m^2 \times 1$ day $\times$ 6 cycles = 12.78 mg

H. $2000\,mg \times mL/180\,mg = 11\,mL$

I. $121\,lb \times \frac{1\,kg}{2.2\,lb} \times 0.25\,mg/kg \times mL/50\,mg = 0.275\,mL = 0.3\,mL$

J. $20\,mEq \times 84\,mg/mEq \times 100\,mL/8400\,mg = 20\,mL$

K. $55\,lb \times \frac{1\,kg}{2.2\,lb} \times 50$ units/kg $\times$ mL/5,000 units = 0.25 mL

L. $50\,\mu g/kg \times \frac{1\,kg}{2.2\,lb} \times 10\,lb \times mL/250\,\mu g = 0.9\,mL$

M. 36 drops/min $\times \frac{1\,mL}{30\text{ drops}} = 1.2\,mL/min$

$1.2\,mL/min \times 180\,min = 216\,mL$

N. $50\,mg/min \times mL/20\,mg = 2.5\,mL/min$

O. drops/min = 1500 mL/24 hr $\times$ hr/60 min $\times$ 25 drops/mL = 26.04 = 26 drops/min.

P. drops/min = 30 drops/mL $\times$ mL/1.25 mg $\times$ 80 mg/hr $\times$ hr/60 min = 32 drops/min.

Q. drops/min = 15 drops/mL $\times$ 150 mL/50 mg $\times$ 25 mg/hr $\times$ hr/60 min = 18.75
= 19 drops/min

R. $275\,mg \times 4\,mL/1000\,mg = 1.1\,mL$

S. $275\,mg/500\,mL \times 1.6\,mL/min \times 60\,min/1\,hr = 52.8\,mg/hr$

T. **(a)** 30 drops/mL × 1000 mL/8 hr × hr/60 min = 62.5 drops/min ≅ 63 drops/min

(b) 1500 units/8hr × hr/60 min × 30 min = 93.75 = 94 units

U. mL/hr = 1.5 mg/min × 100 mL/0.4 g × 1 g/1000 mg × 60 min/hr = 22.5 mL/hr

V. 500 mL × 3.2 mg/mL = 1600 mg in 500 mL

$$\text{mL/min} = 500\,\text{mL}/1600\,\text{mg} \times (20\,\mu\text{g})/\text{kg/min} \times 1\,\text{mg}/1000\,\mu\text{g} \times 154\,\text{lb} \times \frac{1\,\text{kg}}{2.2\,\text{lb}}$$

$$= 0.44\,\text{mL/min}$$

W. **(a)** 1000 mL/8 hr × hr/60 min = 2.08 mL/min

(b) 8 units/100 mL × 1000 mL/8 hr × hr/60 min × 30 min = 5 units

X. $\text{mL}/10{,}000\text{ units} \times 100\text{ units/kg} \times \frac{1\,\text{kg}}{2.2\,\text{lb}} \times 143\,\text{lb} = 0.65\,\text{mL}$

Y. mL/50,000 I.U. × 1 I.U./600 ng × 15,000 μg × 1000 ng/μg = 0.5 mL

Z. 250 mg/2 mL = 300 mg/j

j = 2.4 mL (would need to reconstitute 2 vials to get the dose)

$$\frac{500\text{ mg}}{2.2\text{ mL}} = \frac{300\text{ mg}}{k}$$

k = 1.3 mL

$$\frac{1000\text{ mg}}{3\text{ mL}} = \frac{300\text{ mg}}{z}$$

z = 0.9 mL (the smallest the volume, the best for the patient receiving the IM injection).

CHAPTER 11

PARENTERAL NUTRITION

Learning Objectives: *After completing this chapter the student should be able to:*

1. Calculate body mass index (BMI) as a tool to indicate body fat.
2. Calculate nutritional requirements: energy (caloric), protein, micronutrients and fluid.
3. Perform all calculations related to the *preparation* of a parenteral nutrition formulation.
4. Calculate the volume of a stock solution that will supply a needed quantity of electrolyte, carbohydrate, fat or other nutrient.
5. Determine the caloric content of a specified volume of a carbohydrate or fat preparation.
6. Determine the amount of nitrogen present in a given volume of an amino acid solution.
7. Calculate the ratio between nonprotein calories and nitrogen in a parenteral mixture.

The main importance of nutrition in man is to maintain lean body mass and intact immune system, promote wound healing and prevent diseases and complications. While poor nutritional choices in a healthy subject will contribute to obesity-related chronic conditions and consequences, patients in hospitals and long-term care facilities are often at risk of having a nutritional deficit. Therefore most of today's health care professionals are, in some degree, involved with patient nutrition assessment, support, and counseling. As a result, nutrition support teams (NST), a multidisciplinary group of health care professionals with expertise in nutrition, are frequently in place to provide nutrition support and continuous assessment. Nutrition support provides two basic alternatives: enteral nutrition and parenteral nutrition. We provide a summary of basic concepts related to these specialized nutrition support options in the appendix section and this chapter will focus on calculations involved with each alternative.

ASSESSMENT OF OBESITY: BMI

1. Start by reading the general background information provided in Appendix 5. Nutrition screening uses parameters connected to nutrition-related diseases to identify patients at risk of obesity or malnutrition.

Pharmaceutical Calculations, Fourth Edition, By Joel L. Zatz and Maria Glaucia Teixeira
ISBN 0-471-67623-3

Anthropometrics (body measurements) are frequently used to evaluate body size and proportions. Height and weight are easy to obtain and inexpensive. *Body mass index* (BMI) which uses a ratio of *body weight in kilograms to the height in meters squared*, is strongly correlated to some obesity-associated diseases and total body fat and is used as a clinical standard to assess obesity.

$$\boxed{\text{BMI} = \frac{\text{weight}}{\text{height}^2}}$$

BMI may be easily calculated using the equation provided above or standardized tables. When calculating BMI two intersystem conversions are important: 1 kg = 2.2 lb and 1 m = 39.37 in.

As an example, let us calculate the BMI for a person 5 ft & 1 in in height weighing 120 lb.

$$120\ \text{lb} \times \frac{1\ \text{kg}}{2.2\ \text{lb}} = 54.5\ \text{kg}$$

$$5\ \text{ft}\ 1\text{in} = \left(5\ \text{ft} \times \frac{12\ \text{in}}{\text{ft}}\right) + 1\ \text{in} = 61\ \text{in} \times \frac{1\ \text{m}}{39.37\ \text{in}} = 1.55\ \text{m}$$

$$\text{BMI} = \frac{54.5}{(1.55)^2}$$

BMI = 22.7 kg/m^2. This is a normal weight individual according to the guidelines in Appendix 5 (Table A5.1), which shows the classification recommended by both the NIH (National Institutes of Health) and WHO (World Health Organization).

Calculate the BMI values for the following individuals:

A. An adult weighing 182 lb and measuring 5′5″ in height.

B. A person 5′6″ in height weighing 75 kg.

C. A subject weighing 110 kg and 64 inches in height.

Solutions.

A. 30.4 kg/m^2

B. 26.6 kg/m^2

C. 41.4 kg/m^2

CALCULATIONS

A. $182\ \text{lb} \times \frac{1\ \text{kg}}{2.2\ \text{lb}} = 82.7\ \text{kg}$

$$5'5'' = 65\text{ in} \times \frac{1\text{ m}}{39.37\text{ in}} = 1.65\text{ m}$$

$$\text{BMI} = \frac{82.7}{(1.65)^2} = 30.4\text{ kg/m}^2.$$

B. $$5'6'' = 66\text{ in} \times \frac{1\text{ m}}{39.37\text{ in}} = 1.68\text{ m}$$

$$\text{BMI} = \frac{75}{(1.68)^2} = 26.6\text{ kg/m}^2.$$

C. $$64\text{ in} \times \frac{1\text{ m}}{39.37\text{ in}} = 1.63\text{ m}$$

$$\text{BMI} = \frac{110}{(1.63)^2} = 41.4\text{ kg/m}^2.$$

ASSESSMENT OF MALNUTRITION

2. When nutritional needs are not met by oral intake because the patient does not want to eat or is unable to eat, malnutrition will result unless nutrition is provided in another way.

Critical assessment regarding nutritional requirements and adequacy of nutrition support depends on:

(a) Patient's data related to disease state

(b) Dietary and drug intake history

(c) Laboratory results

Two alternatives are available for nutrition support: *enteral nutrition* (through a feeding tube placed into the stomach or small bowel) or *parenteral nutrition* (nutrition solution administered intravenously). Both methods will require initial nutrition assessment to determine the degree of malnutrition and calculations will be needed to determine specific nutrient requirements.

CALCULATION OF NUTRITIONAL REQUIREMENTS

3. Nutritional needs are primarily focused on caloric needs, as the body needs a constant energy supply to maintain physiological functions. Caloric (nonprotein) requirements vary depending on the patient's physical state, age, height, weight, gender and medical condition, including degree of stress imposed by the disease.

Energy Requirement (Caloric Requirement)

4. Caloric requirement is measured in calories or *kilocalories* (kcal). One *kilocalorie* corresponds to the amount of heat required to raise the temperature of 1 kg of water by 1°C at room temperature. Caloric requirement, also known as estimated *total* *daily calorie* (TDC), total daily energy (TDE) requirement, or anabolic goal (for maintenance of body weight) may be calculated through several methods described in Table 11.1.

As a general rule, glucose, which is administered daily, is used to provide approximately 60–80% of estimated daily caloric requirement for maintenance of body weight. The remaining caloric requirement is complemented by the infusion of fat emulsion 2–3 times a week.

TABLE 11.1 Usual methods to determine caloric requirements

Methods	Description/Equations
Nomogram	Basal metabolic requirements in function of age, sex, height and body weight
Indirect calorimetry	Measured estimate of energy utilization
Ideal body weight (IBW)	IBW rule of thumb or Hamwi equation[a]: *Males* IBW = 106 lb (48 kg) + 6 lb (2.7 kg) for each inch (2.54 cm) over 5 ft (152 cm) *Females* IBW = 100 lb (45 kg) + 5 lb (2.3 kg) for each inch (2.54 cm) over 5 ft (152 cm) **TDC = 45 kcal/kg/day × IBW[b] (kg)**
[c]Harris–Benedict equation (BEE = basal energy expenditure)	*Males* BEE (kcal/day) = $66.5 + (13.75 \times W_{(kg)}) + (5 \times H_{(cm)}) - (6.76 \times A_{(yr)})$ *Females* BEE (kcal/day) = $655 + (9.56 \times W_{(kg)}) + (1.85 \times H_{(cm)}) - (4.68 \times A_{(yr)})$ **TDC = BEE (kcal/day) × 1.25[d] × stress factor[e]**
Rule of thumb	**TDC = 25–35 kcal/kg/day**
General guidelines	**Anabolic goal (TDC) =** **25 kcal/kg/day** (mildly stressed) **30–35 kcal/kg/day** (moderately stressed) **45 kcal/kg/day** (postoperative) **60 kcal/kg/day** (hypercatabolic)[f]

[a] In practice, when the Hamwi equation is used, the resultant weights are rounded to the nearest whole number.

[b] Usually, the smaller value between IBW (ideal body weight) and ABW (actual body weight) is used for calculation of total daily calorie (TDC) or anabolic goal.
ABW is the patient's real weight or the patient's weight at time of consultation. IBW is also known as lean body mass (LBM). An important application of IBW is in the dosing of some drugs that are highly hydrophilic and do not distribute well into fat. For overweight and obese patients, dosing *must* be based on IBW or the patient will be overdosed.

[c] The Harris-Benedict equation determines the Basal Metabolic Expenditure (BME), also known as Basal Energy Expenditure (BEE), Resting Metabolic Energy (RME) or Resting Energy Expenditure (REE).

[d] 1.25 = activity factor (walking, sitting, physical therapy, treatment)

[e] Stress factor = degree of stress imposed by the disease process and may be:
0.85 = simple starvation, hospitalization
1.05–1.15 = elective surgery
1.20–1.40 = sepsis
1.30 = closed head injury
1.40 = multiple trauma
1.50 = systemic inflammatory response syndrome
2.00 = major burn

[f] Hypercatabolic states are sepsis, closed head injury, major trauma, and severe burn.

So Which Method Should I Use to Determine Caloric Requirements?

5. There are no definite rules as to which of the methods in Table 11.1 should be used. The Harris-Benedict equation and the Hamwi equation have traditionally been used and recommended by ASPEN (American Society of Parenteral and Enteral Nutrition) but some institutions consider indirect calorimetry a good guide to have an estimation of energy utilization by the patient. All methods listed above have been used repeatedly in several institutions and have provided relatively similar results (maintenance of patient's body weight) when used as a guide for patient caloric requirement.

---

6. We will now apply the IBW rule of thumb or Hamwi equation to determine the actual body weight (ABW) and ideal body weight (IBW) in kg for a woman of 5′1″ in height and weighing 120 lb. What would be the anabolic goal or total daily calorie requirement (TDC) for this patient?

The patient's actual body weight in kg is:

$$\text{ABW} = 120 \text{ lb} \times \frac{1 \text{ kg}}{2.2 \text{ lb}} = 54.5 \text{ kg}$$

Calculation of IBW in kg is:

$$\text{IBW} = 45 \text{ kg} + (2.3 \times 1 \text{ in.}) = 47.3 = 47 \text{ kg}$$

We will use the smallest body weight (IBW, in this case) for calculation of TDC:

$$\text{TDC} = \frac{45 \text{ kcal}}{\text{kg/day}} \times 47 \text{ kg} = 2115 \text{ kcal/day}$$

Based on IBW, calculate the total daily calorie requirement for a male patient weighing 198 lb and 5′6″ in height.

Solution. 2880 kcal/day

CALCULATIONS

If we consider the patient's actual body weight, it is 198/2.2 = 90 kg. The usual procedure is to calculate TDC using either ABW or IBW, whichever is smaller.

IBW = 106 lb (48 kg) + 6 lb (2.7 kg) for each inch (2.54 cm) over 5 ft (152 cm)

$$\text{IBW} = 106 + (6 \times 6) = 142 \text{ lb} \times \frac{1 \text{ kg}}{2.2 \text{ lb}} = 64 \text{ kg}$$

$$\text{TDC} = \frac{45 \text{ kcal}}{\text{kg/day}} \times 64 \text{ kg} = 2880 \text{ kcal/day}$$

---

7. The weight of a male patient undergoing surgery is 110 lb and his height is 165 cm. What is the total daily energy requirement for maintenance of body weight of this patient? Use the IBW method.

Solution. 2250 kcal/day

CALCULATIONS

$$\text{ABW} = 110 \text{ lb} \times \frac{1 \text{ kg}}{2.2 \text{ lb}} = 50 \text{ kg}$$

$$165 \text{ cm} - 152 \text{ cm} = 13 \text{ cm}$$

$$\text{IBW} = 48 \text{ kg} + (2.7 \text{ kg} \times 13) = 83.1 \text{ kg}$$

$$\text{TDC} = \frac{45 \text{ kcal}}{\text{kg/day}} \times 50 \text{ kg} = 2250 \text{ kcal/day}$$

---

8. Using the Harris-Benedict equation calculate the daily energy requirement for a 38 year old male patient undergoing elective surgery (use a stress factor of 1.10) weighing 132 lb and measuring 5′7″ in height.

Solution. 2043 kcal/day

CALCULATIONS

$$\text{BEE} = 66.5 + (13.75 \times \text{W}_{(\text{kg})}) + (5 \times \text{H}_{(\text{cm})}) - (6.76 \times \text{A}_{(\text{yr})})$$

$$\text{W}_{\text{kg}} = 132 \text{ lb} \times \frac{1 \text{ kg}}{2.2 \text{ lb}} = 60 \text{ kg}$$

$$\text{H}_{\text{cm}} = \left(5 \text{ ft} \times \frac{12 \text{ in}}{\text{ft}}\right) + 7 \text{ in} = 67 \text{ in} \times \frac{2.54 \text{ cm}}{\text{in}} = 170.2 \text{ cm}$$

$$\text{BEE} = 66.5 + (13.75 \times 60) + (5 \times 170.2) - (6.76 \times 38) = 66.5 + 825 + 851 - 256.9$$

$$\text{BEE} = 1486 \text{ kcal/day}$$

$$\text{TDC} = \text{BEE (kcal/day)} \times 1.25 \times \text{stress factor}$$

$$\text{TDC} = 1486 \times 1.25 \times 1.10 = 2043 \text{ kcal/day}$$

9. A 28 year old female patient with closed head injury (stress factor = 1.3) will need nutritional support. Calculate the TDC required for this patient using Harris-Benedict equation. Patient weight and height are 121 lb and 5′2″.

Solution. 2195 kcal/day

CALCULATIONS

$$\text{BEE} = 655 + (9.56 \times W_{(kg)}) + (1.85 \times H_{(cm)}) - (4.68 \times A_{(yr)})$$

$$W_{kg} = 121\ \text{lb} \times \frac{1\ \text{kg}}{2.2\ \text{lb}} = 55\ \text{kg}$$

$$H_{cm} = \left(5\ \text{ft} \times \frac{12\ \text{in}}{\text{ft}}\right) + 2\ \text{in} = 62\ \text{in} \times \frac{2.54\ \text{cm}}{\text{in}} = 157.5\ \text{cm}$$

$$\text{BEE} = 665 + (9.56 \times 55) + (1.85 \times 157.5) - (4.68 \times 28) = 665 + 525.8 + 291.4 - 131 = 1351\ \text{kcal/day}$$

$$\text{TDC} = \text{BEE (kcal/day)} \times 1.25 \times \text{stress factor}$$

$$\text{TDC} = 1351 \times 1.25 \times 1.3 = 2195\ \text{kcal/day}$$

10. Use the Harris-Benedict equation and calculate the caloric requirement for:

A. A 65 year old hospitalized male patient weighing 130 lb and measuring 5′4″ in height. Assume a stress factor of 0.85

B. A 25 year old hospitalized female patient weighing 100 lb and measuring 5′2″ in height. Consider that the patient has multiple trauma (stress factor of 1.4)

Solutions.

A. 1329 kcal/day

B. 2228 kcal/day

CALCULATIONS

A. $BEE = 66.5 + (13.75 \times W_{(kg)}) + (5 \times H_{(cm)}) - (6.76 \times A_{(yr)})$

$$W_{kg} = 130 \text{ lb} \times \frac{1 \text{ kg}}{2.2 \text{ lb}} = 59 \text{ kg}$$

$$H_{cm} = \left(5 \text{ ft} \times \frac{12 \text{ in}}{\text{ft}}\right) + 4 \text{ in} = 64 \text{ in} \times \frac{2.54 \text{ cm}}{\text{in}} = 162.6 \text{ cm}$$

$$BEE = 66.5 + (13.75 \times 59) + (5 \times 162.6) - (6.76 \times 65)$$
$$= 66.5 + 811.3 + 813 - 439.4 = 1251 \text{ kcal/day}$$

$$TDC = BEE \text{ (kcal/day)} \times 1.25 \times \text{stress factor}$$

$$TDC = 1251 \times 1.25 \times 0.85 = 1329 \text{ kcal/day}$$

B. $BEE = 655 + (9.56 \times W_{(kg)}) + (1.85 \times H_{(cm)}) - (4.68 \times A_{(yr)})$

$$W_{kg} = 100 \text{ lb} \times \frac{1 \text{ kg}}{2.2 \text{ lb}} = 45.4 \text{ kg}$$

$$H_{cm} = \left(5 \text{ ft} \times \frac{12 \text{ in}}{\text{ft}}\right) + 2 \text{ in} = 62 \text{ in} \times \frac{2.54 \text{ cm}}{\text{in}} = 157.5 \text{ cm}$$

$$BEE = 665 + (9.56 \times 45.4) + (1.85 \times 157.5) - (4.68 \times 25)$$
$$= 665 + 434 + 291.4 - 117 = 1273 \text{ kcal/day}$$

$$TDC = BEE \text{ (kcal/day)} \times 1.25 \times \text{stress factor}$$

$$TDC = 1273 \times 1.25 \times 1.4 = 2228 \text{ kcal/day}$$

---

11. Determine the TDC for a 32 year old male patient 158 cm in height and weighing 42 kg. Use IBW method, Harris-Benedict method, and rule of thumb method considering the patient is moderately stressed (35 kcal/kg/day or stress factor = 1.05). Analyze the results.

Solutions. See below.

CALCULATIONS

Using body weight:

158 cm − 152 cm = 6 cm

ABW= 42 kg

IBW = 48 kg + (2.7 kg × 6) = 64.2 kg

$$TDC = \frac{45 \text{ kcal}}{\text{kg/day}} \times 42 \text{ kg} = 1890 \text{ kcal/day}$$

Using Harris-Benedict:

$BEE = 66.5 + (13.75 \times W_{(kg)}) + (5 \times H_{(cm)}) - (6.76 \times A_{(yr)})$

$BEE = 66.5 + (13.75 \times 42) + (5 \times 158) - (6.76 \times 32) = 66.5 + 577.5 + 790 - 216.3$
$= 1218 \text{ kcal/day}$

$\text{TDC} = \text{BEE (kcal/day)} \times 1.25 \times \text{stress factor}$

$\text{TDC} = 1218 \times 1.25 \times 1.05 = 1599\,\text{kcal/day}$

Using rule of thumb:

$$\text{TDC} = \frac{35\ \text{kcal}}{\text{kg/day}} \times 42\ \text{kg} = 1470\ \text{kcal/day}$$

These results show that calculations using the Harris-Benedict equation and the rule of thumb produce similar results and seem to be more appropriate than using body weight only because they include a stress factor.

Protein (Nitrogen) Requirement

12. Amino acids, the building blocks of protein, provide the nitrogen necessary for protein synthesis (anabolism). While grams of protein are generally calculated in a nutritional plan, grams of nitrogen are occasionally used as an expression of the amount of protein received by the patient. In a nutrition regimen, amino acids are provided through amino acid solutions. Protein requirements may be reduced in certain disease states (renal failure, hepatic encephalopathy); thus specialized amino acid formulations have been developed for patients whose protein intake is restricted. Recommendations for amino and intake are summarized in Table 11.2.

TABLE 11.2 General recommendations for daily amino acid requirements

Patient Status	Daily Requirement Based on Actual Body Weight (g/kg)
Postoperative (uncomplicated)	0.8–1.0
Postoperative (unable to eat for >10 days)	1.0–1.5
Sepsis and stress	1.2–1.5
Multiple trauma	1.3–1.7
Major burn	1.8–2.5

M.S., 121 lb, will be hospitalized for 2 days after hernia surgery (uncomplicated). What is this patient's daily amino acid requirement? (Assume a daily requirement of 1 g/kg.)

Solution. 55 g

CALCULATIONS

$$121\ \text{lb} \times \frac{1\ \text{kg}}{2.2\ \text{lb}} = 55\ \text{kg}$$

$$55\ \text{kg} \times \frac{1\ \text{g}}{\text{kg}} = 55\ \text{g}$$

13. A patient weighing 165 lb and presenting major burns is hospitalized. The nutrition pharmacist will need to calculate the daily amino acid requirement for this patient. Assuming the patient will receive this nutrient intravenously from a 15% (w/v) amino acid solution, how many milliliters will be required?

Solution. 1250 mL

CALCULATIONS

$$165 \text{ lb} \times \frac{1 \text{ kg}}{2.2 \text{ lb}} \times \frac{2.5 \text{ g}}{\text{kg}} \times \frac{100 \text{ mL}}{15 \text{ g}} = 1250 \text{ mL}$$

---

14. There are three sources of amino acids available in a hospital pharmacy: 3.5%, 7%, and 12.5% amino acid solutions (w/v). Which solution should the pharmacist choose and how many milliliters should be given to a patient weighing 99 lb and presenting sepsis? This patient is under fluid restriction (a patient under fluid restriction should receive the smallest volume of fluid possible).

Solution. 540 mL of the 12.5% solution

CALCULATIONS

Since the patient is under fluid restriction, the most concentrated solution must be chosen to allow the smallest volume. Thus,

$$99 \text{ lb} \times \frac{1 \text{ kg}}{2.2 \text{ lb}} \times \frac{1.5 \text{ g}}{\text{kg}} \times \frac{100 \text{ mL}}{12.5 \text{ g}} = 540 \text{ mL}$$

Micronutrient Requirement

15. The requirements for vitamins, trace minerals and electrolytes (micronutrients) are basically empirical. In general, serum concentrations of electrolytes and trace minerals are measured regularly and adjusted accordingly. Blood concentrations of vitamins are measured in patients with a suspected deficiency. Electrolytes, multivitamins and trace elements are available as multiconcentrates or as individual units to be added to nutrition formulations. Table 11.3 provides guidelines for various micronutrients.

TABLE 11.3 General guidelines for daily electrolyte requirements

Electrolyte	Requirement (mEq/day)
Na (sodium)	80–100
K (potassium)	60–80
Mg (magnesium)	8–16
Ca (calcium)	5–10
PO_4 (phosphate)	15–30
Cl (chloride)	50–100
Acetate	50–100

Daily multivitamin requirement: 10 mL of standard multivitamin injection.
Daily trace elements requirement: 1 mL of standard trace elements injection.

Fluid Requirement

16. Some general recommendations have been developed among practicing nutrition professionals and the choice of one of them for calculation of the total daily fluid (TDF) requirement is based on patient's age, disease state and degree of dehydration. These are shown below in Table 11.4.

TABLE 11.4 Recommendations for fluid intake

Category	Patient Group	Recommendation
I	Adults	35 mL/kg/day
II	Adults with severe dehydration (e.g., burns over a large part of the body)	1500 mL (for first 20 kg) + 20 mL/kg/day (for additional kg > 20 kg)
III	Pediatrics and geriatrics	BW 0–10 kg: 100 mL/kg/day 10 kg < BW < 20 kg: 50 mL/kg/day BW > 20 kg: 20 mL/kg/day

An 8-year-old child was hospitalized after 4 days with diarrhea and vomiting. At consultation, the child weight was 66 lb. What would be the recommended total daily fluid (TDF) for this patient?

Solution. 600 mL

CALCULATIONS

$$66\text{ lb} \times \frac{1\text{ kg}}{2.2\text{ lb}} = 30\text{ kg}$$

Considering that the patient is a child, the best approach is to use category III, BW > 20 kg: 20 mL/kg

$$\text{TDF} = 20\text{ mL/kg} \times 30\text{ kg} = 600\text{ mL}$$

---

17. A 35 year old male patient weighing 110 lb and 5′3″ in height has been admitted with severe burns (stress factor = 2.0). Calculate:

(a) The total daily calorie (TDC) using Harris-Benedict equation.

(b) The daily protein requirement for major burns = 2.5 g/kg. How many milliliters would be administered from a 15% a.a. solution?

(c) The total daily fluid (TDF) using recommendations I or II. Choose the most appropriate and explain your choice.

Solution.

(a) 3169 kcal/day

(b) 833 mL of 15%

(c) use category II recommendation

2100 mL should be used because the patient has major burns and severe dehydration will be present.

CALCULATIONS

(a) $\text{BEE} = 66.5 + (13.75 \times W_{(kg)}) + (5 \times H_{(cm)}) - (6.76 \times A_{(yr)})$

$W_{kg} = 110/2.2 = 50\,\text{kg}$

$$H_{cm} = 5'3'' = 63\text{ in} \times \frac{2.54\text{ cm}}{\text{in}} = 160\text{ cm}$$

$\text{BEE} = 66.5 + (13.75 \times 50) + (5 \times 150) - (6.76 \times 35) = 1267\,\text{kcal/day}$

$\text{TDC} = \text{BEE (kcal/day)} \times 1.25 \times \text{stress factor}$

$\text{TDC} = 1267 \times 1.25 \times 2 = 3169\,\text{kcal/day}$

(b) $$\text{Daily protein} = 110\text{ lb} \times \frac{1\text{ kg}}{2.2\text{ lb}} \times \frac{2.5\text{ g}}{\text{kg}} \times \frac{100\text{ mL}}{15\text{ g}} = 833\text{ mL of }15\%\text{ a.a. solution.}$$

(c) $$\text{TDF} = \frac{35\text{ kcal}}{\text{kg/day}} = \frac{35\text{ mL}}{\text{kg}} \times 50\text{ kg} = 170\text{ mL}$$

$\text{TDF} = 1500\,\text{mL (for 20 kg BW)} + 20\,\text{mL/kg (additional kg} > 20\,\text{kg)}$

$\text{TDF} = 1500 + (20 \times 30) = 2100\,\text{mL}$

---

18. An adult patient weighing 60 kg was admitted for an elective surgery. Considering the patient as mildly stressed calculate TDC (general guidelines = 25 kcal/kg/day).

Solution. 1500 kcal/day

CALCULATIONS

$$\text{TDC} = \frac{25\ \text{kcal}}{\text{kg/day}} \times 60\ \text{kg} = 1500\ \text{kcal/day}$$

PARENTERAL NUTRITION

19. The term *TPN* or *PN*, which stands for *total parenteral nutrition* or simply *parenteral nutrition*, is frequently used to describe an intravenous infusion mixture that is expected to supply all needed fluid, electrolytes, calories, essential fatty acids, and vitamins: in short, everything needed to sustain life.

Calculations Related to Preparation or Compounding of Parenteral Nutrition Formulations

20. In preparing solutions for parenteral nutrition, the pharmacist makes extensive use of commercial products. Check Table A5.2 (Appendix 5) for sources, strengths and dosage/contents of solutions generally used in the preparation of parenteral nutrition.

The major fluid contribution in a PN will come from amino acid solution, dextrose solution and fat emulsion (and occasionally added water). Electrolytes (micronutrients) are in general, calculated in an individual basis, depending on the patient's needs.

We will now concentrate on many of the calculations that have to be performed in connection with the preparation or compounding of parenteral nutrition formulations.

Calculation of Electrolytes

21. A TPN solution is to contain sodium, 60 mEq/L and calcium, 9 mEq/L. A sodium chloride solution with a concentration of 4 mEq/mL is available. How much of this solution is needed to contribute the sodium ion in preparing 2 L of the TPN solution?

Solution. 30 mL

CALCULATIONS

$j \times 4\ \text{mEq/mL} = 2\ \text{L} \times 60\ \text{mEq/L}$

$j = 30\ \text{mL}$

---

22. Referring to the previous example, how many milliliters of a calcium gluconate solution containing 0.45 mEq/mL of calcium ion should be used to prepare 2 L of the TPN solution?

Solution. 40 mL

CALCULATIONS

$j \times 0.45\,\text{mEq/mL} = 2\,\text{L} \times 9\,\text{mEq/L}$
$j = 40\,\text{mL}$

---

23. Calculate the quantity of KCl solution, 2 mEq/mL, needed to supply potassium ion in 1.75 L of a solution to contain 40 mEq of potassium ion/liter.

Solution. 35 mL

CALCULATIONS

$j \times 2\,\text{mEq/mL} = 1.75\,\text{L} \times 40\,\text{mEq/L}$
$j = 35\,\text{mL}$

---

24. One liter of a parenteral nutrition mixture is to contain 8 mEq of calcium ion and 100 mEq of chloride. Other electrolyte solutions that have been used have already supplied 95 mEq of chloride. Our problem is to decide what combination of calcium gluconate, 0.45 mEq/mL, and calcium chloride, 1.36 mEq/mL should be used.

First, let us recall that *each electrolyte yields the same number of mEq of cation and anion.* Therefore, the calcium chloride solution contains 1.36 mEq/mL of chloride also.

Next, we have to decide which ion, Ca^{+2} or Cl^-, is needed in smallest quantity. If it is the calcium, we can get this entire ion from the calcium chloride stock solution which will also supply some of the chloride, and use another salt to contribute the remaining chloride. If the chloride is needed in smallest quantity, then we add enough calcium chloride stock to supply the chloride and just part of the calcium. The gluconate stock solution is used to contribute the remaining calcium ion. In this case, we need 8 mEq of Ca^{+2} and 5 mEq of Cl^-. Can you determine what quantity of each stock solution to add?

Solution. 3.7 mL of $CaCl_2$ stock; 6.7 mL of gluconate stock

CALCULATIONS

For chloride, 100 mEq − 95 mEq = 5 mEq remain
$j \times 1.36\,\text{mEq/mL} = 5\,\text{mEq}$
$j = 3.68\,\text{mL} = 3.7\,\text{mL}$ of calcium chloride stock

This stock solution contributes 5 mEq of chloride ion and 5 mEq of calcium ion. That means that we still need 3 mEq of calcium, which will come from the gluconate stock:

$j \times 0.45\,\text{mEq/mL} = 3\,\text{mEq}$

$j = 6.67\,\text{mL} = 6.7\,\text{mL}$ of calcium gluconate stock

---

25. One liter of a TPN solution is to contain 80 mEq of sodium and 30 mEq of acetate. Available stock solutions contain 4 mEq/mL sodium as sodium chloride and 2 mEq/mL sodium as sodium acetate. What quantity of each solution will supply the needed quantities of both ions?

Solution. 15 mL of sodium acetate; 12.5 mL of sodium chloride

CALCULATIONS

Acetate is the ion needed in smallest quantity (30 mEq of acetate vs. 80 mEq of sodium). Therefore, we start with sodium acetate.

$j \times 2\,\text{mEq/mL} = 30\,\text{mEq}$

$j = 15\,\text{mL}$ of sodium acetate solution

This also supplies 30 mEq of sodium ion. We still need 50 mEq of Na^+; it will come from the sodium chloride solution.

$j \times 4\,\text{mEq/mL} = 50\,\text{mEq}$

$j = 12.5\,\text{mL}$ of sodium chloride solution

---

26. Most inorganic components of parenteral nutrition mixtures are specified in terms of mEq. An exception is the *phosphorus* content, which is frequently prescribed as *millimoles* of phosphate or occasionally as milligrams of elemental phosphorus. The reason for this is that in the body, phosphorus exists mainly as a mixture of monobasic and dibasic forms, which differ in valence. The balance between these depends on the pH. In healthy persons, the blood pH falls within a narrow range so a calculation based on mEq is possible. However, the pH may vary from normal values in many individuals who are candidates for parenteral nutrition, making a calculation based on milliequivalents ambiguous. Using millimoles, which are mass units to describe phosphorus content, avoids these uncertainties.

One source of phosphorus is a potassium phosphate injection, which contains a mixture of dibasic potassium phosphate (K_2HPO_4) and monobasic potassium phosphate (KH_2PO_4). Each milliliter of the injection contains 65.2 mg of elemental P which is equivalent to 3 mmol of P as phosphate.

What quantity of this potassium phosphate injection should be used in the preparation of 5 L of a solution that is to contain 12 mmol of phosphorus in each 100 mL?

Solution. 200 mL

CALCULATIONS

P content of the injection is 3 mmol/mL

$j \times 3\,\text{mmol/mL} = 5000\,\text{mL} \times 12\,\text{mmol/100 mL}$

$j = 200\,\text{mL}$

---

27. A TPN mixture is to contain 0.22 mg of P per milliliter. How many milliliters of a sodium phosphate injection containing 93 mg of P per milliliter should be used in the preparation of 3 L?

Solution. 7.1 mL

CALCULATIONS

$j \times 93\,\text{mg/mL} = 3000\,\text{mL} \times 0.22\,\text{mg/mL}$

$j = 7.1\,\text{mL}$

Calculation of Non-Nitrogen Calories from Dextrose Solutions and Fat Emulsions

28. Dextrose is a simple sugar commonly used almost exclusively as the source of carbohydrate calories in parenteral nutrition. Sterile dextrose solutions in several concentrations are available to serve as a sugar source. The calculations are similar to those we have already seen.

A TPN formula calls for dextrose 20%. What volume of 50% dextrose injection should be utilized for each liter of the TPN formula?

Solution. 400 mL

CALCULATIONS

$j \times 50\,\text{g/100 mL} = 1000\,\text{mL} \times 20\,\text{g/100 mL}$

$j = 400\,\text{mL}$

---

29. A TPN formula for 2 L is to contain 25% dextrose. What volume of a 70% dextrose injection will supply the needed sugar?

Solution. 714 mL

CALCULATIONS

$j \times 70\,g/100\,mL = 2000\,mL \times 25\,g/100\,mL$

$j = 714\,mL$

---

30. Each gram of dextrose monohydrate (the usual form of this compound utilized in TPN solutions) supplies 3.4 kcal of energy. How many kilocalories would be supplied by 2 L of a 25% dextrose monohydrate solution?

Solution. 1700 kcal

CALCULATIONS

$$2000\text{ mL} \times \frac{25\text{ g}}{100\text{ mL}} \times 3.4\text{ kcal/g} = 1700\text{ kcal}$$

---

31. Glycerin is sometimes used as an alternative energy source in some parenteral nutrition formulations. It produces 4.32 kcal of energy per gram.

A commercial stock solution contains various amino acids, electrolytes and 3% w/v glycerin. What is the caloric contribution of the glycerin in 100 mL of this stock solution?

Solution. 13 kcal

CALCULATIONS

$$100\text{ mL} \times \frac{3\text{ g}}{100\text{ mL}} \times 4.32\text{ kcal/g} = 13\text{ kcal}$$

Calculation of Nitrogen Calories, Grams of Nitrogen, and Calorie to Nitrogen Ratio

32. As an example, a TPN solution is to contain 4% protein. The source is a commercial product that contains 10 g of amino acids per 100 mL (10%).

Let's say that we wish to prepare 1 L of the TPN solution. We have to calculate the amount of source product to use. A simple approach is to begin with a mass balance equation. In the equation, *aa* stands for amino acids.

Let j = volume of amino acid source to be used amount of aa in source = amount of aa in TPN

$$j \times 10\,\text{g}/100\,\text{mL} = 1000\,\text{mL} \times 4\,\text{g}/100\,\text{mL}$$
$$j = 400\,\text{mL}$$

---

33. How many milliliters of the same amino acid source product would be needed to prepare 2.5 L of a TPN solution containing 5.2% amino acids?

Solution. 1300 mL

CALCULATIONS

$j \times 10\,\text{g}/100\,\text{mL} = 2500\,\text{mL} \times 5.2\,\text{g}/100\,\text{mL}$

$j = 1300\,\text{mL}$

---

34. The amount of nitrogen is frequently used as a general indication of amino acids available. *1 g of nitrogen is equivalent to 6–6.5 g of amino acids* (aa), depending on the preparation used. One commercial preparation that contains amino acids equivalent to 10% protein delivers 1.53 g of nitrogen/100 mL. For this product,

$$\frac{10\ \text{g aa}/100\ \text{mL}}{1.53\ \text{g N}/100\ \text{mL}} = \frac{6.5\ \text{g aa}}{1\ \text{g N}}$$

An amino acid formulation contains 23.8 g of nitrogen per liter. If the ratio of amino acids to nitrogen is 6.3, what is the percentage concentration of amino acids?

Solution. 15%

CALCULATIONS

$23.8\,\text{g/L} = 2.38\,\text{g}/100\,\text{mL}$

$2.38\,\text{g N}/100\,\text{mL} \times 6.3\,\text{g aa/g N} = 15\,\text{g aa}/100\,\text{mL} = 15\%$

---

35. A male patient weighing 80 kg is to receive 2.5 L of a parenteral mixture every 24 hours. If his nitrogen requirement is 0.3 g/kg/day, how many milliliters of an amino acid solution containing 14 grams of nitrogen per liter should be used to prepare the mixture?

Solution. 1710 mL

CALCULATIONS

$80\,kg \times 0.3\,gN/kg/d = 24\,gN/d$

$24\,g = j \times 14\,g/1000\,mL$

$j = 1710\,mL$

---

36. Referring to previous example, calculate the amount of a 10% amino acid solution that should be used to prepare the parenteral mixture. Assume 1 gN = 6.25 g amino acids for this source.

Solution. 1500 mL

CALCULATIONS

$24\,gN \times 6.25\,g\,aa/gN = 150\,g\,aa$

$$\frac{150\text{ g}}{10\text{ g}/100\text{ mL}} = 1500\text{ mL}$$

---

37. Although protein can supply calories to the body, it is most efficiently utilized for building tissue when other energy sources, namely carbohydrate and fat, are administered. This "spares" the nitrogen. *The ratio of nonprotein kcal to grams of nitrogen is sometimes used as a basis for estimating protein needs in patients.*

Each liter of a parenteral nutrition solution supplies 260 nonprotein kcal and 16 g of amino acids. Assuming that each gram of nitrogen is equivalent to 6.25 g of amino acids, the nitrogen content is

$$\frac{16\text{ g aa}}{6.25\text{ g aa}/\text{g N}} = 2.56\text{ g N}$$

and the energy : nitrogen ratio is

$$\frac{260\text{ kcal}}{2.56\text{ g N}} = 102\text{ kcal}/\text{g N}$$

What is the nonprotein kcal : nitrogen ratio for a parenteral solution that contains 10% amino acids and 30% dextrose? In this solution, 1 gN = 6 g amino acids.

Solution. 61 kcal/gN

CALCULATIONS

Dextrose: 30 g/100 mL × 3.4 kcal/g = 1.02 kcal/mL

$$\text{N: } 10 \text{ g aa}/100 \text{ mL} \times \frac{1 \text{ g N}}{6 \text{ g aa}} = 0.0167 \text{ g N}/\text{mL}$$

$$\frac{1.02 \text{ kcal}}{0.0167 \text{ g N}} = 61 \text{ kcal}/\text{g N}$$

---

38. A pharmacist combines 300 mL of a 70% dextrose solution with 700 mL of a solution containing 10% amino acids. The combination is to be administered at the same time as 400 mL of an emulsion containing 10% fat. For the emulsion, each milliliter delivers 1.1 kcal. For the amino acid solution, 1 g N = 6.25 g amino acids. Calculate the ratio of nonprotein calories to nitrogen overall.

Solution. 103 kcal/g N

CALCULATIONS

Fat: 400 mL × 1.1 kcal/mL = 440 kcal

Dextrose: 300 mL × 70 g/100 mL × 3.4 kcal/g = 714 kcal

Total nonprotein calories = 440 kcal + 714 kcal = 1154 kcal

$$\text{N: } 700 \text{ mL} \times 10 \text{ g}/100 \text{ mL} \times \frac{1 \text{ g N}}{6.25 \text{ g aa}} = 11.2 \text{ g}$$

$$\frac{1154 \text{ kcal}}{11.2 \text{ g N}} = 103 \text{ kcal}/\text{g N}$$

TOTAL NUTRITION ADMIXTURE (TNA)

39. TPN solutions that contain fat as an extra non-nitrogen energy source are known as *TNA (total nutrition admixture) or 3-in-1 solutions.* The fat is supplied in the form of an emulsion, a liquid preparation in which microscopic oil globules are dispersed in a water medium. IV Fat emulsions are made from either soybean oil or a mixture of safflower and soybean oils, and contain egg phosphatides as an emulsifier and glycerin to adjust the osmolarity. Products currently available contain 10% fat, which supplies 1.1 kcal/mL, 20% fat, which supplies 2 kcal/mL and 30% fat with 3 kcal/mL.

Clinical experience with parenteral nutrition has shown that a combination of carbohydrate and fat is better tolerated and utilized than a larger quantity of either component by itself. In addition to calories, the fat emulsion contributes essential fatty acids.

If 500 mL of a 10% fat emulsion (1.1 kcal/mL) were used to prepare 1 L of a TNA solution how many kilocalories were contributed by the emulsion?

Solution. 550 kcal

CALCULATIONS

$500\,\text{mL} \times 1.1\,\text{kcal/mL} = 550\,\text{kcal}$

---

40. Assuming that another component of the solution prepared above is 1000 mL of 24% dextrose. What percentage of the caloric value of the two energy sources is supplied by the fat emulsion?

Solution. 40%

CALCULATIONS

Calories contributed by dextrose:

$$1000\text{ mL} \times \frac{24\text{ g}}{100\text{ mL}} \times 3.4\text{ kcal/g} = 816\text{ kcal}$$

From previous problem, the fat emulsion contributes 550 kcal.

$$\frac{550}{550+816} = 0.40 = 40\%$$

---

41. One component of a TNA formula is a 20% fat emulsion, which contributes 2 kcal/mL.

(a) What is the caloric content of 225 mL of this emulsion?

(b) How many milliliters should be used to supply 800 kcal?

Solution. 450 kcal; 400 mL

CALCULATIONS

(a) $225\,\text{mL} \times 2\,\text{kcal/mL} = 450\,\text{kcal}$

(b) $800\,\text{kcal} \times \text{mL}/2\,\text{kcal} = 400\,\text{mL}$

42. The formula for a parenteral nutrition mixture calls for 750 mL of 25% dextrose. How many milliliters of a 10% fat emulsion (1.1 kcal/mL) are needed to supply the same number of calories as the dextrose?

Solution. 580 mL

CALCULATIONS

For the dextrose:

$$750\text{ mL} \times \frac{25\text{ g}}{100\text{ mL}} \times 3.4\text{ kcal/g} = 638\text{ kcal}$$

For the fat emulsion:

$638\text{ kcal} = 1.1\text{ kcal/mL} \times j$

$j = 580\text{ mL}$

43. Here are some more advanced calculations related to parenteral nutrition.

A. ZN, a 68-year old female is receiving the following TPN:

FreAmine 8.5%	450 mL
D 45 W	550 mL

Rate: 85 mL/h

Liposyn II 20% (fat emulsion 20%), 250 mL/day, IV infusion at 20 mL/h
Calculate the total daily calorie intake by ZN.

B. Use the sources indicated below to calculate the amount of each component required in preparing 1000 cc of the following parenteral nutrition solution:

Amino acids	4.5%
Dextrose	15 %
NaCl	30 mEq
Ca gluc.	2.5 mEq
Insulin	15 units
Heparin	2500 units
SWFI	q.s. to 1000 mL

Available in the pharmacy:

15 % AA injection (500 cc); 50% dextrose (500 mL); 15% NaCl sol. (20 mL); 4.6 mEq/10 mL Ca gluconate (10 mL); U-100 Insulin (10 mL); 1000 units/mL heparin sodium (5 mL); SWFI (500 mL).

C. A hospital pharmacy received the following medication order:

Maria Spencer   Rm. 101   Dr. John Holt
Wt. = 10 kg

Central TPN

50 ml/kg over 24 hr

Amino acids	1.2 g/kg/24 h
Dextrose	15%
Sodium	1.2 mEq/kg/24 h
Potassium	4.5 mEq/kg/24 h
Chlorine	1.5 mEq/kg/24 h
Phosphorus	2.5 mmol/kg/24 h
Acetate	0.58 mEq/kg/24 h
Calcium	0.45 mEq/kg/24 h

Based on the following sources available in the pharmacy, calculate what is asked for in the questions that follow.

D50 = dextrose solution. 50%: AA sol. 15%: contains 12.5 mmol P/liter and 15 mEq Na/liter

K phosphate: 3 mmol P/mL and 4.4 mEq K/mL

KCl (MW = 74.5): 2 mEq/mL

NaCl (MW = 58.5): 4 mEq/mL

Na acetate (MW = 141): 2 mEq/mL

Ca gluconate (MW = 430): 4.65 mEq/10 mL

SWFI

C.1 The volumes of AA and Dextrose needed to prepare the PN are:

(a) 80 mL 15% AA and 24 mL D50
(b) 80 mL 15% AA and 150 mL D50
(c) 80 mL 15% AA and 220 mL D50
(d) 375 mL 15% AA and 150 mL D50
(e) 375 mL 15% AA and 24 mL D50

C.2 The phosphorus requirement will be fulfilled by:

(a) 8.3 mL K phosphate
(b) 8 mL K phosphate
(c) 4.15 mL K phosphate
(d) 4 mL K phosphate
(e) 5.7 mL K phosphate

C.3 The potassium requirement will be completed by adding:

(a) 4.9 mL KCl
(b) 4.24 mL KCl
(c) 22.5 mL KCl
(d) 11.25 mL KCl
(e) No KCl is needed

C.4 A volume of ____ mL of NaCl will take care of chlorine requirement:

(a) 3.8 mL
(b) 2.8 mL
(c) 3.3 mL
(d) 1.3 mL
(e) 2.3 mL

C.5 The sodium requirement will be fulfilled by:

(a) 3.4 mL of Na acetate
(b) 5.4 mL of Na acetate
(c) 2.8 mL of Na acetate
(d) 3.8 mL of Na acetate
(e) No Na acetate is needed.

C.6 The calcium amount will be provided by:

(a) 9.7 mL Ca gluconate
(b) 19.4 mL Ca gluconate
(c) 4.9 mL Ca gluconate
(d) 1 mL Ca gluconate
(e) None of the above

C.7 Sterile water for injection (SWFI) to adjust to the required final volume of PN is:

(a) 243 mL
(b) 270 mL
(c) 200 mL
(d) 240 mL
(e) 280 mL

C.8 The flow rate of this TPN would be:

(a) 0.5 mL/min
(b) 42 mL/hr
(c) 0.7 mL/min
(d) 21 mL/hr
(e) 2.1 mL/hr

C.9 The baby will receive from this TPN a total of:

(a) 303 kcal

(b) 89 kcal

(c) 480 kcal

(d) 422 kcal

(e) 266 kcal

Solutions.

A. 1495 kcal

B. 300 mL of 15% amino acid solution; 300 mL of 50% dextrose solution; 11.7 mL of NaCl; 5.4 mL Ca gluconate; 0.15 mL insulin U-100; 2.5 mL heparin 1,000 Units/mL; 380 mL SWFI.

C. 80 mL of 15% amino acid solution and 150 mL of dextrose 50%; 8 mL potassium phosphates; 4.9 mL KCl; 1.3 mL of NaCl; 2.8 mL of Na acetate; 9.7 mL of Ca gluconate; 243 mL SWFI; 21 mL/hr; 303 kcal.

CALCULATIONS

A. $$\text{Amino acids} = 450\text{ mL} \times \frac{8.5\text{ g}}{100\text{ mL}} \times 4\text{ kcal/g} = 153\text{ kcal}$$

$$\text{Dextrose} = 550\text{ mL} \times \frac{45\text{ g}}{100\text{ mL}} \times 3.4\text{ kcal/g} = 814.5\text{ kcal}$$

20% fat emulsion = 2 kcal/mL

Fat = 250 mL × 2 kcal/mL = 500 kcal

Total daily calorie intake = 153 + 841.5 + 500 = 1494.5 = 1495 kcal

B. Amino acids: 1000 mL × 4.5% = z × 15%

z = 300 mL of 15% amino acids solution

Dextrose: 1000 mL × 15% = y × 50%

y = 300 mL of 50% Dextrose solution

NaCl: $$30\text{ mEq} \times \frac{\text{mmol}}{\text{mEq}} \times \frac{58.5\text{ mg}}{1\text{ mmol}} \times \frac{100\text{ mL}}{15\text{ g}} \times \frac{1\text{ g}}{1000\text{ mg}} = 11.7\text{ mL}$$

Ca gluconate: $$2.5\text{ mEq} \times \frac{10\text{ mL}}{4.6\text{ mEq}} = 5.4\text{ mL}$$

Insulin: $$15\text{ units} \times \frac{1\text{ mL}}{100\text{ units}} = 0.15\text{ mL}$$

Heparin: $$2500\text{ units} \times \frac{1\text{ mL}}{1000\text{ units}} = 2.5\text{ mL}$$

SWFI: 1000 mL − (300 + 300 + 11.7 + 5.4 + 0.15 + 2.5) = 380 mL

C.1 $\frac{50\text{ mL}}{\text{kg}} \times 10\text{ kg} = 500\text{ mL} = \text{Total volume of PN to be prepared}$

$\frac{1.2\text{ g}}{\text{kg}} \times 10\text{ kg} = 12\text{ g a.a needed}$

$12\text{ g} \times \frac{100\text{ mL}}{15\text{ g}} = 80\text{ mL of 15\% AA solution}$

$500\,\text{mL} \times 15\% = z \times 50\%$

$z = 150\,\text{mL of dextrose 50\%}$

C.2 Phosphorus:

Need: $2.5\,\text{mmol/kg} \times 10\,\text{kg} = 25\,\text{mmol}$

Have already: $80\text{ ml} \times \frac{12.5\text{ mmol}}{1000\text{ mL}} = 1\text{ mmol}$

Need to add: $25 - 1 = 24\,\text{mmol}$

$24\text{ mmol} \times \frac{1\text{ mL}}{3\text{ mmol}} = 8\text{ mL postassium phosphates}$

C.3 Potassium:

Need: $\frac{4.5\text{ mEq}}{\text{kg}} \times 10\text{ kg} = 45\text{ mEq}$

Have already: $8\,\text{mL (KPhos)} \times 4.4\,\text{mEq/mL} = 35.2\,\text{mEq}$

Need to add: $45 - 35.2 = 9.8\,\text{mEq}$

$9.8\text{ mEq} \times \frac{1\text{ mL}}{2\text{ mEq}} = 4.9\text{ mL KCI}$

C.4 Chlorine:

Need: $\frac{1.5\text{ mEq}}{\text{kg}} \times 10\text{ kg} = 15\text{ mEq}$

Have already: 9.8 mEq (from KCl)

Need to add: $15 - 9.8 = 5.2\,\text{mEq}$

$5.2\text{ mEq} \times \frac{1\text{ mL}}{4\text{ mEq}} = 1.3\text{ mL of NaCl}$

C.5 Sodium:

Need: $\frac{1.2\text{ mEq}}{\text{kg}} \times 10\text{ kg} = 12\text{ mEq}$

Have already:

5.2 mEq (from NaCl)

$80\text{ mL} \times \frac{15\text{ mEq}}{1000\text{ mL}} = 1.2\text{ mEq (from a.a solution)}$

Need to add: $12 - (5.2 + 1.2) = 5.6\,\text{mEq}$

$5.6\text{ mEq} \times \frac{1\text{ mL}}{2\text{ mEq}} = 2.8\text{ mL of Na acetate}$

C.6 Calcium:

$$\text{Need: } \frac{0.45\ \text{mEq}}{\text{kg}} \times 10\ \text{kg} = 4.5\ \text{mEq}$$

$$4.5\ \text{mEq} \times \frac{1\ \text{mL}}{0.465\ \text{mEq}} = 9.67 = 9.7\ \text{mL of Ca gluconate}$$

C.7 Total TPN = 500 mL

Base solution = 230 mL

Additives = 8 + 4.9 + 1.3 + 2.8 + 9.7 = 26.7 mL = 27 mL

SWFI = 500 − (230 + 26.7) = 243.3 mL = 243 mL

C.8 Flow rate:

$$\frac{500\ \text{mL}}{24\ \text{hr}} = 20.8 = 21\ \text{mL/hr}$$

C.9 Total calories:

$$80\ \text{mL} \times \frac{15\ \text{g}}{100\ \text{mL}} \times \frac{4\ \text{kcal}}{\text{g}} = 48\ \text{kcal}$$

$$150\ \text{mL} \times \frac{50\ \text{g}}{100\ \text{mL}} \times 3.4\ \text{kcal/g} = 255\ \text{kcal}$$

Total kcal = 48 + 255 = 303 kcal

Calculations Related to Design of a Parenteral Nutrition Solution

44. When parenteral nutrition is required for a given patient, the pharmacist is responsible for all the dosing and compounding calculations. It is also the pharmacist's duty to remember that, TPN is a very complex admixture containing several individual chemicals, all bringing their own physicochemical characteristics that will influence their behavior when combined in a single container. Compatibility and stability of all components of a PN admixture are prerequisites to the safe use of PNs.

While all calculations of nutrients provided by different solutions to supply the needed requirements of electrolyte, carbohydrate, fat and other nutrients are the pharmacists' responsibility, the design of a parenteral nutrition formula depends on several nutrition professionals, who will discuss and decide, as a team, the best regimen for the patient. If you are interested in more detail on this subject, check Appendix 5. We provide also one practical example of calculations related to the design of a parenteral nutrition formula.

ENTERAL NUTRITION

45. Enteral nutrition, the most natural method for nutrition support, needs to accommodate the needs and lifestyles of patients. Implementation, monitoring and management of enteral feeding belong to a very specialized group of health professionals. Some basics are provided in Appendix 5.

REVIEW PROBLEMS

A. A parenteral solution is to contain 3% amino acids. How many milliliters of a 10% amino acid stock solution should be used to prepare 4 liters of the parenteral solution?

B. A TPN solution is to contain 35 mEq of sodium ion per liter. How many milliliters of a stock solution containing 2.5 mEq of sodium (as NaCl) per mL should be used to prepare 1 L?

C. How many milliliters of a calcium gluconate solution, 0.45 mEq/mL should be used to supply 20 mEq of calcium ion to a TPN solution?

D. A TPN solution is to contain the following quantities in each liter: 60 mEq of Na^+; 60 mEq of K^+; 100 mEq of acetate; 60 mEq of Cl^-; 10 mEq of Mg^{2+}. How many mL of a potassium acetate stock solution, 5 mEq/mL, will supply the potassium needed for 2 L of the TPN solution? How many mEq of acetate will the stock solution provide?

E. Referring to the previous problem, how many mL of a magnesium sulfate ($MgSO_4$, FW = 120.4) injection, 4 mEq/mL, will supply the magnesium needed?

F. A parenteral nutrition formula calls for 400 mg of P. How many milliliters of a potassium phosphates injection containing 2.1 mmol/mL phosphate should be used?

G. A TPN formula includes 22% dextrose. How many mL of a 70% dextrose injection should be used for each liter of the TPN mixture?

H. How many kilocalories will be provided by each liter of the solution made in problem G?

I. How many grams of glycerin will provide 100 kcal? Each gram of glycerin produces 4.32 kcal of energy.

J. A 20% fat emulsion yields 2 kcal/mL. How many mL will provide 1200 kilocalories?

K. A total of 800 mL of the emulsion described in problem J are administered over the same time period as 2000 mL of a 20% dextrose solution. What is the ratio of calories from fat to total calories?

L. An amino acid preparation contains 40 g of amino acids per liter, equivalent to 6.5 g of nitrogen per liter. How many grams of amino acids are equivalent to each gram of nitrogen?

M. A stock preparation for parenteral nutrition contains 10 grams of amino acids per 100 mL. How many grams of nitrogen would be contained in 2.5 L if each g N = 6.25 g amino acids?

N. A patient is to receive 0.5 g/kg of nitrogen. How many mL of a 15% amino acid solution will supply the amount of nitrogen needed by a 65 kg woman if 1 g N = 6.25 g amino acids?

O. A parenteral solution contains 2% amino acids and 25% dextrose. Assuming that 1 gN = 6.25 g amino acids, calculate the ratio of nonprotein calories to nitrogen.

P. A TNA mixture consists of 750 mL of a solution containing 4.2% amino acids and 25% dextrose and 500 mL of a 10% fat emulsion. Calculate the ratio of nonprotein calories to nitrogen if each mL of the fat emulsion = 1.1 kcal and 6.25 g amino acids = 1 gN.

Q. KM, a 72-year old male is receiving the following TPN. Calculate the total daily calorie intake by KM.

FreAmine 8.5% 500 mL

D 25 W 500 mL

Rate: 110 mL/h

Intralipid 10% (10% fat emulsion), 200 mL/day, IV infusion, 20 mL/h

R. A five-foot female patient, formerly weighing 121 lb., has lost 20 lb. in the past four weeks and will require IV nutrition prior to a planned major operation. Her doctor calculated a nutritional plan based on her ideal weight (110 lb). How many total kcal will this patient receive if the nutritional plan designed for her includes 500 mL of 10% fat emulsion, 1000 mL dextrose 20% and the protein requirement is 0.75 g/kg/day?

S. The following TPN order needs to be prepared. Do all calculations required to compound it.

Margaret Raymond Rm. ICU 505 Dr. Paul Preston

Central TPN ______________ 2000 cc daily @ 100 mL/hr

2.8% Amino acids and 20% Dextrose

Sodium chloride: 45 mmols

Potassium phosphate: 24 mmols
Magnesium sulfate: 16 mmols
Calcium gluconate: 5 mmols
Insulin R: 100 Units
Trace minerals: 1 mL
MVI-12: 10 mL
Lipids: 20%, 500 mL daily (piggyback)

Solutions available in the pharmacy:
Aminosyn sol., 7%
Dextrose sol., 50%
SWFI
NaCl (MW= 58.5), 50 mL vial of 15% solution
K Phosphate sol., 3 mmol P/mL and 4.4 mEq K/mL
$MgSO_4$ (MW = 120), 20 mL 50% solution
Ca gluconate ($C_{12}H_{22}CaO_{14}$, MW = 430), 0.465 mEq/mL solution
Insulin regular U-100, 10 mL vial

T. Based on the previous question, calculate the total daily calories (TDC or TDE), in kcal, that this patient will receive (TPN + fat emulsion).

U. A medication order is received for 1 liter of PN with a final concentration of 15% dextrose. The physician specifies that *500 mL is to be amino acid solution*. The pharmacy has 50% dextrose injection in 500-mL bottles and 500-mL bags of SWFI. How would you prepare this order? Show calculations.

V. The following pediatric TPN needs to be prepared. Do all calculations required to prepare it.

James Kearney   Rm. 205 PEDS   Dr. Brian Moore

Wt. = 5 kg

Peripheral TPN – 10 mL/kg q 8 h daily

Amino acids: 1.5 g/kg/24 hr

Dextrose: 7%

Sodium: 1.3 mEq/kg/24 hr

Potassium: 3 mEq/kg/24 hr

Chlorine: 2 mEq/kg/24 hr

Phosphorus: 1.5 mmol/kg/24hr

Calcium: 0.25 mEq/kg/24 hr

Syringe pump @ 10 mL/hr

Available solutions:

SWFI

A.A sol., 10% (contains 4 mmol/L sodium)

Dextrose sol., 50%

K Phosphates sol. = 3 mmol P/mL and 4.4 mEq K/mL

KCl sol. = 2 mEq/mL

NaCl sol. = 4 mEq/mL

Na acet. sol. = 2 mEq/mL

Ca gluconate sol. = 0.2325 mmol/mL

W. The following tailored TPN order has to be prepared. Show calculations and selected solutions. Remember that the easiest way to prepare a PN is by combining premeasured volumes or volumes that will be read from graduations on the sides of a large container.

1000 mL 3.5% AA and 15% dextrose q 24 hr—Central Line

Additives:

Sodium ....20 mmol
Potassium ....15 mmol
Chloride...............................18.5 mmol
Calcium ....4 mmol
Magnesium...........................3 mmol
Phosphorus ....9 mmol
MVI.....................................10 mL
MTE ....1 mL
Insulin R..............................90 units

Run at 80 mL/hr through central line

Available solutions:

Aminosyn 7% (250 mL, 500 mL, 1000 mL): has 3.2 mmol *potassium*/liter and 3 mmol *sodium*/liter

Aminosyn 10% (250 mL, 500 mL, 1000 mL): has 3.2 mmol *potassium*/liter and 3 mmol *sodium*/liter

Dextrose 50% (250 mL, 500 mL, 1000 mL)

Dextrose 70% (250 mL, 500 mL, 1000 mL)

SWFI (250 mL, 500 mL, 1000 mL)

Potassium phosphates, 3 mM/mL P and 4.4 mEq/mL K

Calcium gluconate, 0.465 mEq/mL

Magnesium sulfate, 4 mEq/mL

Sodium chloride, 2.5 mEq/mL

Multivitamins Infusion (MVI), standard solution

Multitrace elements (MTE), standard solution

Humulin R, U-100

Solutions to Review Problems

A. 1200 mL

B. 14 mL

C. 44.4 mL

D. 24 mL; 120 mEq of acetate
E. 5 mL
F. 6.14 mL
G. 314 mL
H. 748 kcal
I. 23.1 g
J. 600 mL
K. 54%
L. 6.15 g aa/g N
M. 40 g
N. 1350 mL
O. 265 kcal/g N
P. 236 kcal/g N
Q. 815 kcal
R. 1380 kcal
S. See calculations.
T. 1984 kcal
U. Add 500 mL AA + 300 mL 50% Dextrose + 200 mL SWFI to an empty 1-liter container.
V. See calculations.
W. See calculations.

Calculations to Review Problems

A. $j \times \frac{10 \text{ g}}{100 \text{ mL}} = 4000 \text{ mL} \times 3 \text{ g}/100 \text{ mL}$

$j = 1200 \text{ mL}$

B. $j \times \frac{2.5 \text{ mEq}}{\text{mL}} = 1 \text{ L} \times 35 \text{ mEq}/\text{L}$

$j = 14 \text{ mL}$

C. $j \times \frac{0.45 \text{ mEq}}{\text{mL}} = 20 \text{ mEq}$

$j = 44.4 \text{ mL}$

D. $j \times \frac{5 \text{ mEq}}{\text{mL}} = 2 \text{ L} \times 60 \text{ mEq}/\text{L}$

$j = 24 \text{ mL}$

The potassium acetate solution will contain the same number of millimoles of acetate, 120 mEq.

E. $j \times \frac{4 \text{ mEq}}{\text{mL}} = 2 \text{ L} \times 10 \text{ mEq}/\text{L}$

$j = 5 \text{ mL}$

F. $\frac{400 \text{ mg}}{31 \text{ mg/mmol}} = 12.9 \text{ mmol P}$

$j \times \frac{2.1 \text{ mmol}}{\text{mL}} = 12.9 \text{ mmol}$

$j = 6.14 \text{ mL}$

G. $j \times \frac{70 \text{ g}}{100 \text{ mL}} = 1000 \text{ mL} \times 22 \text{ g/100 mL}$

$j = 314 \text{ mL}$

H. $1000 \text{ mL} \times \frac{22 \text{ g}}{100 \text{ mL}} \times 3.4 \text{ kcal/g} = 748 \text{ kcal}$

I. $\frac{100 \text{ kcal}}{4.32 \text{ kcal/g}} = 23.1 \text{ g}$

J. 1200 kcal × mL/2 kcal = 600 mL

K. 800 mL × 2 kcal/mL = 1600 kcal from fat

$2000 \text{ mL} \times \frac{20 \text{ g}}{100 \text{ mL}} \times 3.4 \text{ kcal/g} = 1360 \text{ kcal}$

$\frac{1600 \text{ mg}}{1600 + 1360} = 0.54 = 54\%$

L. $\frac{40 \text{ g aa}}{6.5 \text{ g N}} = 6.15 \text{ g aa/g N}$

M. $2500 \text{ mL} \times \frac{10 \text{ g}}{100 \text{ mL}} \times \frac{1 \text{ g N}}{6.25 \text{ g aa}} = 40 \text{ g N}$

N. $0.5 \text{ g N/kg} \times 65 \text{ kg} \times 6.25 \text{ g aa/g N} = j \times \frac{15 \text{ g}}{100 \text{ mL}}$

$j = 1350 \text{ mL}$

O. $\frac{25 \text{ g/100 mL} \times 3.4 \text{ kcal/g}}{2 \text{ g aa/100 mL} \times \frac{1 \text{ g N}}{6.25 \text{ g aa}}} = \frac{0.85 \text{ kcal}}{0.0032 \text{ g N}} = 265 \text{ kcal/g N}$

P. $\frac{(750 \text{ mL} \times 25 \text{ g/100 mL} \times 3.4 \text{ kcal/g}) + (500 \text{ mL} \times 1.1 \text{ kcal/mL})}{750 \text{ mL} \times 4.2 \text{ g aa/100 mL} \times \frac{1 \text{ g N}}{6.25 \text{ g aa}}}$

$= \frac{1188 \text{ kcal}}{5.04 \text{ g N}} = 236 \text{ kcal/g N}$

Q. $\text{A.A.} = 500 \text{ mL} \times \frac{8.5 \text{ g}}{100 \text{ mL}} \times 4 \text{ kcal/g} = 170 \text{ kcal}$

$\text{Dextrose} = 500 \text{ mL} \times \frac{25 \text{ g}}{100 \text{ mL}} \times 3.4 \text{ kcal/g} = 425 \text{ kcal}$

10% fat emulsion = 1.1 kcal/mL

Fat = 200 mL × 1.1 kcal/mL = 220 kcal

Total daily calorie intake = 170 + 425 + 220 = 815 kcal

R. 500 mL × 1.1 kcal/mL= 550 kcal

$$1000 \text{ mL} \times \frac{20 \text{ g}}{100 \text{ mL}} \times 3.4 \text{ kcal/g} = 680 \text{ kcal}$$

$$\frac{0.75 \text{ g}}{\text{kg}} \times \frac{1 \text{ kg}}{2.2 \text{ lb}} \times 110 \text{ lb} \times 4 \text{ kcal/g} = 150 \text{ kcal}$$

Total kcal = 1380 kcal

S. *Amino acids*: 2.8% × 2000 mL = 7% × z

z = 800 mL of 7% A.A. solution

Dextrose: 20% × 2000 mL = 50% × y

Y = 800 mL of 50% dextrose solution

SWFI: 2000 − 1600 = 400 mL SWFI

Sodium chloride:

Need: 45 mmol

Have: 15% solution

$$15 \text{ g/100 mL} = \frac{150 \text{ mg}}{\text{ml}} \times \frac{1 \text{ mmol}}{58.5 \text{ mg}} = 2.56 \text{ mmol/mL}$$

$$\text{Need to add: } 45 \text{ mmol} \times \frac{1 \text{ mL}}{2.56 \text{ mmol}} = 17.6 \text{ mL of } 15\% \text{ NaCl}$$

Potassium phosphates:

Need: 24 mmol

Have: 3 mmol/mL

$$\text{Need to add: } 24 \text{ mmol} \times \frac{1 \text{ mL}}{3 \text{ mmol}} = 8 \text{ mL of KPhos. sol.}$$

Magnesium sulfate:

Need: 16 mmol

Have: 50% solution

$$50 \text{ g/mL} = \frac{500 \text{ mg}}{\text{mL}} \times \frac{1 \text{ mmol}}{120 \text{ mg}} = 4.17 \text{ mmol/mL}$$

$$\text{Need to add: } 16 \text{ mmol} \times \frac{1 \text{ mL}}{4.17 \text{ mmol}} = 3.8 \text{ mL of } 50\% \text{ MgSO}_4$$

Calcium gluconate:

Need: 5 mmol

Have: 0.465 mEq/mL = 0.2325 mmol/mL

$$\text{Need to add: } 5 \text{ mmol} \times \frac{1 \text{ mL}}{0.2325 \text{ mmol}} = 21.5 \text{ mL of CaGluc. sol.}$$

Insulin:

Need: 100 units

Have: U-100 = 100 units/mL

Need to add: 1 mL of U-100 insulin

Trace minerals: 1 mL

Multivitamin infusion (MVI): 10 mL

Total TPN:

Base solution: 2000 mL

Additives: 17.6 + 8 + 3.8 + 21.5 + 1 + 1 + 10 = 63 mL

Final TPN volume = 2000 + 63 = 2063 mL

T. A.A.: $800 \text{ mL} \times \frac{7 \text{ g}}{100 \text{ mL}} \times 4 \text{ kcal/g} = 224 \text{ kcal}$

Dextrose: $800 \text{ mL} \times \frac{50 \text{ g}}{100 \text{ mL}} \times 3.4 \text{ kcal/g} = 1360 \text{ kcal}$

Fat: 200 mL × 2 kcal/mL = 400 kcal

TDC: 224 + 1360 + 400 = 1984 kcal

U. A.A.: 500 mL

Dextrose: 1000 mL × 15% = z × 50%

Z = 300 mL of 50% dextrose solution

SWFI: 1000 mL − (500 + 300) = 200 mL SWFI

Final product is 1 liter of 15% dextrose including 500 mL A.A. solution.

V. *Amino acid:*

$\frac{1.5 \text{ g}}{\text{kg}} \times 5 \text{ kg} = 7.5 \text{ g} \times \frac{100 \text{ mL}}{10 \text{ g}} = 75 \text{ mL of } 10\% \text{ A.A. solution}$

Dextrose:

$\frac{10 \text{ mL}}{\text{kg}} \times 5 \text{ kg} = 50 \text{ mL q } 8 \text{ hr} = 50 \times 3 = 150 \text{ mL total PN}$

150 mL × 7% = z × 50%

z = 21 mL of 50% dextrose solution

Phosphorus:

Need: $\frac{1.5 \text{ mmol}}{\text{kg}} \times 5 \text{ kg} = 7.5 \text{ mmol}$

Add: $7.5 \text{ mmol} \times \frac{1 \text{ mL}}{3 \text{ mmol}} = 2.5 \text{ mL}$ of potassium phosphates solution

Potassium:

Need: $\frac{3 \text{ mEq}}{\text{kg}} \times 5 \text{ kg} = 15 \text{ mEq}$

Have: $2.5 \text{ mL} \times \frac{4.4 \text{ mEq}}{\text{mL}} = 11 \text{ mEq}$

Add: $15 - 11 = 4 \text{ mEq} \times \frac{1 \text{ mL}}{2 \text{ mEq}} = 2 \text{ mL}$ of potassium chloride solution

Chloride:

Need: $\frac{2 \text{ mEq}}{\text{kg}} \times 5 \text{ kg} = 10 \text{ mEq}$

Have: 4 mEq (from KCl)

Add: $10 - 4 = 6 \text{ mEq} \times \frac{1 \text{ mL}}{4 \text{ mEq}} = 1.5 \text{ mL}$ of sodium chloride solution

Sodium:

Need: $\frac{1.3 \text{ mEq}}{\text{kg}} \times 5 \text{ kg} = 6.5 \text{ mEq}$

Have: 6 mEq (from 1.5 mL NaCl)

$$75 \text{ mL(A.A.)} \times \frac{4 \text{ mmol}}{1000 \text{ mL}} = 0.3 \text{ mmol} = 0.3 \text{ mEq}$$

Add: $6.5 - (6 + 0.3) = 0.2 \text{ mEq} \times \frac{1 \text{ mL}}{2 \text{ mEq}} = 0.1$ mL of sodium acetate solution

Calcium:

Need: $\frac{0.25 \text{ mEq}}{\text{kg}} \times 5 \text{ kg} = 1.25 \text{ mEq}$

Add: $1.25 \text{ mEq} \times \frac{1 \text{ mL}}{0.465 \text{ mEq}} = 2.7$ mL of calcium gluconate solution

Total PN: 150 mL

Base solution: 75 + 21 = 96 mL

Additives: 2.5 + 2 + 1.5 + 0.1 + 2.7 = 8.8 mL

SWFI: 150 − (96 + 8.8) = 150 − 105 = 45 mL of SWFI

W. *Amino acid:*

1000 mL × 3.5% = a × 7%

a = 500 mL of 7% A.A. solution

1000 mL × 3.5% = b × 10%

b = 350 mL of 7% A.A. solution

Use 500 mL of 7% A.A. solution

Dextrose:

1000 mL × 15% = c × 50%

c = 300 mL of 50% dextrose

1000 mL × 15% = d × 70%

d = 214.3 mL of 70% dextrose

Use 300 mL of 50% dextrose

Total base solution = 1000 mL

SWFI = 1000 − (500 + 300) = 200 mL

Phosphorus:

$9 \text{ mmol} \times \frac{1 \text{ mL}}{3 \text{ mmol}} = 3$ mL potassium phosphates solution

Potassium:

Need: 15 mmol

Have: $3 \text{ mL} \times \frac{4.4 \text{ mEq}}{\text{mL}} = 13.2 \text{ mEq} = 13.2 \text{ mmol}$

$$\frac{3.2 \text{ mmol}}{1000 \text{ mL}} \times 500 \text{ mL (A.A. sol)} = 1.6 \text{ mmol}$$

Add: 13.2 + 1.6 = 14.8 mmol (No need to add; <10% of prescribed amount)

Calcium:

$$4 \text{ mmol} \times \frac{1 \text{ mL}}{0.2325 \text{ mmol}} = 17.2 \text{ mL of calcium gluconate solution}$$

Magnesium:

$$3 \text{ mmol} \times \frac{1 \text{ mL}}{2 \text{ mmol}} = 1.5 \text{ mL of magnesium sulfate solution}$$

Sodium:

Need: 20 mmol

$$\text{Have: } \frac{3 \text{ mmol}}{1000 \text{ mL}} \times 500 \text{ mL (A.A. sol)} = 1.5 \text{ mmol}$$

Add: 20 − 1.5 = 18.5 mmol

$$18.5 \text{ mmol} \times \frac{1 \text{ mL}}{2.5 \text{ mmol}} = 7.4 \text{ mL of sodium chloride solution}$$

Chloride:

Got 18.5 mmol (from NaCl)

Insulin:

$$90 \text{ units} \times \frac{1 \text{ mL}}{100 \text{ units}} = 0.9 \text{ mL U-100 Humulin R}$$

Trace elements: 1 mL

Multivitamins: 10 mL

Final volume: 1000 mL (base solution) + 41 mL (additives) = 1041 mL

CHAPTER 12

BIOLOGICS FOR IMMUNIZATION

***Learning Objectives:** After completing this chapter the student should be able to:*

1. Describe the pharmacist's role in immunizations.
2. Recognize different types of vaccines and their strengths.
3. Calculate doses for the administration of bacterial and viral vaccines, and toxoids.

One of the greatest accomplishments of the medical field has been the ability to protect individuals from common diseases (e.g. influenza) and life-threatening ones (e.g. rabies) through the administration of immunizing agents or antibody-containing preparations. Inexpensive childhood immunizations have decreased considerably the morbidity and mortality of several infectious diseases and their consequences. Adulthood immunization (against e.g. hepatitis A and B, influenza), needs attention since many adults do not recognize the need for immunization throughout their lifetime. In the USA, pharmacists have worked with allied health practitioners to encourage immunizations and there are 41 states where pharmacists are actually permitted to administer vaccines. Regardless of the practice setting, pharmacists are involved in the design, implementation, patient screening, counseling, and logistics of vaccination programs. Because they provide patient education and care in a variety of health care settings, pharmacists are in a good position to develop immunization programs. In addition, since most pharmacies are in residential neighborhoods, people not only do not have to travel so far to receive immunizations, but each pharmacy may provide the optimal site for immunizations since the majority are open during the day and on weekends.

Immunity, defined as resistance to a specified disease, may be natural or acquired. Natural immunity depends on several factors, including species, race, and individual ability to resist diseases. Acquired immunity is achieved through *vaccination*, which is the use of biologics for immunization to allow the body to build up its own defenses. *Vaccines* have prophylactic action in the development of immunity and may contain fractions or whole, living, attenuated, or killed microorganisms. *Toxoids* are modified and detoxified bacterial toxins, often combined with adjuvants (e.g. aluminum hydroxide, aluminum phosphate) to enhance their antigenicity. *Immunoglobulins* provide temporary prophylaxis (1–2 weeks) to susceptible individuals, or protection during a critical period of exposure (e.g. antivenins for bites from poisonous snakes or patients exposed to hepatitis).

Pharmaceutical Calculations, Fourth Edition, By Joel L. Zatz and Maria Glaucia Teixeira
ISBN 0-471-67623-3

Biologics are substances used therapeutically in humans and produced by a living organism. They include hormones, antibiotics, vitamins, vaccines, and antivenins. In a broad sense, *biologics for immunization* (FDA) or *immunobiologics* (CDC-ACIP, Center for Disease Control—Advisory Committee on Immunization Practices) are biologic products designed to help develop any type of immunity and may be a virus, a therapeutic serum, toxin, antitoxin, or analogous product.

The 2003 Childhood & Adolescent Immunization Schedule, catch-up schedule, and recommended adult immunizations, may be obtained from the Center for Disease Control and Prevention (www.cdc.gov/nip) and will not be discussed in this text. This chapter will emphasize different types of vaccines, their strengths, and calculations related to their preparation and administration. Check Table A.6 in the appendix for some examples of vaccines and toxoids and their main characteristics.

VACCINES

Bacterial Vaccines

1. Bacteria are grown in a suitable broth under ideal conditions and processed to provide two types of vaccines: attenuated (live microorganisms) or inactivated (killed microorganisms). Attenuated vaccines may also be produced by removing or altering several base pairs of DNA within a key region of the gene structure of the pathogenic organism. As a result, the organism is allowed to survive and multiply but not cause the disease. Another method used to prepare bacterial vaccines is recombinant DNA technology, which introduces the genes that code for the desired antigen into nonpathogenic organisms. The vaccine product may contain one single immunogen (monovalent vaccine) or several immunogens (polyvalent vaccine) to stimulate immunity against the same disease.

The strength of bacterial vaccines may be expressed as:

- Total protective units (PU) per milliliter (or per dose): quantity of bacteria capable of promoting immunity against the organism.
- Total organisms or colony forming units: CFU.
- Micrograms of immunogen per milliliter (or in each dose): μg/mL.

Viral Vaccines

2. Viruses are grown and propagated in one of several animate media (e.g. embryonic egg, skin of living calves, chick embryo cell culture) then purified to contain one single immunogen (monovalent vaccine) or several immunogens (polyvalent vaccine) to elicit immunity against the same disease. The vaccine product may also be a mixed vaccine with several viral immunogens for several different viral diseases (e.g. MMR vaccine = measles, mumps, rubella).

The strengths of viral vaccines are expressed as:

- Tissue culture infectious doses ($TCID_{50}$): quantity of virus estimated to infect 50% of inoculated cultures.
- Micrograms of immunogen per milliliter (or per dose): μg/mL.
- International units (IU): units of activity based on tests in animals.

- Plaque-forming units (PFU).
- D-antigen units: units based on immunodiffusion method.
- ELISA units (ELu): based on antigen activity assayed in reference to a standard using enzyme-linked immunosorbent assay (ELISA). (e.g. hepatitis A vaccine (HAV).)

Cancer Vaccines

3. Several vaccines have been developed for melanoma, colorectal cancer, renal cell carcinoma, lung cancer, cervical cancer, and breast and ovarian cancers. Most are still under clinical trials but their main goal is to increase antigen awareness to the immune cells or increase co-stimulatory signals that induce immune response (e.g., stimulation of T cells, LAK cells = lymphokine-activated killer cells, NK cells = natural killer cells, all with antitumor activity).

The development of some cancer vaccines has two basic targets:

1. Prevention of cancer in high-risk patients due to family history.
2. Therapy in combination with surgery, radiation therapy, and chemotherapy.

TOXOIDS

4. Toxoids are detoxified toxins from bacteria. The bacterial toxin, detoxified with formaldehyde, may be plain or contain an aluminum salt as adjuvant. Some products contain a single immunogen (e.g. tetanus toxoid), multiple immunogens or mixed immunogens (e.g. tetanus and diphtheria toxoids). Other products combine toxoids and vaccines in a single preparation (DPT = diphtheria and tetanus toxoids + pertussis vaccine). This type of product has the main advantage of providing broad immunization coverage in the same injection.

The strengths of toxoids are expressed as:

- Flocculating unit (L*f* unit): the smallest amount of toxin capable of flocculating (or clumping) 1 unit of standard antitoxin.

OTHER IMMUNOBIOLOGICS

5. Passive immunity may be provided by human immune sera and globulins (homologous sera) or animal immune sera (heterologous sera). These products are developed from blood of humans who have had the specific disease or after having been immunized against it with a specific biologic product. They may also be developed in animals (usually horse). Immunity is generally of brief duration (a few weeks only) but of great value for treatment of acute disease (immediate supply of immunoglobulin) or prevention of illness when immediate protection is needed.

PHARMACEUTICAL CHARACTERISTICS AND COMPOUNDING DILUTIONS

6. Immunologic drugs are available in numerous dosage forms but most frequently they are solutions or suspensions for injection. Some products are joined with devices (e.g. multiple-puncture devices). Suspensions must always be thoroughly agitated before withdrawal of each dose to assure uniformity. Powders must be reconstituted with a specific quantity of recommended diluent to avoid physical or chemical incompatibility and consequent inactivation of the drug.

For standard 10-fold serial dilution, compound the most concentrated vial first then compound each dilution by withdrawing a 1 mL aliquot and adding it to a 9 mL vial of suitable diluent. Alternately, add 0.2 mL of concentrate to 1.8 mL diluent or add 0.5 mL concentrate to 4.5 mL diluent. Pre-filled vials with 9, 4.5, and 1.8 mL are commercially available from allergen-extract manufacturers. If a five-fold dilution is needed, add 2 mL concentrate to 8 mL diluent, 1 mL to 4 mL, 0.5 mL to 2 mL, or 0.4 mL to 1.6 mL.

Accidentally challenging a hypersensitive patient directly with a full-strength antigen may result in anaphylaxis and death. Error prevention should be emphasized by double checking each compounding step, to make sure transposition (reversing) of vials or labels has not occurred. The most concentrated label should appear on the vial whose contents have the darkest color.

CALCULATIONS INVOLVING BIOLOGICS FOR IMMUNIZATION

7. Despite the fact that vaccines and toxoids are expressed in unusual concentrations, most calculations are straightforward and can be solved by a simple proportion or by dimensional analysis. Some sample problems are shown to make you feel comfortable when the time comes to calculate and prepare doses for immunization.

One-half the adult immunizing dose of BCG vaccine is usually given to children 1 month of age. If a BCG vaccine product contains an average of 4×10^8 CFU (colony-forming units) per milliliter and the adult dose is 0.2 mL, how many CFU would a month-old child receive?

Solution. 4×10^7 CFU

CALCULATION

$$\frac{4 \times 10^8 \text{ CFU}}{1 \text{ ml}} \times 0.1 \text{ ml} = 4 \times 10^7 \text{ CFU} = 40 \text{ million colony-forming units}$$

---

8. A 5 mL vial of DTP vaccine (diphtheria, tetanus, and pertussis) contains 40 PU (protective units) of pertussis. If the standard schedule of administration is a total of 16 PU of pertussis administered at 2, 4, 6, and 18 months of age, how many milliliters would be needed for each dose?

Solution. 0.5 mL

CALCULATIONS

$$\frac{16 \text{ PU}}{4 \text{ injections}} = 4\text{PU / dose}$$

$$4 \text{ PU} \times \frac{5 \text{ mL}}{40 \text{ PU}} = 0.5 \text{ mL}$$

---

9. The concentration of a typhoid vaccine product is 8 units/mL. For adults and children older than 10 years of age, the full dose is two 0.5 mL SC 4 weeks apart. How many individuals can be fully vaccinated with a 10 mL vial?

Solution. 10 individuals

CALCULATIONS

$$\frac{8 \text{ units}}{1 \text{ mL}} \times 1 \text{ mL} = 8 \text{ units / full dose}$$

$$\frac{8 \text{ units}}{1 \text{ mL}} \times 10 \text{ mL(vial)} = 80 \text{ units/vial}$$

$$80 \text{ units} \times \frac{\text{dose}}{8 \text{ units}} = 10 \text{ doses}$$

---

10. Measles virus vaccine live contains not less than 1000 $TCID_{50}$ per 0.5 mL dose. How many doses can be administered from a 10 mL multidose vial?

Solution. 20 doses

CALCULATIONS

$$10 \text{ mL} \times \frac{\text{dose}}{0.5 \text{ mL}} = 20 \text{ doses}$$

---

11. Dosage recommendations by CDC and FDA of a hepatitis B vaccine product for both pre- and post-exposure prophylaxis of adults is 20 micrograms by IM injection at 0, 1, and 6 months. How many patients will be completely immunized from a 10 mL multidose vial containing 200 µg?

Solution. 3 patients

CALCULATIONS

$$20\ \mu\text{g} \times 3 \text{ doses} = 60\ \mu\text{g (per patient)}$$

$$60\ \mu\text{g (patient)} \times \frac{10 \text{ mL}}{200\ \mu\text{g}} = 3 \text{ patients}$$

---

12. A poliovirus vaccine inactivated product contains 40, 8, and 32 D-antigen units per 0.5 mL dose of poliovirus types 1, 2, and 3, respectively. How many antigen units of each poliovirus type would a child receive from a complete immunization that includes 3 doses?

Solution. 120, 24, and 96 antigen units of poliovirus types 1, 2, and 3, respectively.

CALCULATIONS

$$\frac{40 \text{ antigen units}}{0.5 \text{ mL}} \times \frac{0.5 \text{ mL}}{\text{dose}} \times 3 \text{ doses} = 120 \text{ units of Type 1}$$

$$\frac{8 \text{ antigen units}}{0.5 \text{ mL}} \times \frac{0.5 \text{ mL}}{\text{dose}} \times 3 \text{ doses} = 24 \text{ units of Type 2}$$

$$\frac{32 \text{ antigen units}}{0.5 \text{ mL}} \times \frac{0.5 \text{ mL}}{\text{dose}} \times 3 \text{ doses} = 96 \text{ units of Type 3}$$

13. A traveler, visiting a foreign area of enzootic rabies, was prescribed to receive vaccination for pre-exposure prophylaxis. Primary vaccination schedule was 1 mL IM or 0.1 mL ID of human diploid cell rabies vaccine on days 0, 7, and 21. Imovax® Rabies Vaccine is available in products for intramuscular (IM) injection, containing not less than 2.5 IU/mL, and for intradermal (ID) injection, containing 0.25 IU/0.1 mL. If this patient receives all three doses as scheduled, how many international units (IU) he will receive from either IM or ID type of vaccination schedule?

Solution. 7.5 international units by IM and 0.75 IU by ID

CALCULATIONS

IM vaccination: $\frac{1\text{ mL}}{\text{dose}} \times 3\text{ doses} \times \frac{2.5\text{ IU}}{\text{mL}} = 7.5\text{ IU}$

ID vaccination: $\frac{0.1\text{ mL}}{\text{dose}} \times 3\text{ doses} \times \frac{0.25\text{ IU}}{0.1\text{ mL}} = 0.75\text{ IU}$

14. A test of hypersensitivity for Yellow Fever vaccine requires injection of 0.02 mL intradermally. If a yellow fever vaccine product contains 0.7 plaque-forming units (PFU) per 0.5 mL, how many PFU will be injected?

Solution. 0.028 PFU

CALCULATIONS

$$\frac{0.7\text{ PFU}}{0.5\text{ mL}} \times 0.02\text{ mL} = 0.028\text{ PFU}$$

15. Hepatitis A Vaccine (HAV) *pediatric* is provided in a concentration of 360 ELu/0.5 mL (ELISA units/mL). How many ELISA units will a child receive after the prescribed immunization regimen of 360 ELu on days 0 and 30 and a single booster 12 months later?

Solution. 1080 ELu

CALCULATIONS

$$\frac{360\ \text{ELu}}{0.5\ \text{mL}} \times \frac{0.5\ \text{mL}}{\text{dose}} \times 3\text{doses} = 1080\ \text{ELu}$$

---

16. A tetanus and diphtheria toxoids (Td) adsorbed product for adult use has a concentration of 5 L*f* units tetanus toxoid and 2 L*f* units diphtheria toxoid per 0.5 mL. Complete immunization is achieved by administering two 0.5 mL doses at an interval of 4–8 weeks and a third reinforcing dose 12 months later. How many L*f* units of each toxoid would a patient receive after complete immunization?

Solution. 15 L*f* units of tetanus toxoid and 6 L*f* units of diphtheria toxoid

CALCULATIONS

$$\frac{5\ \text{L}f\text{units}}{0.5\ \text{mL}} \times \frac{0.5\ \text{mL}}{\text{dose}} \times 3\ \text{doses} = 15\ \text{L}f\text{units}$$

$$\frac{2\ \text{L}f\text{units}}{0.5\ \text{mL}} \times \frac{0.5\ \text{mL}}{\text{dose}} \times 3\ \text{doses} = 6\ \text{L}f\text{units}$$

---

17. The concentration of measles, mumps and rubella virus vaccine live is not less than 1000 measles $TCID_{50}$ (tissue culture infectious dose), not less than 20,000 mumps $TCID_{50}$ and not less than 1000 $TCID_{50}$ rubella per 0.5 mL dose. If a child received the primary dose and a booster dose after 12 months, what is the total $TCID_{50}$ of each virus received?

Solution. 2000, 40,000, and 2000 $TCID_{50}$ of measles, mumps and rubella, respectively.

CALCULATIONS

$$2\ \text{doses} \times \frac{0.5\ \text{mL}}{\text{dose}} = 1\ \text{ml}$$

$$\frac{1000\ TCID_{50}}{0.5\ \text{mL}} \times 1\ \text{mL} = 2000\ TCID_{50}\ \text{measles virus}$$

$$\frac{20\,000\ \text{TCID}_{50}}{0.5\ \text{mL}} \times 1\ \text{mL} = 40{,}000\ \text{TCID}_{50}\ \text{mumps virus}$$

$$\frac{1000\ \text{TCID}_{50}}{0.5\ \text{mL}} \times 1\ \text{mL} = 2000\ \text{TCID}_{50}\ \text{rubella virus}$$

---

18. Poliovirus vaccine inactivated (e-IVP) is reserved for vaccination of adults and immunocompromised persons and their contacts. If a patient receives the primary series that consists of two 0.5 mL doses at 1–2 months interval with a third dose 6–12 months later, how many milligrams each preservative will the patient receive, considering that 0.5% 2-phenoxyethanol and 0.02% formaldehyde are present in the dosage form?

Solution. 7.5 mg 2-phenoxyethanol and 0.3 mg formaldehyde

CALCULATIONS

$$3\ \text{doses} \times \frac{0.5\ \text{mL}}{\text{dose}} \times \frac{0.5\ \text{g}}{100\ \text{mL}} \times 1000\ \text{mg/g} = 7.5\ \text{mg}\ \text{2-phenoxyethanol}$$

$$3\ \text{doses} \times \frac{0.5\ \text{mL}}{\text{dose}} \times \frac{0.02\ \text{g}}{100\ \text{mL}} \times 1000\ \text{mg/g} = 0.3\ \text{mg}\ \text{formaldehyde}$$

REVIEW PROBLEMS

Now, do the following problems and check your answers at the end.

A. An immunobiologic product for pediatric use contains 50 L*f* units of tetanus toxoid in each 5 mL. If a pediatric patient is to receive a reinforcing dose of 5 L*f* units, how many milliliters of product should be administered?

B. If an anthrax vaccine product contains 0.02% formaldehyde as preservative and a patient receives a dose of 0.5 mL by SC route, how many micrograms of preservative will the patient receive?

C. A mumps virus vaccine live is prepared to contain 20,000 $TCID_{50}$ per 0.5 mL dose. What is the $TCID_{50}$ content of a 30 mL multidose vial?

D. How many micrograms of 23 polysaccharides are there in a package of 100 doses of pneumococcal vaccine 23-valent, based on a concentration of 25 μg/0.5 mL dose?

E. Primary immunizing series for DTP (diphtheria, tetanus, and pertussis) vaccine includes four 0.5 mL doses. If a DTP product contains 6.5 L*f* units of diphtheria toxoid, 5 L*f* units of tetanus toxoid and 4 L*f* units of pertussis vaccine per 0.5 mL, how many units of each antigen will be received by a patient at completion of series?

F. How many colony forming units (CFU) will a patient receive when 0.25 mL BCG vaccine is applied percutaneously from a BCG vaccine product that contains 5×10^8 CFU/mL?

G. Cholera vaccine has 0.5% phenol as a preservative. If a 4-year-old patient received 0.2 mL by ID injection as a booster dose, how many milligrams of phenol did the patient receive?

H. A tetanus toxoid adsorbed product contains 5 L*f* units per 0.5 mL. For children immunized during the first year of life, the primary immunizing series consists of three 0.5 mL doses 4–8 weeks apart, followed by a 4th reinforcing 0.5 mL dose 6–12 months after the third dose. How many total units are given at the end of primary series?

I. Diphtheria toxoid adsorbed for pediatric use is available as 15 L*f* units per 0.5 mL. If a primary immunizing series includes three 0.5 mL doses, how many children would be vaccinated from a 5 mL multidose vial?

J. The strength of rubella virus vaccine live is 1000 $TCID_{50}$ per 0.5 mL dose. How many $TCID_{50}$ are there in a 50-dose vial?

K. If a poliovirus vaccine contains 40 D-antigen units of poliovirus type 1 per 0.5 mL dose, how many D-antigen units would be present in a 20 mL multidose vial?

L. A rabies vaccine product contains 0.01% thimerosal. How many micrograms of preservative will be received in 1 mL intradermal pre-exposure prophylaxis dose?

Solutions to Review Problems

A. 0.5 mL

B. 100 µg

C. 1,200,000 $TCID_{50}$

D. 2500 µg

E. 26 L*f* units of diphtheria toxoid; 20 L*f* units of tetanus toxoid; 16 L*f* units of pertussis vaccine

F. 125×10^6 CFU

G. 1 mg

H. 2 mL; 20 L*f* units

I. 3 children

J. 50,000 $TCID_{50}$

K. 1600 D-antigen units

L. 100 µg of thimerosal

Calculations for Review Problems

A. $5 \text{ L}\mathit{f}\text{units} \times \dfrac{5 \text{ mL}}{50 \text{ L}\mathit{f}\text{units}} = 0.5 \text{ mL}$

B. $0.02 \text{ g}/100 \text{ mL} = 20 \text{ mg}/100 \text{ mL} = \dfrac{20\,000 \text{ µg}}{100 \text{ mL}} \times 0.5 \text{ mL} = 100 \text{ µg}$

C. $$\frac{20\,000\ \text{TCID}_{50}}{0.5\ \text{mL}} = \frac{z}{30\ \text{mL}}$$

$$z = 1{,}200{,}000\ \text{TCID}_{50}$$

D. $$\frac{25\ \mu\text{g}}{\text{dose}} \times 100\ \text{doses} = 2500\ \mu\text{g}$$

E. $$\frac{6.5\ \text{L}\mathit{f}\text{units}}{\text{dose}} \times 4\ \text{doses} = 26\ \text{L}\mathit{f}\text{units of diphtheria toxoid}$$

$$\frac{5\ \text{L}\mathit{f}\text{units}}{\text{dose}} \times 4\ \text{doses} = 20\ \text{L}\mathit{f}\text{units of tetanus toxoid}$$

$$\frac{4\ \text{L}\mathit{f}\text{units}}{\text{dose}} \times 4\ \text{doses} = 16\ \text{L}\mathit{f}\text{units of pertussis vaccine}$$

F. $$\frac{5\times10^{8}\ \text{CFU}}{\text{mL}} \times 0.25\ \text{mL} = 1.25\times10^{8} = 125\times10^{6}\ \text{CFU}$$

G. $$\frac{0.5\ \text{g}}{100\ \text{mL}} \times \frac{1000\ \text{mg}}{1\ \text{g}} \times 0.2\ \text{mL} = 1\ \text{mg}$$

H. $$4\ \text{doses} \times \frac{0.5\ \text{mL}}{\text{dose}} = 2\ \text{mL total}$$

$$\frac{5\ \text{L}\mathit{f}\text{units}}{0.5\ \text{mL}} \times 2\ \text{mL} = 20\ \text{L}\mathit{f}\text{units}$$

I. $$5\ \text{mL} \times \frac{\text{child}}{1.5\ \text{mL}} = 3.3 = 3\ \text{children}$$

J. $$\frac{1000\ \text{TCID}_{50}}{\text{dose}} \times 50\ \text{doses} = 50{,}000\ \text{TCID}_{50}$$

K. $$\frac{40\ \text{D-antigen units}}{0.5\ \text{mL}} \times 20\ \text{mL} = 1600\ \text{D-antigen units}$$

D-antigen units

L. $$\frac{0.01\ \text{g}}{100\ \text{mL}} \times \frac{1\,000\,000\ \mu\text{g}}{1\ \text{g}} \times 1\ \text{mL} = 100\ \mu\text{g of thimerosal}$$

CHAPTER 13

RADIOACTIVE DECAY

***Learning Objectives:** After completing this chapter the student should be able to:*

1. State the units for radioactive disintegration and convert from one set of units to another.
2. Write the equations for radioactive decay.
3. Calculate the half-life of an isotope from its decay constant (or the other way around).
4. Determine the amount of radioactive isotope remaining after an indicated period of time from a knowledge of the amount present initially and the half-life.

Radioactive isotopes undergo changes in nuclear structure with the emission of energy. This process, called *radioactive decay*, continues until a stable isotope is formed. The rate at which nuclear change occurs is an important characteristic of a particular isotope.

Specialized units are utilized to express radioactivity and its effect on living tissues. Dosage—that is, the amount of radioactivity—is a crucial issue for radioactive substances used for diagnostic and therapeutic purposes. Because of radioactive decay, the potency of a radioactive isotope is a function of time, so that the length of time elapsed following manufacture must be taken into consideration when calculating therapeutic dosage.

---

1. When identifying a particular isotope of an atom, the usual symbol is preceded by the atomic weight of the isotope. Thus tritium (a radioactive isotope of hydrogen) whose atomic weight is 3, would be written ^{3}H while the most common hydrogen isotope is ^{1}H. The symbol for carbon, atomic weight 14, is ^{14}C.

Write the symbol for the tin isotope with atomic weight 113.

Solution. ^{113}Sn

---

2. When the nucleus of a radioactive atom undergoes a change, the atom is said to *decay* (to produce a different atom). Another name for this event is *radioactive disintegration*. Different atoms undergo decay by different processes, so the energy produced as a result of the process depends on the isotope involved.

Pharmaceutical Calculations, Fourth Edition, By Joel L. Zatz and Maria Glaucia Teixeira
ISBN 0-471-67623-3

One way of describing the quantity of a particular radioactive material is through the number of disintegrations that take place in unit time, usually seconds. The reason is that the number of disintegrations is proportional to the amount of material present. The fundamental unit is the Becquerel (Bq), which represents one disintegration per second (dps).

$$1\ \text{Bq} = 1\ \text{dps}$$

The becquerel is usually too small a unit for practical purposes, so kilo- or megabecquerels are more commonly used.

$$10^3\ \text{Bq} = 1\ \text{kilobecquerel (kBq)}$$

$$10^6\ \text{Bq} = 1\ \text{megabecquerel (MBq)} = 10^3\ \text{kBq}$$

A. If the radioactivity of a material is 1.40×10^2 kBq, how many dps does that represent?

B. Express the radioactivity of a substance with 6.33×10^7 disintegrations per second in terms of megabecquerels.

Solutions.

A. 1.4×10^5 dps

B. 63.3 MBq

CALCULATIONS

A. $1.40 \times 10^2\ \text{kBq} \times 1000\ \text{dps/kBq} = 1.4 \times 10^5\ \text{dps}$

B. $6.33 \times 10^7\ \text{dps} \times \dfrac{1\ \text{MBq}}{10^6\ \text{dps}} = 63.3\ \text{MBq}$

---

3. A second unit, older but still widely used, is the curie (Ci), defined as 3.7×10^{10} dps. As with other units, prefixes are used to scale the unit size and bring numerical values into a convenient range. For small quantities of radiation, millicuries or microcuries are commonly used.

$$1\ \text{Ci} = 1000\ \text{mCi} = 1 \times 10^6\ \mu\text{Ci} = 3.7 \times 10^{10}\ \text{dps}$$

A. How many disintegrations per second are represented by a material whose activity is 9.26 mCi?

B. If an isotope undergoes 10^7 disintegrations per minute, how many microcuries does this represent?

C. How many kBq are equivalent to $20\,\mu$Ci of a radioactive substance?

Solutions.

A. 3.43×10^8 dps

B. $4.51\,\mu$Ci

C. 740 kBq

CALCULATIONS

A. $$9.26 \text{ mCi} \times \frac{1 \text{ Ci}}{10^3 \text{ mCi}} \times \frac{3.7 \times 10^{10} \text{ dps}}{1 \text{ Ci}} = 3.43 \times 10^8 \text{ dps}$$

B. $$\frac{10^7 \text{ dpm}}{1 \text{ min}} \times \frac{1 \text{ min}}{60 \text{ s}} = 1.67 \times 10^5 \text{ dps}$$

$$1.67 \times 10^5 \text{ dps} \times \frac{1 \text{ Ci}}{3.7 \times 10^{10} \text{ dps}} \times 10^6\ \mu\text{Ci/Ci} = 4.51\,\mu\text{Ci}$$

C. $$20\,\mu\text{Ci} \times \frac{1 \text{ Ci}}{10^6\,\mu\text{Ci}} \times 3.7 \times 10^{10} \text{ dps/Ci} \times \frac{1 \text{ kBq}}{1000 \text{dps}} = 740 \text{ kBq}$$

---

4. Disintegrations per second observed for a radioisotopes sample is proportional to the number of radioactive molecules present. The following equation applies to an individual isotope:

$$\text{rate of decay} = \lambda N$$

In this equation, N is the number of radioactive molecules at any time, t, and λ (lambda) is a rate constant that depends on the identity of the particular substance. Using the notation of differential calculus, the rate of decay can be represented as $-dN/dt$. This expression describes the change in N with time; the negative sign in front of the expression indicates that as time goes on (increases), N decreases. In the absence of the negative sign, the expression would say that the amount of radioactive substance should grow over time rather than decay.

Which of the following statements are true?

A. The number of disintegrations per unit time is proportional to the amount of a given radioactive substance present.

B. The number of molecules of a given isotope decreases as time goes on.

Solution. Both statements are true.

---

5. Calculate λ, the decay rate constant, for an isotope if 100 kBq decays instantaneously at a rate of 0.023 MBq/year.

Solution. 0.23 year^{-1}

CALCULATIONS

rate of decay = λN

$$\lambda = \frac{\text{rate of decay}}{N} = \frac{0.023\ \text{MBq/year}}{0.1\ \text{MBq}} = 0.23\ \text{year}^{-1}$$

---

6. In frame 5 we saw that radioactive decay could be described by this equation:

$$-\frac{dN}{dt} = \lambda N$$

This equation can be rearranged to

$$\frac{dN}{N} = -\lambda dt$$

and solved by integration to yield

$$N = N_0 e^{-\lambda t}$$

N_0 is defined as the initial number of radioactive atoms present; e is the base of natural logarithms, 2.718. Using this equation and a calculator or log table, it is possible to calculate the amount of radioactivity at any time if we know the original activity and the rate constant.

But before we do any calculations of that type, let us explore some properties of this equation by dividing both sides by N_0:

$$\frac{N}{N_0} = e^{-\lambda t}$$

The product of λ and t in this equation has to be dimensionless. Therefore the units of λ are reciprocal time (1/time). For example, if the unit of time used is seconds (s), the units of λ are expressed as 1/s usually written s^{-1}.

Write the units for λ when time is in years.

Solution. $\frac{1}{\text{year}}$ or year^{-1}

---

7. We saw that the decay equation could be written

$$\frac{N}{N_0} = e^{-\lambda t}$$

Notice that the left-hand side of this equation is a ratio of numerical values. This ratio represents the fraction of radioactive atoms remaining after decay has proceeded for a given period of time, t. The right-hand sign of the equation contains a constant, λ in addition to t. If we choose any arbitrary time period, say 1 h, the value of $e^{-\lambda t}$ will always be the same and so will N/N_0, regardless of the amount of radioactivity originally present. In other words, during any uniform time period, a constant fraction (or percentage) of the amount present at the beginning of the time period will decay.

As an illustration, let us assume that we have 100 units of a radioactive substance with a value of λ of $0.1\,h^{-1}$. The value of N at various times is shown in Table 13.1, along with the ratio of N values for each 10-h period of time.

TABLE 13.1 Calculated values of *N* over time: N_0 = 100 units; $\lambda = 0.1\,h^{-1}$

Time (h)	*N* (arbitrary units)	Ratio*
0	100.00	
10	36.79	0.368
20	13.53	0.368
30	4.98	0.368
40	1.83	0.368

* Number of molecules at end of 10-h period/Number of molecules at beginning of 10-h period

As you can see from the table, the amount of decay from 0 to 10h is much greater than the decay from 10 to 20h, and so on. As time goes on, the change in N becomes smaller and smaller. However, as is shown in the third column, the fraction remaining after any 10-h period is the same regardless of the value of N at the beginning of the period.

Rank the following starting quantities of ^{226}Ra in terms of the *percent* of original activity remaining after 14 weeks of storage. (Or would they all be equal?)

1. 1 kBQ
2. 2 kBQ
3. 3 kBQ

Solution. Equal percentages of activity remain.

---

8. A radioactive substance loses 4.80% of its activity in 3.0 months. If the initial activity were 6.19 kBQ, predict how many kBQ will remain at the end of 6.0 months.

Solution. 5.61 kBQ

CALCULATIONS

After 3 months, 6.19 kBQ × 0.952 = 5.89 kBQ remain. 5.89 kBQ is the starting amount for the second 3-month period. At the end of that time, the amount remaining should be

5.89 kBQ × 0.952 = 5.61 kBQ

---

9. ^{90}Sr loses half its activity in 25 days. In how many days will the activity of 16 MBq decay to 1 MBq?

Solution. 100 days

CALCULATION

In 25 days, activity will be 8 MBq; in 50 days, 4 MBq; in 75 days, 2 MBq; in 100 days, 1 MBq.

---

10. The decay equation can be expressed in several formats. Here are three ways of writing this equation:

$$N = N_0 e^{-\lambda t} \tag{1}$$

$$\ln N = \ln N_0 - \lambda t \tag{2}$$

$$\log N = \log N_0 - [\lambda/2.3]t \tag{3}$$

In Eq. 2, "ln" means *natural log* (to the base e). In Eq. (3), "log" is the logarithm to base 10.

Using a calculator or log table, it is possible to predict the activity of a radioisotope after any length of time from a knowledge of the initial activity and the value of λ.

The *half-life* is a useful parameter that comes from these equations. The half-life is defined the length of time required for the activity to drop to one-half its initial value. At this time, $t_{0.5}$, $N = 1/2\ N_0$. If we make these substitutions in one of the equations, the following result is obtained:

$$t_{0.5} = 0.693/\lambda$$

From this equation we see that the half-life depends only on the value of λ. It is independent of the initial activity.

Calculate the half-life of an isotope whose λ value is 0.023 day^{-1}.

Solution. 30.1 days

CALCULATION

$$t_{0.5} = \frac{0.693}{0.023\ \text{day}^{-1}} = 30.1\ \text{days}$$

11. What is the value of the decay rate constant, λ , for an isotope whose half-life is 16 h?

Solution. $0.043\,h^{-1}$

CALCULATION

$t_{0.5} = 0.693/\lambda$

$\lambda = 0.693/t_{0.5}$

$\lambda = 0.693/16\,h = 0.043\,h^{-1}$

12. If the amount of a radioactive isotope drops to half its initial value in $t_{0.5}$ h, then in another $t_{0.5}$ hours the amount will drop in half again, or in other words, to one-fourth the initial value. During the next $t_{0.5}$ hours, the amount will drop to half once more, so that one-eighth the original amount remains. It follows that the isotope will persist for an exceedingly long period of time.

The half-life of a radioisotope is 60 days. How many days will it take for 64 μCi of the isotope to drop in activity to 8 μCi?

Solution. 180 days

CALCULATION

Three half-lives are required; 3×60 days = 180 days.

13. Try these problems that review the concept and calculations involving half-life:

A. The decay rate constant for ^{230}Th is 9.1×10^{-6} year^{-1}. Calculate its half-life.

B. The half-life of ^{222}Rn (radon) is 3.82 days. Calculate its decay rate constant in h^{-1}.

C. The half-life of ^{210}Bi is 5 days. What percentage of a quantity of this material will remain after 10 days?

D. The half-life of ^{210}Pb is 22 years. How long would it take for 48 μCi of ^{210}Pb to decay to 1.5 μCi?

Solutions.

A. 7.6×10^4 years

B. 7.56×10^{-3} h

C. 25% (50% remains after 1 half-life; 25% after 2 half-lives)

D. 110 years

CALCULATIONS

A. $t_{0.5} = \dfrac{0.693}{\lambda} = \dfrac{0.693}{9.1 \times 10^{-6}\,\text{year}^{-1}} = 7.6 \times 10^4\,\text{years}$

B. $\lambda = \dfrac{0.693}{3.82\,\text{days}} \times \dfrac{1\,\text{day}}{24\,\text{h}} = 7.56 \times 10^{-3}\,\text{h}$

C. $0.5 \times 0.5 = 0.25 = 25\%$

D. 5 half-lives would bring the quantity down to 1/32 of the original value, from 48 μCi to 1.5 μCi.

5×22 years = 110 years

---

14. When considering time periods equal in magnitude to the half-life, it is easy to calculate the amount of radioactivity remaining. For other lengths of time, the equations describing radioactive decay can be employed:

$$N = N_0 e^{-\lambda t} \tag{1}$$

$$\ln N = \ln N_0 - \lambda t \tag{2}$$

$$\log N = \log N_0 - [\lambda/2.3]t \tag{3}$$

As an example, a sample of a radioisotope whose decay rate constant is 0.02 year^{-1} has an activity of 0.5 MBq. Let us calculate the activity remaining after the sample stands for 10 years.

Using Eq. (1), $N = 0.5\,\text{MBq} \times e^{-[(0.02)(10)]} = 0.41\,\text{MBq}$.

Using Eq. (2), $\ln N = \ln(0.5\,\text{MBq}) - (0.02 \times 10)$.

$\ln N = -0.693 - 0.2 = -0.893$

$N = 0.41\,\text{MBq}$

Using Eq. (3), $\log N = \log (0.5\,\text{MBq}) - \frac{0.02 \times 10}{2.3}$.

$\log N = -0.301 - 0.087 = -0.388$

$N = 0.41\,\text{MBq}$

Use any appropriate equation to predict the activity of a sample of ^{195}Au, whose half-life is 183 days, after storage for 300 days. Initially, 206 μCi are present.

Solution. 66.1 μCi

CALCULATIONS

$\lambda = 0.693/183\ \text{days} = 0.00379\ \text{day}^{-1}$

$N = 206\,\mu\text{Ci} \times e^{-[(0.00379)(300)]} = 66.1\,\mu\text{Ci}$

---

15. A radioisotope whose half-life is 82 h is stored for 6 days. If the initial activity is 72 kBq, predict the activity at the end of the storage period.

Solution. 21.3 kBq

CALCULATIONS

$\frac{0.693}{82\ \text{h}} = 0.00845\ \text{h}^{-1}$

$N = 72\ \text{kBq} \times e^{-[(0.00845)(144)]} = 21.3\ \text{kBq}$

REVIEW PROBLEMS

Do all of the following problems before checking your answers:

A. Write the symbol for cobalt isotope with an atomic weight of 57.

B. 625 μCi of an isotope are present in a container. How many MBq does this quantity represent?

C. How many millicuries are represented by 950 kBq?

D. Calculate the instantaneous rate of decay, in μCi/day, for 120 mCi of an isotope whose decay rate constant is 0.006 day^{-1}.

E. What is the decay rate constant of an isotope with a half-life of 14.3 days?

F. Calculate the half-life, in seconds, of an isotope for which $\lambda = 0.0725\,h^{-1}$.

G. The half-life of radioactive sodium (atomic weight 22) is 2.60 years. How long will it take for 10 mCi of this isotope to decay to 2.5 mCi?

H. Referring to the isotope in question G, how many microcuries will be left after 25 years?

I. The half-life of an isotope is 6.0 days. To what quantity will 280 kBq decay in 36 days?

J. The half-life of an isotope is 122 days. If 1 mCi of this substance is stored for 1 year, how many μCi will remain?

Solutions to Review Problems

A. ^{57}Co

B. 23.1 MBq

C. 0.0256 mCi

D. 0.72 mCi/day

E. 0.0485 day^{-1}

F. 34,400 s

G. 5.20 years

H. 12.6 μCi

I. 4.38 kBq

J. 126 μCi

Calculations for Review Problems

A. As above.

B. $625\ \mu\text{Ci} \times \dfrac{1\text{Ci}}{10^6\ \mu\text{Ci}} \times 3.7 \times 10^{10}\ \text{dps/Ci} = 23.1 \times 10^6\ \text{dps} = 23.1\ \text{MBq}$

C. $950\ \text{kBq} = 950 \times 10^3\ \text{dps}$

$$950 \times 10^3\ \text{dps} \times \frac{1\ \text{Ci}}{3.7 \times 10^{10}\ \text{dps}} \times \frac{10^3\ \text{mCi}}{1\ \text{Ci}} = 0.0256\ \text{mCi}$$

D. $\text{Rate} = \lambda N = 0.006\ \text{day}^{-1} \times 120\,\text{mCi} = 0.72\,\text{mCi/day}$

E. $\lambda = \dfrac{0.693}{14.3\ \text{days}} = 0.0485\ \text{day}^{-1}$

F. $t_{0.5} = \dfrac{0.693}{0.725\ \text{h}^{-1}} = 9.56\ \text{h}$

G. Since the final value is 1/4 the initial value, 2 half-lives must pass.
2.60 years × 2 = 5.20 years

H. $N = N_0 e^{-\lambda t}$

$$\lambda = \frac{0.693}{2.60\ \text{years}} = 0.267\ \text{year}^{-1}$$

$$N = 10\ \text{mCi} \times e^{-[(0.26)(25)]} = 0.0126\ \text{mCi} = 12.6\ \mu\text{Ci}$$

I. 36 days represents 6 half-lives. During that time, the amount will drop to 1/64 the initial quantity.

$$\frac{280\ \text{kBq}}{64} = 4.38\ \text{kBq}$$

J. $\lambda = \dfrac{0.693}{122\ \text{days}} = 0.00568$

$$N = 1000\ \mu\text{Ci} \times e^{-[(0.00568)(365)]} = 126\ \mu\text{Ci}$$

CHAPTER 14

RATE LAWS AND SHELF LIFE

LEARNING OBJECTIVES: After completing this chapter the student should be able to:

1. Define what is meant by the order of a rate reaction and state the order of the two rate laws most useful in describing drug degradation.
2. Calculate the half-life of a drug given the reaction order, a rate constant, and for zero-order processes, an initial concentration.
3. Calculate the time required for 10% loss of a drug given the reaction order, a rate constant, and for zero-order processes, an initial concentration.
4. Calculate the concentration of drug remaining at a given time from a knowledge of the reaction order, initial concentration, and either half-life or rate constant.
5. Given a rate constant at a specific temperature and the activation energy, calculate the rate constant at another temperature.
6. Given the shelf life based on drug content at one temperature, calculate the shelf life at another temperature.

Over the years, a number of tests have been developed to confirm the identity and purity of the many drug substances that are used to treat disease. Even if a drug is 100% pure when it is used in the preparation of a product, chemical changes typically occur over a period of time. The substances produced by these processes differ from the starting material in potency and toxicity. The rate at which drug loss occurs is of great concern; at some point in time, the residual drug concentration will drop to values below allowable limits and the product will no longer meet acceptable standards for drug content. *Shelf life* is the name given to the time period during which drug content and other characteristics remain within acceptable limits.

RATE LAWS

1. The process of drug breakdown (also known as *drug degradation*) occurs by a variety of mechanisms that depend on the chemistry of the drug and its environment. In this and succeeding frames, we explore two simple rate laws with wide applicability.

The names of these laws come from a fundamental rate relation, which takes the form

$$\text{rate} \propto C^P$$

This expression states that the rate is proportional to the drug concentration (C) raised to some power, P. The value of P identifies the *order* of the equation. Its origin in the case

Pharmaceutical Calculations, Fourth Edition, By Joel L. Zatz and Maria Glaucia Teixeira
ISBN 0-471-67623-3

of relatively simple reactions is the number of molecules of a given reactant that participate in the reaction. In a practical sense, the assignment of order is empirical, based on the way a system behaves rather than fundamental mechanisms.

Thus a first-order rate expression has the form

$$\text{rate} \propto C^1 \text{ (which is the same as rate} \propto C)$$

while a second order equation would be

$$\text{rate} \propto C^2$$

Write the form for a zero-order rate expression.

Solution. Rate $\propto C^0$

---

2. The two rate laws of interest to us are zero- and first-order. Let us rewrite the first-order expression appearing in the previous frame.

$$\text{rate} \propto C^1$$
$$\text{rate} \propto k_1 C^1 \propto k_1 C$$

The subscript 1 is used with k to emphasize that this equation is first-order. Write the equation (including a proportionality constant, k_0) to describe the rate of a zero-order process.

Solution. Rate $= k_0 C^0 = k_0$ (any number raised to the zero power has a value of one)

---

3. We can make these rate expressions more informative by substituting a mathematical expression for the word "rate." Using the notation of differential calculus, the rate can be represented as $-\mathrm{d}C/\mathrm{d}t$. This expression represents the change in C with time; the negative sign in front of the expression indicates that as time goes on (increases), C decreases. In the absence of the negative sign, the expression would say that drug concentration grows over time rather than dropping as degradation occurs.

Let us start with the zero-order rate equation:

$$-\frac{\mathrm{d}C}{\mathrm{d}t} = k_0$$

Integration leads to the following solution:

$$C = C_0 - k_0 t$$

C_0 is the initial concentration. C and C_0 are in the same units, which might, for example, be g/100 mL or mg/mL or mol/L, while t can be any valid time units (i.e., seconds, minutes, hours, days, etc.). The product of k_0 and t must have the same units as C_0, to allow one term in the equation to be subtracted from the other. k_0 therefore has the units of concentration divided by time. If C_0 is in mg/mL and t is in hours, the units of k_0 are mg/mL/h.

Another way of writing this is $mg\,mL^{-1}\,h^{-1}$.
What units would be used for k_0 if drug concentration is expressed in moles/liter and time in days?

Solution. $mol\,L^{-1}\,day^{-1}$

---

4. A degradation process that follows the zero-order rate law is called a zero-order (or apparent zero-order) process. We could also say that it follows zero-order kinetics. (The word *kinetics* refers to things in motion; in chemistry, the same word describes rates of chemical change.)

As t changes from zero (the time at which the solution is prepared), drug concentration drops from C_0; the equation allows us to calculate the value of C corresponding to any value of t. As you can see from the equation, the concentration of drug remaining in solution decreases over time in a linear fashion. This is the defining characteristic of a zero-order relationship.

If a plot of drug concentration vs. time is a straight line, what is the order of the process?

(a) zero

(b) first

(c) second

Solution. Zero is correct. Linear degradation rate is synonymous with a zero-order rate law.

---

5. A drug degrades by a zero-order process with a rate constant of $0.040\,mg\,mL^{-1}\,year^{-1}$ at 25 °C. If a 0.25% w/v solution is prepared and stored at 25 °C, what concentration will remain after 2 years?

Solution. 2.42 mg/mL (or 0.242%)

CALCULATIONS

$$0.25\%\ \text{w/v} = \frac{250\ \text{mg}}{100\ \text{mL}} = 2.5\ \text{mg/mL}$$

$$C = C_0 - k_0 t = 2.5\,\text{mg/mL} - (0.040\,\text{mg mL}^{-1}\,\text{year}^{-1} \times 2\ \text{years}) = 2.42\,\text{mg/mL}$$

---

6. Many degradation processes appear to follow first-order kinetics. Recall that a first-order rate law means that degradation is proportional to drug concentration.

$$-\frac{dC}{dt} = k_1 C$$

This expression should look familiar to you. It has the same form as the equation for radioactive decay. All of the mathematical properties that we studied in Chapter 13 apply here.

The three variations of the integrated equation for first-order degradation are:

$$C = C_0 e^{-k_1 t} \tag{1}$$

$$\ln C = \ln C_0 - k_1 t \tag{2}$$

$$\log C = \log C_0 - \frac{1}{2.3} t \tag{3}$$

Recall that "ln" refers to the natural logarithm (to the base *e*), while "log" refers to the logarithm to the base 10. The other symbols were defined earlier. The units of k_1 are reciprocal time s^{-1}, h^{-1}, $days^{-1}$, and so on. Dividing both sides in Eq. (1) by C_0 leads to

$$\frac{C}{C_0} = e^{-k_1 t}$$

The left-hand side of the equation is now a ratio of the concentration values. This ratio represents the fraction of the original concentration remaining after degradation has proceeded for a given period of time, *t*. Over any time period chosen, the ratio C/C_0 is constant regardless of what the starting concentration, C_0, was. Over that same time period, the fraction of unchanged drug remaining will always be the same.

We have two flasks containing different concentrations of the same drug. The concentration in flask A is 0.11 molar; the concentration in flask B is 0.35 molar. If the degradation process is first-order and both flasks are stored at the same temperature for 6 months, in which will the highest percentage of initial drug remain?

Flask A

Flask B

They will be the same

Solution. They will each have the same percentage of initial concentration because the degradation process is first-order.

---

7. An aqueous solution containing 1.5 mg/mL of a drug loses 6.0% of its activity in 3 months. If degradation occurs by first-order kinetics, predict how much of the drug will remain at the end of 6 months. (*Hint*: What is the concentration after 3 months? Knowing the percent retained, repeat the calculation for another 3 months.)

Solution. 1.33 mg/mL

CALCULATION

The concentration remaining at the end of 3 months is 1.5 mg/mL × 0.94 = 1.41 mg/mL. 1.41 mg/mL is the starting concentration for the second 3-month period. At the end of that time, the concentration remaining should be 1.41 mg/mL × 0.94 = 1.33 mg/mL.

---

8. The concentration of a drug in solution drops by 20% in 25 days when stored at 25 °C. What concentration will remain if a 0.16 molar solution of the drug is stored under the same conditions for 75 days?

Solution. 0.082 molar

CALCULATION

The drug retains 80% of its initial concentration in each 25-day period. The concentration remaining after three such periods is 0.16 molar × 0.8 × 0.8 × 0.8 = 0.082 molar.

---

9. We have seen that the fraction of drug remaining after a given time is constant when first-order kinetics apply. One way of using this property is to ask how much time has to pass in order to bring the concentration down to a specified fraction of the original value.

One standard way of expressing this idea is in terms of a half-life or half-time for degradation. This is the time required to bring the concentration down to 50% of its initial value. The symbol used is $t_{1/2}$, $t_{0.5}$, or $t_{50\%}$.

The second fraction of great interest, $t_{0.9}$, relates to drug stability in manufactured products. This is the time required for the concentration to drop to 90% of its original value. For most products, 90% of original content is the point at which the end of the shelf life is reached. Thus, $t_{0.9}$ is equivalent to the shelf life.

We can obtain both of these values by making the appropriate substitution in any of the rate equations. $C = 0.5\ C_0$ when $t = t_{0.5}$ and $C = 0.9\ C_0$ when $t = t_{0.9}$. When we do so, the following results are obtained:

$$t_{0.5} = \frac{0.693}{k_1}$$

$$t_{0.9} = \frac{0.105}{k_1}$$

Notice that neither equation contains a concentration term; $t_{0.5}$ and $t_{0.9}$ for a first-order process are independent of the starting concentration.

A solution containing 0.636 g/100 mL of drug is stored in a closed container at 20 °C. It degrades by a first-order process and the rate constant at this temperature is 0.00215 day^{-1}.

Calculate the time required for the concentration to drop to

(a) 90%

(b) 50% of its initial value.

Solutions.

(a) 48.8 days

(b) 322 days

CALCULATIONS

$$t_{0.9} = \frac{0.105}{k_1} = \frac{0.105}{0.00215\ \text{day}^{-1}} = 48.8\ \text{days}$$

$$t_{0.5} = \frac{0.693}{k_1} = \frac{0.693}{0.00215\ \text{day}^{-1}} = 322\ \text{days}$$

---

10. A drug in a clear solution is present at a concentration of 1.0 mg/mL after manufacture. If the half-life of the drug is 800 days, is the product's shelf life at least six months?

Solution. No. $t_{0.9}$ = 121 days.

CALCULATIONS

$$k_1 = \frac{0.693}{t_{0.5}} = \frac{0.693}{800\ \text{days}} = 0.000866\ \text{day}^{-1}$$

$$t_{0.9} = \frac{0.105}{k_1} = \frac{0.105}{0.000866\ \text{day}^{-1}} = 121\ \text{days}$$

---

11. In the previous frames we have been exploring some outstanding properties of zero- and first-order rate laws. Let us now step back and review these characteristics. Table 14.1 summarizes the important mathematical relationships.

TABLE 14.1 Summary of equations describing zero- and first-order rate processes

Property	Zero-order	First-order
Rate equation	$-\frac{dC}{dt} = k_0$	$-\frac{dC}{dt} = k_1 C$
Integrated equation(s)	$C = C_0 - k_0 t$	$C = C_0 e^{-k_1 t}$ $\ln C = \ln C_0 - k_1 t$ $\log C = \log C_0 - \frac{k_1}{2.3} t$
$t_{0.5}$	$\frac{C_0}{2k_0}$	$t_{0.5} = \frac{0.693}{k_1}$
$t_{0.9}$	$\frac{C_0}{10k_0}$	$t_{0.9} = \frac{0.105}{k_1}$

Review the relationships in Table 14.1 and make sure that you are familiar with them. The only new items in the table are equations for $t_{0.5}$ and $t_{0.9}$ for a zero-order process. Note that both quantities are dependent on the initial concentration, in contrast to the corresponding first-order equations.

A drug degrades in a zero-order process with a k_0 value of $1\,\mu g\,mL^{-1}\,day^{-1}$. A solution containing 2.68 mg/mL is prepared. Calculate the half-life and the time at which 90% of the initial concentration will remain.

Solutions. $t_{0.5} = 1340$ days; $= t_{0.9} = 268$ days

CALCULATIONS

$$t_{0.5} = \frac{C_0}{2k_0} = \frac{2680\,\mu g/mL}{(2)(1\,\mu g\,mL^{-1}\,day^{-1})} = 1340 \text{ days}$$

$$t_{0.9} = \frac{C_0}{10k_0} = \frac{2680\,\mu g/mL}{(10)(1\,\mu g\,mL^{-1}\,day^{-1})} = 268 \text{ days}$$

---

12. Some numerical examples will further illustrate the patterns involved. Assume that we have a drug solution whose concentration is 1.50 mg/mL. Let us use the equations to calculate the concentrations remaining at various times for both rate processes. Table 14.2 shows values calculated from the appropriate equations while, Figure 14.1 contains plots of the same data.

Note that, as expected, the zero-order plot is linear with time. At some point the concentration will drop to zero. The first-order plot is not linear. Its slope decreases over time, so that the curve drops more and more slowly toward zero as time goes on.

TABLE 14.2 Concentrations over time for two drugs whose degradation follows different kinetic patterns

	Concentration (mg/mL)	
Time (days)	Zero-order	First-order
0	1.50	1.50
10	1.30	1.23
20	1.10	1.01
30	0.90	0.823
40	0.70	0.674
50	0.50	0.552
60	0.30	0.452
70	0.10	0.370

$C_0 = 1.5$ mg/mL; $k_0 = 0.02$ mg/mL/day; $k_1 = 0.02$ day^{-1}

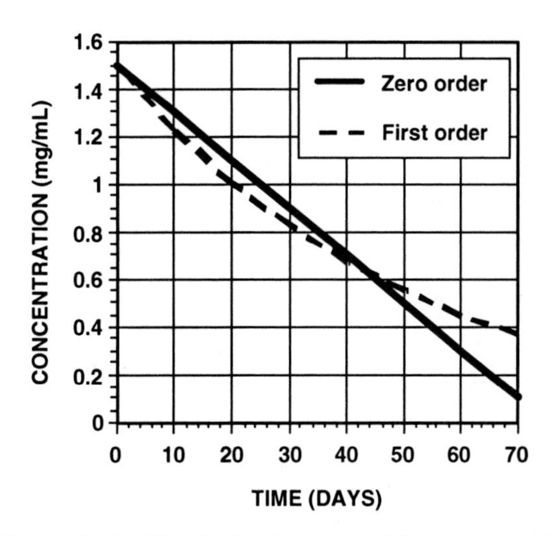

Figure 14.1 Kinetic data for zero and first-order reaction.

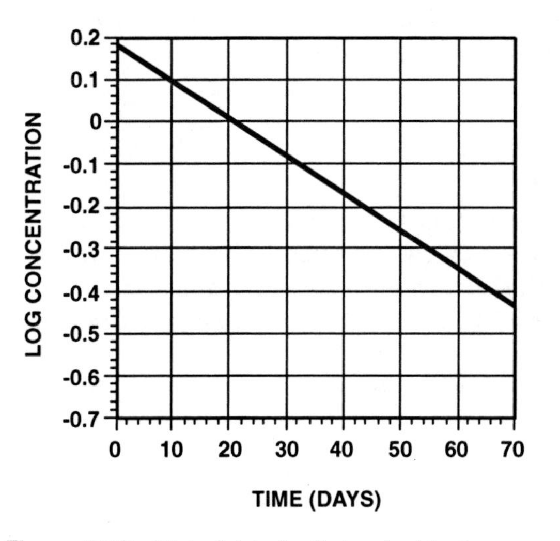

Figure 14.2 Plot of data for first-order kinetic process.

However, the same data for a first-order process can be re-plotted by taking the logarithm (base 10) of concentration and plotting it against time, as suggested by the third form of the integrated first-order equation:

$$\log C = \log C_0 - \frac{k_1}{2.3}t$$

According to this equation, when $\log C$ is plotted against t, the result will be a straight line with a slope of $k_1/2.3$ and an intercept on the vertical axis of $\log C_0$. Figure 14.2, based on the data in Table 14.2, illustrates these points.

---

13. Chlordiazepoxide degrades in acid solution at 80 °C by a first-order process. The half-life is 7 h. A 1% solution of this drug is prepared; calculate the concentration remaining (in mg/mL) after 1 h of storage at the same temperature.

Solution. 9.06 mg/mL

CALCULATIONS

1% = 1 g/100 mL = 10 mg/mL

$$k_1 = \frac{0.693}{t_{0.5}} = \frac{0.693}{7\ \text{h}} = 0.099\ \text{h}^{-1}$$

$$C = C_0 e^{-k_1 t} = (10\ \text{mg/mL})\ e^{-[(0.099)(1)]} = 9.06\ \text{mg/mL}$$

---

14. The rate constant, and therefore the reaction rate, are a function of several factors, notably pH and temperature. The *Arrhenius equation* shows the relationship between rate constant and temperature:

$$k = A\exp\left|-\frac{E_a}{RT}\right|$$

The notation "exp" in this equation indicates that the mathematical factor e is raised to the power that follows (in brackets). E_a is an activation energy, a parameter related to the reaction mechanism. It is assumed to be constant. R is the gas constant and T the absolute temperature. (Use of upper case distinguishes temperature from the symbol for "time.")

If the rate constant (k) at one temperature, T_1, and E_a are known, the rate constant at a second temperature can be determined from the following form of the Arrhenius equation:

$$\log k_2 = \log k + \frac{E_a(T_2 - T)}{2.3RTT_2}$$

where k_2 is the rate constant at the new temperature, T_2.

Let's use this equation in an example. There are a couple of things to review first. The absolute temperature, in °K, can be found by adding 273.16 to the temperature in °C. The units of E_a are typically calories/mol or joules/mol. These units may be scaled, for example to kcal/mol. The value and units of R depend on the units for E_a. If E_a is in cal/mol, R is 1.987 cal/deg/mol (cal deg^{-1} mol^{-1}); if E_a is in J/mol, R is 8.314 J/deg/mol (J deg^{-1} mol^{-1}).

The first-order degradation rate constant for a drug at 35 °C is 0.092 h^{-1} (at pH 5). Calculate the rate constant at the same pH at 45 °C if the activation energy is 18.0 kcal/mol.

Solution. 0.232 h^{-1}

CALCULATIONS

$$\log k_2 = \log k + \frac{E_a(T_2 - T)}{2.3RTT_2}$$

$$= \log(0.092\ \text{h}^{-1}) + \frac{(18000\ \text{cal/mol})(10\ \text{deg})}{(2.3)(1.987\ \text{cal deg}^{-1}\ \text{mol}^{-1})(308.16\ \text{deg})(318.16\ \text{deg})}$$

$$\log k_2 = -1.036 + 0.402$$

$$k_2 = 0.232\ \text{h}^{-1}$$

---

15. Calculate $t_{0.5}$ at 5 °C for the same compound at the same pH. (5 °C approximates refrigerator temperature.)

Solution. 180 h

CALCULATIONS

$$\log k_2 = \log k + \frac{E_a(T_2 - T)}{2.3RTT_2} = \log(0.092) + \frac{(18,000)(-30)}{(2.3)(1.987)(308.16)(278.16)}$$

$$= -1.036 - 1.378$$

$$k_2 = 0.00385\ \text{h}^{-1}$$

$$t_{0.5} = \frac{0.693}{k_1} = \frac{0.693}{0.00385\ \text{h}^{-1}} = 180\ \text{h}$$

16. The rate constant for first-order degradation of a drug at 60 °C is $1.67 \times 10^{-3}\,h^{-1}$. If the activation energy is 20.1 kcal/mol, calculate the amount remaining after storage for 200 h at 25 °C. The initial concentration was 5.0 mg/mL.

Solution. 4.95 mg/mL

CALCULATIONS

$$\log k_2 = \log k + \frac{E_a(T_2 - T)}{2.3\,RTT_2} = \log(0.00167) + \frac{(20{,}100)(-35)}{(2.3)(1.987)(333.16)(298.16)}$$

$$= -2.777 - 1.550$$

$$k_2 = 471 \times 10^{-5}\,h^{-1}$$

This is the first-order rate constant at 25 °C.

$$C = C_0 e^{-kt} = (5\,mg/mL)\,e^{-[(0.0000471)(200)]} = 4.95\,mg/mL$$

SHELF LIFE

17. The shelf life determines the expiration date that is affixed to drug products. One way to estimate shelf life is to store samples of a test batch under expected environmental conditions and determine the drug concentration periodically. When its value drops below the set specification (typically, 90% of the labeled value), the end of product life has been reached. For relatively stable formulations, this process can take several years.

We can get a good idea of the shelf life at room temperature by conducting stability tests at elevated temperatures. If exposure to higher temperatures does not introduce a new reaction, the only effect of temperature will be to increase reaction rate. Because reactions proceed much more rapidly at high temperature, the rate constant can be determined in a relatively short period of time. By determining the rate constant at several temperatures, the activation energy can be calculated. This allows extrapolation of a rate constant from an elevated temperature to the corresponding value at room temperature (or even refrigerator temperature) so that the shelf life under normal storage conditions can be projected.

As an example, the shelf life ($t_{0.9}$) for a drug at 50 °C is 93 days. The activation energy for the first-order degradation process is 80,000 J/mol. To estimate the shelf life at 25 °C, first calculate the rate constant at 50 °C, then use the Arrhenius equation to obtain the rate constant at 25 °C, and then calculate $t_{0.9}$ at that temperature.

$$k = \frac{0.105}{t_{0.9}} = \frac{0.105}{93\text{ days}} = 0.00113\text{ day}^{-1}$$

$$\log k_2 = \log k + \frac{E_a(T_2 - T)}{2.3\ RTT_2} = \log(0.00113) + \frac{(80{,}000)(-25)}{(2.3)(8.314)(323.16)(298.16)}$$

$$= -2.947 - 1.085 = -4.032$$

$$k_2 = 9.27 \times 10^{-5}\ \text{day}^{-1}$$

$$t_{0.9} = \frac{0.105}{9.27 \times 10^{-5}\text{day}^{-1}} = 1130 \text{ days (about 3 years)}$$

The number of steps can be reduced by substituting $0.105/t_{0.9}$ for k in the Arrhenius equation. The result is

$$\log t_{0.9(2)} = \log t_{0.9(1)} - \frac{E_a(T_2 - T)}{2.3\ RTT_2}$$

where $t_{0.9(2)}$ is the shelf life at the new temperature and $t_{0.9(1)}$ is the shelf life at the original temperature. You can verify that this equation is independent of the reaction order. In other words, the same equation applies to first- and zero-order reactions.

Let's try the same calculation again, using this equation:

$$\log t_{0.9(2)} = \log t_{0.9(1)} - \frac{Ea(T_2 - T)}{2.3\ RTT_2}$$

$$\log t_{0.9(2)} = \log(94) - \frac{(80{,}000)(-25)}{(2.3)(8.314)(323.16)(298.16)} = 1.973 - (-1.085) = 3.058$$

$$t_{0.9(2)} = 1140 \text{ days}$$

Now, calculate the shelf life at 5 °C.

Solution. 11,700 days

CALCULATIONS

$$\log t_{0.9(2)} = \log t_{0.9(1)} - \frac{Ea(T_2 - T)}{2.3RTT_2}$$

$$\log t_{0.9(2)} = \log(94) - \frac{(80{,}000)(-45)}{(2.3)(8.314)(323.16)(278.16)} = 1.973 - (-2.094) = 4.067$$

$$t_{0.9(2)} = 11{,}700 \text{ days}$$

18. The value of $t_{0.9}$ for a drug is 35 minutes at 35 °C. What is the shelf life at 5 °C? The activation energy is 15.5 kcal/mol.

Solution. 538 minutes

CALCULATIONS

$$\log t_{0.9(2)} = \log t_{0.9(1)} - \frac{Ea(T_2 - T)}{2.3RTT_2}$$

$$\log t_{0.9(2)} = \log(35) - \frac{(15,500)(-30)}{(2.3)(1.987)(308.16)(278.16)} = 1.544 - (-1.187) = 2.731$$

$$t_{0.9(2)} = 538\,\text{m}$$

REVIEW PROBLEMS

Here are some questions that review the material in this chapter. Try doing all of them before checking your answers.

A. The rate of a degradation of a drug is proportional to the concentration of that drug. What is the order of the reaction?

B. Write the rate equation for a degradation process that proceeds at a constant rate.

C. The rate constant for zero-order degradation of a drug is $0.860\,\mu g\,mL^{-1}\,h^{-1}$ at 20 °C. A 3.3% solution is prepared and stored at that temperature for 1 year. Predict the drug concentration that will be found at the end of that time.

D. What is its shelf life of the drug in question C?

E. The rate constant for digoxin degradation at 25 °C is $2.08 \times 10^{-4}\,s^{-1}$ at a certain acid pH. The process is first-order. Calculate the half-life and $t_{0.9}$ (both in minutes) of a solution at the same pH and temperature.

F. The half-life of a drug in aqueous solution at 25 °C is 820 days. If degradation is a first-order process and a solution containing 0.750 g/L is prepared, calculate the concentration remaining after 1 year at 25 °C.

G. Referring to the drug in question F, calculate the concentration remaining if the same solution were stored for 1 year at 10 °C. The activation energy for this drug is 17.4 kcal/mol.

H. A solution contains 0.25% w/v drug when freshly prepared. If the half-life (first-order degradation process) of the drug in this solution at room temperature is 12 days, predict the drug concentration, in mg/mL, after storage for 24 days.

I. Meperidine has a shelf life of 220 days at a pH of 3.7 and a temperature of 25 °C. Assuming an activation energy of 19 kcal/mol, predict the shelf life at 5 °C at the same pH.

J. The degradation rate constant for a drug is $4.1 \times 10^{-4}\,s^{-1}$ at 70 °C. Assuming degradation to be apparent first-order, calculate the shelf life at this temperature.

K. Assuming the activation energy of the drug in question J is 22 kcal/mol, predict the shelf life at 25 °C, in hours.

L. The half-life of a drug in solution that degrades by an apparent first-order process is 23 days. How long will it take for 10% of the drug to degrade at that temperature?

M. Referring to the drug in question L, what percentage of the initial drug present will remain after storage for 4 h?

N. The shelf life of a drug in a solution whose pH matches that of gastric fluid is 8.8 h at 5 °C. An oral dose is 2 µg in 5 mL. Assuming an activation energy of 14 kcal/mol, predict the number of minutes required for unchanged drug to drop to 1.8 µg/5 mL after exposure of a dose to gastric fluid at 37 °C.

Solutions to Review Problems

A. First-order

B. rate $\propto C^0$ or rate $= k_0$

C. 25,500 µg/mL = 25.5 mg/mL = 2.55%

D. 3840 h

E. $t_{0.5} = 55.5$ min; $t_{0.9} = 8.4$ min

F. 0.551 g/L

G. 0.703 g/L

H. 0.625 mg/mL

I. 2210 days

J. 256 s

K. 9.31 h

L. 3.49 days

M. 88.6%

N. 39 min

Calculations for Review Problems

A. As above.

B. As above.

C. For consistency of units, concentration is expressed in μg/mL and time in hours.

3.3% = 3300 mg/100 mL = 33 mg/mL = 33,000 μg/mL

1 year = 365 days/year × 24 h/day = 8760 h

$C = C_0 - k_0 t = 33{,}000\,\mu\text{g/mL} - (0.860\,\mu\text{g mL}^{-1}\text{h}^{-1} \times 8760\,\text{h}) = 25{,}500\,\mu\text{g/mL}$

D. $$t_{0.9} = \frac{C_0}{10\ \text{k}_0} = \frac{33{,}000\ \mu\text{g/mL}}{(10)(0.860\ \mu\text{g/mL h}^{-1})} = 3840\ \text{h}$$

E. $$t_{0.5} = \frac{0.693}{k_1} = \frac{0.693}{2.08\times10^{-4}\ \text{s}^{-1}} = 3330\ \text{s} = 55.5\ \text{min}$$

$$t_{0.9} = \frac{0.105}{k_1} = \frac{0.105}{2.08\times10^{-4}\ \text{s}^{-1}} = 505\ \text{s} = 8.4\ \text{m}$$

F. $$k_1 = \frac{0.693}{t_{0.5}} = \frac{0.693}{820\ \text{days}} = 8.45\times10^{-4}\ \text{day}^{-1}$$

$C = C_0 e^{-k_1 t} = (0.75\,\text{g/L})\ e^{-[(0.000845)(365)]} = 0.551\,\text{g/L}$

G. $$\log k_2 = \log k + \frac{Ea(T_2 - T)}{2.3\ RTT_2}$$

$$= \log(8.45\times10^{-4}\text{day}^{-1}) + \frac{(17{,}400\ \text{cal/mol})(-15\ \text{deg})}{(2.3)(1.987\ \text{cal deg}^{-1}\ \text{mol}^{-1})(298.16\ \text{deg})(283.16\ \text{deg})}$$

$= -3.073 - 0.676$

$k_2 = 0.000178\ \text{day}^{-1}$

$C = (0.75\,\text{g/L})\ e^{-[(0.000178)(365)]} = 0.703\,\text{g/L}$

H. 24 days is 2 half-lives; the concentration remaining will be

2.5 mg/mL × 0.5 × 0.5 = 0.625 mg/mL

I. $$\log t_{0.9(2)} = \log t_{0.9(1)} - \frac{Ea(T_2 - T)}{2.3\ RTT_2}$$

$$\log_{0.9(2)} = \log(220) - \frac{(19{,}000)(-20)}{(2.3)(1.987)(298.16)(278.16)} = 2.342 - (-1.003) = 3.345$$

$\log\ t_{0.9(2)} = 2210$ days

J. $$t_{0.9} = \frac{0.105}{k_1} = \frac{0.105}{0.00041\ \text{s}^{-1}} = 256\ \text{s}$$

K. $$\log t_{0.9(2)} = \log(256) - \frac{(22{,}000)(-45)}{(2.3)(1.987)(343.16)(298.16)} = 2.408 - (-2.117) = 4.525$$

$t_{0.9(2)} = 33{,}500\,\text{s} = 9.31\,\text{h}$

L. $$k_1 = \frac{0.693}{t_{0.5}} = \frac{0.693}{23\text{ days}} = 0.0301\text{ day}^{-1}$$

$$t_{0.9} = \frac{0.105}{k_1} = \frac{0.105}{0.0301\text{ day}^{-1}} = 3.49\text{ days}$$

M. $C = C_0 e^{-k_1 t} = (100\%)\, e^{-[(0.0301)(4)]} = 88.6\%$

N. The time required for the concentration to drop from 2 µg/5 mL to 1.8 µg/5 mL defines $t_{0.9}$.

$$\log t_{0.9(2)} = \log(8.8) - \frac{(14,000)(32)}{(2.3)(1.987)(278.16)(310.16)} = 0.944 - (1.136) = -0.192$$

$$\log t_{0.9(2)} = 0.643\text{ h} = 39\text{ min}$$

APPENDIXES

These appendixes contain

- Support information and references for pharmaceutical and clinical calculations.
- Other calculations of practical importance.

Pharmaceutical Calculations, Fourth Edition, By Joel L. Zatz and Maria Glaucia Teixeira
ISBN 0-471-67623-3

APPENDIX 1: SYSTEMS OF MEASUREMENT

Metric System

- Official system of measurement used in pharmacy and medicine
- Basic units: liter (L) for volume, gram (g) for weight, meter (m) for length
- Prefixes applied to the basic unit specify a power of 10, which is multiplied times that unit (see Table A1.1).

TABLE A1.1 Metric measures and multiples of basic units

Prefix	Power-of-10	Symbol
Giga	10^9	G
Mega	10^6	M
Kilo	10^3	k
Hecto	10^2	h
Deka	10	da
Basic unit	*1*	—
Deci	10^{-1}	d
Centi	10^{-2}	c
Milli	10^{-3}	m
Micro	10^{-6}	μ
Nano	10^{-9}	n
Pico	10^{-12}	p

Common Systems: Avoirdupois and Apothecary

- *Avoirdupois*: system used in commerce in USA; pharmaceuticals are purchased by the ounce and pound.
- *Apothecary*: traditional system of measurement used in pharmacy; obsolete at present.

Important measures and symbols in use in the health field:

Avoirdupois

1 pound (lb) = 16 ounces (oz)

Apothecary

1 gallon (gal) = 4 quarts (qt)

1 quart (qt) = 2 pints (pt)

1 pint (pt) = 16 fluidounces (f ℥)

Practical Intersystem Conversions and Household Approximate Measures Used in Pharmacy and Medical Practice

Length

1 meter (m) = 39.37 inches (in or ″)

1 inch (in) = 2.54 centimeters (cm)

1 foot (ft or') = 12 inches (in or ″)

Volume

1 gallon (gal) = 3785 milliliters (mL)

1 pint (pt) = 473 milliliters (mL)

1 pint (pt) = 2 cups

1 cup = 8 fluidounces (f ℥) = 240 milliliters (mL)

1 tumblerful = 8 fluidounces (f ℥) = 240 milliliters (mL)

1 fluidounce (f ℥) = 29.6 milliliters (mL)

1 fluidounce (f ℥) = 2 tablespoonfuls (T or tbsp)

1 tablespoonful (T or tbsp) = $\frac{1}{2}$ fluidounce (f ℥) = 15 milliliters (mL)

1 tablespoonful (T or tbsp) = 3 teaspoonfuls (tsp)

1 teaspoonful (tsp) = 5 milliliters (mL)

Weight

1 pound (lb) = 16 ounces (oz) (avoirdupois)

1 pound (lb) (avoirdupois) = 454 grams (g)

1 kilogram (kg) = 2.2 pounds (lb) (avoirdupois)

1 ounce (oz) (avoirdupois) = 28.4 grams (g)

Approximate Equivalents: Apothecary to Metric System

- Not to be used in prescription compounding or for the conversion of specific quantities in converting pharmaceutical formulas
- Used when a prefabricated dosage form is prescribed in one system of units but is available in strengths given in a different system (metric and apothecary)

1 qt ≈ 1000 mL

1 pt ≈ 500 mL

8 f ℥ ≈ 250 mL

1 f ℥ ≈ 30 mL

1 grain (gr) ≈ 60 mg

$1\frac{1}{2}$ gr ≈ 100 mg

1 ounce (℥) ≈ 30 g

For a complete table of metric doses with approximate apothecary equivalents, consult *Remington: The Science and Practice of Pharmacy*.

APPENDIX 2: CALIBRATION OF MEDICINAL DROPPER

- Used when doses of liquids for oral administration are too small to be measured with a teaspoon
- By knowing the volume of each drop, the number of drops of a liquid necessary to make up the desired dose can be calculated
- The volume of a drop depends upon the size of the dropper orifice and the viscosity and surface tension of the liquid. Personal factors (variation in hand pressure, speed of dropping, and angle at which the dropper is held) can also contribute to variability in dropper volume
- A medicinal dropper is calibrated for use only with the specific liquid formulation that is dispensed, and will provide a statement of equivalence between milliliters and "drops" as a measure of volume
- Calibration: slowly drop the formulation into a small graduated cylinder (10 mL) and count the number of drops that exactly equal a definite volume (≥2.0 mL)
- The United States Pharmacopeia and National Formulary describes medicine droppers in section <1101>

For example, a solution for administration to a child was prepared. The dose of the solution was 1.2 mL. (A dose measuring 1/4 teaspoonful would be a very inaccurate way of administering the solution.) A dropper calibrated with the solution delivered 2.0 mL in 44 drops. We may therefore write

$$2.0\,\text{mL} = 44\ \text{drops}$$

or

$$1.0\,\text{mL} = 22\ \text{drops}$$

The number of drops equivalent to 1.2 mL is

$$1.2\text{mL} \times \frac{22\ \text{drops}}{1.0\text{mL}} = 26\ \text{drops}$$

A dropper was found to deliver 3.0 mL of saline solution in 75 drops. If 14 drops were put in a test tube, how many cubic centimeters of saline solution did the test tube contain?

Solution. $0.56\,\text{cm}^3$

CALCULATIONS

$3.0\,\text{mL} = 75\ \text{drops}$

$$14\ \text{drops} \times \frac{3.0\text{mL}}{75\ \text{drops}} = 0.56\text{mL} = 0.56\text{cm}^3$$

APPENDIX 3: MEDICAL ABBREVIATIONS

The summarized list of abbreviations presented in Table A3.1 has been compiled to assist health professionals in reading and transcribing prescriptions, medical records, and related communications. However, it is important to highlight that confusion caused by abbreviations has made a requirement by the Joint Commission on Accreditation of Healthcare Organizations to all hospitals to prepare and publish a comprehensive list of abbreviations approved for use in the institution. Thus, some abbreviations that are standard in one specific institution may not be in another.

TABLE A3.1 Most commonly used medical abbreviations

Abbreviation	Meaning	Abbreviation	Meaning
		Related to dosage forms	
amp	ampul	oint	ointment
cap, caps	capsule	pulv, pulvis	powder
crm, crem	cream	sol or soln	solution
chart	powder in a paper	supp	suppository
elix	elixir	susp	suspension
gtt	drops	syr	syrup
inj	injection	tab, tabs	tablet
liq	liquid	troch	troche, lozenge
MDI	metered dose inhaler	ung, unguentum	ointment
neb	nebulizer		
		Related to dosage regimen	
a.c.	before meals	pm, PM	afternoon, evening
ad lib	at pleasure, freely	prn, p.r.n.	as needed, when required
am, AM	in the morning	Pt.	patient
ATC	around the clock	q	every
bid, b.i.d.	twice a day	q12h	every 12 hours
c., c̄	with	qAM	every morning
cc	cubic centimeter (= milliliter)	qd	every day, once daily
d	day	qh	every hour
e.o.d	every other day	qid	four times a day
gl. aq.	glass of water	qod	every other day
h, hr.	hour	qPM	every evening
hs	at bedtime	rep.	repeat
i.c.	between meals	s., ś	without
KVO	keep vein open	s.o.s.	if there is need
min	minute	stat	immediately
noct	night	t.a.t	until all taken
om	every morning	tbsp, TBSP, T	tablespoonful
on	every night	t.i.d., tid, TID	three times a day
p.c.	after meals	t.i.w., TIW	three times a week
postop	postoperatively	tsp, TSP	teaspoonful

TABLE A3.1 Most commonly used medical abbreviations *(Continued)*

Abbreviation	Meaning	Abbreviation	Meaning
		Related to route of administration	
a.d., AD	right ear	ol, OL, os	left eye
a.s., a.l., AL	left ear	OU, O_2	each eye, both eyes
au	each ear	po, p.o.	by mouth, orally, swallowed
ID	intradermal	PR	suppository (when written after drug name)
IM	intramuscular	PV	vaginally (when written after drug name)
IV	intravenous	rect	rectally
IVP	intravenous push	SC, SQ, subQ	subcutaneously
IVB	intravenous bolus	SL	sublingually (under the tongue)
IVPB	intravenous piggyback	s/s, S/S	swish and spit
NPO	nothing by mouth	top	topically (on the skin)
od, OD	right eye		
		In directions to the pharmacist	
Aa, āā	of each	max.	maximum
ad	up to	min.	minimum, minutes
aq	water	No., #	number
aq. pur.	purified water	NTE	not to exceed
disp	dispense	pt	pint
d.t.d., DTD	give of such doses	q.s.	up to
div	divide	qt	quart
e.m.p.	in the manner prescribed	S.A., secundum artem	according to art, use your skill and judgment
F, ft.	make		
gal	gallon		
i	one	v	five
ii	two	WA	while awake
iv	four	x, X	ten, times
M	mix	x2	two times
m. ft.	mix to make		
		Miscellaneous	
Δ	change	HCTZ	hydrochlorothiazide
Alb.	albumin	HR	heart rate
APAP	acetaminophen (N-acetyl-para-aminophenol)	HT	hypertension
		MI	myocardial infarction
BP	blood pressure	MVI	multivitamins infusion
BS	blood sugar	NS	normal saline; 0.9% sodium chloride
BSA	body surface area	½ NS	half strength normal saline; 0.45% NaCl sol.
BUN	blood urea nitrogen		
CBC	complete blood count	PN	parenteral Nutrition
CHF	congestive heart failure	RL	ringer's Lactated
Cr	creatinine	temp	temperature
D5W	dextrose 5% in water	TNA; 3-in-1	total Nutrient admixture
Hb	hemoglobin	TPN	total parenteral nutrition

APPENDIX 4: NAMES, ATOMIC WEIGHTS AND SYMBOLS OF SOME CHEMICAL ELEMENTS WITH PHARMACEUTICAL IMPORTANCE

Name	Symbol	Atomic Weight used in Pharmaceutical Calculations
Aluminum	Al	27
Antimony	Sb	122
Argon	Ar	40
Arsenic	As	75
Barium	Ba	137
Bismuth	Bi	209
Boron	B	11
Bromine	Br	80
Cadmium	Cd	112
Calcium	Ca	40
Carbon	C	12
Cesium	Cs	133
Chlorine	Cl	35.5
Chromium	Cr	52
Cobalt	Co	59
Copper	Cu	64
Fluorine	F	19
Gold	Au	197
Helium	He	4
Hydrogen	H	1
Indium	In	115
Iodine	I	127
Iron	Fe	56
Lead	Pb	207
Lithium	Li	7
Magnesium	Mg	24
Manganese	Mn	55
Mercury	Hg	201
Nickel	Ni	59
Nitrogen	N	14
Oxygen	O	16
Phosphorus	P	31
Platinum	Pt	195
Plutonium	Pu	244
Potassium	K	39
Radium	Ra	226
Radon	Rn	222
Selenium	Se	79
Silicon	Si	28
Silver	Ag	108
Sodium	Na	23
Strontium	Sr	88
Sulfur	S	32
Technetium	Tc	98
Tin	Sn	119
Titanium	Ti	48
Tungsten	W	184
Uranium	U	238
Vanadium	V	51
Zinc	Zn	65

APPENDIX 5: PARENTERAL AND ENTERAL NUTRITION

Screening and Assessing Nutritional Needs

Increased interest in specialized nutrition support (Enteral and Parenteral) and the recognition that these therapies are both life-saving and life-threatening have led to creation of interdisciplinary organizations (e.g. ASPEN, American Society for Parenteral and Enteral Nutrition) dedicated to assure the best nutrition care to patients. As a consequence, professions involved with nutrition support (medicine, nursing, pharmacy, and nutrition professionals) have developed certification processes. Board-certified nutrition support pharmacists are now active members of nutrition support teams throughout USA (check www.bpsweb.org).

Nutrition screening uses parameters connected to nutrition-related diseases to identify patients at risk of obesity or malnutrition.

Management of excessive weight is fundamental to prevent greater risk of chronic degenerative diseases, disease complications, and mortality. Obesity, a heterogeneous disease, involves numerous factors including genetic, environmental, and psychological. However, despite its etiology, obesity results when the amount of energy expended over time is smaller than the energy intake. Body weight, a practical tool for nutrition calculations and medical management of obese patients, may be modified by several disease states that cause increases in fluid retention. Therefore, body mass index (BMI) is now considered as the most useful clinical standard for assessing excessive body weight because it utilizes a ratio of weight to height. Body mass index is defined as the weight of the patient in kg divided by the height in meters squared (kg/m^2).

Anthropometrics (body measurements) are frequently used to evaluate body size and proportions. Height and weight are easy to obtain and inexpensive. Several classifications of obesity have been proposed based on different health studies. Table A5.1 shows the classification recommended by both the NIH (National Institutes of Health) and WHO (World Health Organization).

In the severely malnourished patient, the increased morbidity and mortality will more than justify the risks, costs, and time involved with correcting the nutritional deficit. Evaluation of clinical and nutritional status of a patient should be performed before implementation of any effective feeding program. Critical assessment regarding nutritional requirements and adequacy of nutrition support depends on:

(a) Patient's data related to disease state

(b) Dietary and drug intake history

(c) Laboratory results

TABLE A5.1 BMI and weight classification

BMI (kg/m^2)	Weight category
<18.5	Underweight
18.5–24.9	Healthy weight
25–29.9	Overweight (mild obesity)
30–39.9	Obesity
≥40	Severe obesity

General nutrition practices relevant to most types of patients are well known but some adult (obese and disease-specific) and pediatric patients frequently require expertise in management of associated comorbidities, such as illnesses (type-2 diabetes mellitus, dyslipidemia, cardiovascular and respiratory defects), other medications and/or physical disability.

Sound nutritional status is vital for total health care. Unfortunately, many acutely and chronically ill patients, who are malnourished, often do not want to eat or are unable to eat spontaneously. Proper nutrition is crucial in the therapy of multiple trauma, cancer, and neurologically impaired patients. Patients who are unable to meet their nutritional requirements through *oral* intake will need supplementation by other routes. The *enteral route* (bypass of upper gastrointestinal tract and placement of a feeding tube into the stomach or small bowel) is generally preferred, whenever possible, because it is associated with improved outcome, fewer complications, and lower costs. The *parenteral route* (bypass of all gastrointestinal tract using intravenous feeding) is reserved for patients whose gastrointestinal tract is not functioning, since it involves more risks and higher costs.

Parenteral Nutrition

The term TPN or PN, which stands for total parenteral nutrition or simply parenteral nutrition, is frequently used to describe a mixture that is expected to supply all needed fluid, electrolytes, calories, essential fatty acids, and vitamins—in short, everything needed to sustain life.

Parenteral nutrition (PN) may be administered through one of the large, central veins (e.g. subclavian vein), a smaller peripheral vein, or both. The patient sometimes is totally reliant on intravenous administration for nutrition.

TPN design and calculations involve *four* primary components:

(i) Carbohydrates: primary energy substrate

(ii) Proteins: source of amino acids (nitrogen) for new protein synthesis

(iii) Fat: source of essential fatty acids and non-protein calories

(iv) Electrolytes: essential nutrients that perform many critical physiologic functions, including homeostasis and acid-base balance.

Other components of PN admixtures are:

Vitamins: standardized solutions provide most hydrosoluble and liposoluble vitamins as a fixed dose, while other vitamins (K, A, C, B_{12}, folic acid) are added when a deficiency state exists or when needs are increased due to a disease or clinical condition.

Trace minerals: essential to normal metabolism and growth, are used in a daily recommended dose from standard solutions or, as supplemental replacement doses for increased requirements (e.g. burn injury) or increased losses (large-volume fistula output), or restricted (e.g. copper retention with biliary disease).

Sterile water: added to adjust the total volume of fluid needed for a 24-h fluid intake or to create a specific concentration when concentrated dextrose (e.g. 70%), amino acids (e.g. 15%) and lipid (e.g. 30%) are used as the base products.

These are typically solutions or dispersions in which each unit of volume contains known amounts of essential nutrient components. Table A5.2 contains sources of various components of parenteral nutrition formulations.

TABLE A5.2 Solutions used in parenteral nutrition

Sources	Solutions	Typical Concentration	Dosage/Content
Non-nitrogen sources	Dextrose solutions	Range from 5 to 70%	3.4 kcal/g
	Lipid emulsions	10%, 20%, 30%	1.1, 2.0, and 3.0 kcal/mL respect.
Nitrogen source	Amino acid solutions	Range from 3 to 15%	4 kcal/g
Electrolytes	Multiple-electrolyte formulation	Standard	Standard
	Single-entity electrolyte solution	Variable	Dependent on compatibility
Vitamins	Standard solutions	Adult or Pediatric	Fixed dose (10 mL/day)
	Individual vitamin solutions	Variable	Supplemental dose
Trace minerals	Standard solutions	Adult or pediatric	Fixed dose (1 mL/day)
	Single-entity solutions	Variable	Supplemental dose
Other additives	Insulin	U-100, U-500	1 unit/100 cc TPN (max 100 units/liter)
	Vit.K, Vit.B_{12}, Iron, H_2 blockers, other dugs	Variable	As needed, dependent on compatibility

As the pharmacy profession gets more involved in clinical practice, pharmacists are frequently called on to be part of nutrition support teams. This multidisciplinary collaborative team ensures that every patient receives optimal nutrition care. When parenteral nutrition is required for a given patient, the pharmacist is responsible for all the dosing and compounding calculations. It is the pharmacist's responsibility to remember that, TPN is a very complex admixture containing several individual chemicals, all bringing their own physicochemical characteristics that will influence their behavior when combined in a single container. Compatibility and stability of all components of a PN admixture are prerequisites to the safe use of PNs.

While the calculations of nutrients provided by different solutions to supply a needed requirement of electrolyte, carbohydrate, fat, and other nutrients are mainly on pharmacists' shoulders, the design and establishment of a PN formula depends on several nutrition professionals. These professionals will discuss and decide, as a team, the best regimen for each specific patient.

General guidelines for calculations related to establishing a TPN regimen are:

1. The target total daily energy, fluid, and protein requirements (or range) should be determined. These are done through the calculation of TDE, TDF, and daily protein requirement considering the patient's stress level.
2. From available solutions, the volumes of protein, dextrose, and fat that the patient will receive are calculated.
3. The doses of electrolytes ordered for the patient based on laboratory analysis and the volumes of each electrolyte solution that will be needed for the patient are calculated.

We will analyze the steps involved in defining a parenteral nutrition regimen and will discuss in detail one example.

Let us consider the following practical case. MP, an 80-year-old female, wt. 52 kg, ht. 5′2″, has been vomiting for 5 days each time she ingested food. After admission to High Country Hospital, screening and assessment lead to the diagnosis of pancreatitis. The consulting surgeon suggested placing her on parenteral nutrition.

What would be the suggested TPN regimen for this patient?

Calculate the IBW to compare to the patient's ABW: use the smallest

IBW (female) = 45 kg + (2.3 kg × additional in.>5 ft)

IBW = 45 + (2.3 × 2) = 49.6 = 50 kg

ABW = 52 kg

Calculate the *total daily energy*:

TDE = 50 kg × 35 kcal/kg/day (moderately stressed) = 1750 kcal ~ 1800 kcal

Calculate the *total daily fluid*:

TDF = 1500 mL (BW = 20 kg) + 20 mL/kg (additional kg >20 kg)

TDF = 1500 + (20 × 30) = 2100 mL

Calculate the *daily protein* (amino acids) requirement:

Sepsis and stress: 1.2–1.5 g/kg (pancreatitis = stress)

Daily protein = 52 kg × 1.2 = 62.4 = 62 g/d

Amino acid solution 6.2% 1000 mL/day would provide 62 g/day

As a general rule, *lipid* emulsion (Intralipid or Liposyn, 10% or 20%) is administered to prevent essential fatty acid deficiency.

We will suggest 200 mL of 20% fat emulsion three times a week.

Calculate calories from fat, amino acids, and dextrose/day:

Calories from fat = 200 mL × 2 kcal/mL = 400 kcal

Calories from amino acid (protein) and dextrose = 1800 − 400 = 1400 kcal

Calories from protein = 62 g × 4 kcal/g = 248 = 250 kcal

Calories from dextrose = 1400 − 250 = 1150 kcal

Grams of dextrose needed to provide 1150 kcal/d = 1150 kcal × 1 g/3.4 kcal = 338 = 340 g/day

Dextrose 34% 1000 mL/day would provide dextrose 340 g/day

The starting regimen would be:

Base solution:

Dextrose 34% 500 mL + amino acid 6.2% 500 mL = 1 L

2 L of base solution would provide 340 g of dextrose and 62 g of amino acids.

Patient's required fluid = 2100 mL/d

Rate of infusion of TPN base solution would be = 2100 mL/24 h = 87.5 = 88 mL/hr

In general, the base solution is started with a low flow rate (e.g., 40–50 mL/hr) and is gradually increased until reaching the goal rate (88 mL/hr).

Fat emulsion 20% 200 mL will be infused at 20 mL/hr (lipids should be infused over at least 8 hr each time). Lipid injection is also administered as a continuous infusion (over 24 h), which causes less suppression on reticuloendothelial system (RES) with optimal metabolic lipid utilization.

Electrolytes/vitamins/trace elements (in 1 L of base solution): added based on patient's baseline electrolytes and the recommended electrolyte contents per day or per 1 L of TPN.

Multivitamins infusion (MVI) 10 mL/day

Trace elements 1 mL/day

In some situations, the patient cannot tolerate high concentrations of glucose. The original concentration of dextrose proposed for the base solution in the case above (34%) could be reduced to 20%, for example. A higher flow rate would then be used to achieve similar calories per day but total daily fluid would be larger than the calculated 2100 ml.

The total calories intake/day for this patient with the modified regimen would then be calculated as follows:

Caloric intake provided by 1 L of base solution would be:

Dextrose 20% 500 mL = 20 g/100 mL × 500 mL × 3.4 kcal/g = 340 kcal

Amino acids 6.2% 500 mL = 6.2 g/100 mL × 500 mL × 4 kcal/g = 124 kcal

Total calories = 340 + 124 = 464 kcal

Flow rate = 100 mL/hr

Total volume/day = 100 mL/hr × 24 hr = 2400 mL or 2.4 L

2.4 L would provide = 464 kcal/L × 2.4 L = 1113.6 = 1100 kcal, plus 200 mL of lipid emulsion 20% will give a total of 1100 + 400 = 1500 kcal/d (on the day that lipid is given)

On the day that the patient does not receive lipid emulsion, the rate of infusion of TPN base solution could be advanced to 110–120 mL/h, which would provide ~ 1225–1336 kcal/day with consequent increased total daily fluid intake (2640 mL and 2880 mL)

TNA (Total Nutrient Admixture)

Total nutrient admixture (TNA) or 3-in-1 nutrition solution includes fat emulsion in the PN container. The presence of lipid in a TNA alters the physicochemical character of the PN admixture, making it inherently less stable than a PN solution. However, TNA is ideal for stable patients, has a possible protective effect on hepatic microsomal enzyme function, and brings the advantage of a continuous infusion of lipids.

Tapering Off of Parenteral Nutrition

When a patient on PN recovers GI function, *tapering off* of parenteral nutrition must be predicted to prevent hypoglycemia. Steps for tapering off a TPN:

1. Gradually reduce the rate of infusion over 18–24 hr before discontinuation.
2. May reduce the rate 20–40 mL every 4–6 hr until the final rate reaches approximately 40 mL/hr.

3. May rapidly taper off by decreasing the rate to 40 mL/h, run this rate for 2–4 hr then discontinue.

Example:
BJ is on TPN at 80 mL/hr. The physician orders to have it discontinued but the last liter of TPN base solution is still needed. How should it be discontinued?

Run at 80 mL/hr for 4 hr

Volume = 320 mL

Reduce rate to 60 mL/hr for 6 hr

Volume = 360 mL

Reduce rate to 40 mL/hr for 8 hr

Volume = 320 mL

Discontinue

320 + 360 + 320 = 1000 mL

Enteral Nutrition

Patients who are unable to meet their nutritional needs through oral intake, but have a functional GI tract, may need supplementation or complete nutrition through the enteral feeding route. The length of time feeding will define the type of feeding tube and the multiple access routes. Enteral feeding may be short-term (<3 weeks) or long-term (>3 weeks). Methods of tube insertion include blind passage (e.g. nasogastric tube), operative or laparoscopic (e.g. gastrostomy, jejunostomy tube), and radiological (e.g. nasoenteric tube) placement.

Guidelines and Calculations for Enteral Feeding

The basic guidelines and calculations are similar to those used for parenteral nutrition. Adequate individualized regimen has to be defined. The total daily calorie, total daily protein, and total daily fluid requirements need to be calculated using the same methods described previously for parenteral nutrition. Electrolytes, multivitamins, and other necessary elements also need to be considered. There are several commercial products (formulas) available with known characteristics and composition to facilitate optimal formula selection.

Enteral Formulations

The choice depends on the degree of digestive function. Two major types of formulas are available:

1. Polymeric, for patients with significant degree of digestive function.
2. Elemental (or oligomeric), for patients with little or no digestive function.

Several other *disease-specific formulas* are available. As examples, Pulmocare, a high fat and low carbohydrate formula, is used to lower CO_2 production and retention, and Hepatic-Aid, a formula containing an increased branched chain amino acid-to-aromatic amino acid ratio (AAA-to-BCAA ratio), is intended to improve outcomes in patients with encephalopathy.

In general formula characteristics are:

- Protein content: 11–15% of total calories/can (250 mL)
- Caloric density: 1–2 kcal/mL
- Osmolality: Polymeric: 300–550 mOsmol/kg of water
  Elemental: 550–700 mOsmol/kg of water

Delivery Methods

Gravity controlled or pump-assisted methods are recommended depending on where the patient is located (home, ambulatory, or hospital). Delivery may be *continuous* or without interruption (the patient is fed through 24 hr), *intermittent*, when nutrition is carried out periodically (for example, a bag with formula is administered every 6 hr), as a *bolus*, which means the patient is fed through a single dose, or *cyclic*, which occurs periodically, depending on the course of the symptoms experienced. Independently of the method used, assessment of GI tolerance and gastric residuals should be performed to evaluate gastric emptying of enteral feedings.

APPENDIX 6: BIOLOGICS FOR IMMUNIZATION

TABLE A6.1 Some examples of vaccines and toxoids and main characteristics

Biologics for Immunization	Product Concentration	Route	Microorganism	Viability	Antigenic Form	Disease Transmission
			Bacterial vaccines and toxoids			
BCG (Bacillus Calmette Guerin) vaccine	1–8 × 10^8 CFU (50 mg) per mL	Percutaneous	Mycobacterium bovis	Live, attenuated	Whole bacterium	Coughing, singing, sneezing from infected persons
Cholera vaccine	8 units (4 billion vibrios) of each type per mL	ID (≥5 yrs) SC or IM	Vibrio colerae	Inactivated	Whole bacterium	Contaminated water or food
DTP (Diphtheria, Tetanus, and Pertussis) vaccine	6.5–12.5 L*f* units Diphtheria toxoid 5 L*f* units tetanus toxoid 4 units pertussis vaccine per 0.5 mL product	IM	Corynebacterium diphtheriae Clostridium tetani Bordetella pertussis	Inactivated	Toxoid mixture + whole-cell pertussis bacteria	D = contact with carrier T = spores introduced into the body P = airborne droplets from infected persons
Tetanus toxoid, adsorbed (TT)	5–10 L*f* units/0.5 mL	IM or jet injection	Clostridium tetani	Inactivated	Toxoid	Spores introduced into body
Haemophilus influenzae type B conjugate vaccine	10–25 mcg polysaccharide per 0.5 mL	IM	Haemophilus influenzae	Inactivated	Capsular polysaccharide fragments	Droplets from nose and throat during infectious period
Plague vaccine	1.8–2.2 × 10^9 bacteria/mL	IM or jet injection	Yersinia pestis	Inactivated	Whole bacterium	Infected rodent and their fleas
Typhoid vaccine	8 units/mL	SC ID	Salmonella typhi	Inactivated	Whole bacterium	Contaminated water or food

TABLE A6.1 *(Continued)*

Biologics for Immunization	Product Concentration	Route	Microorganism	Viability	Antigenic Form	Disease Transmission
			Viral vaccines			
Hepatitis A vaccine (HAV)	1440 ELu/mL	IM	Single-stranded RNA Human Enterovirus 72	Inactivated	Lysed whole viruses	Person-to-person by fecal–oral route
Hepatitis B vaccine (HBV)	5, 10 and 40 mcg/mL	IM ID	Double-stranded DNA virus, family Hepadnaviridae	Inactivated	Purified antigen	Blood, saliva, semen, vaginal fluids
Influenza virus Trivalent A & B vaccine	3 hemagglutinin antigens, each 15 mcg/0.5 mL	IM or jet injection	Single-stranded RNA Influenzavirus genus	Inactivated	Split- or whole-virus	Airborne or droplet spread
MMR (Measles, Mumps, and Rubella) vaccine live	NLT 1000 measles $TCID_{50}$ NLT 20,000 mumps $TCID_{50}$ NLT 1000 rubella $TCID_{50}$ per 0.5 mL	SC	Measles virus, g. Mor-billivirus Mumps virus, g. Para-myxovirus Rubella virus, g. Rubivirus	Live, attenuated	Whole viruses	Airborne or droplet spread
Poliovirus trivalent (IPV, e-IPV) vaccine, subcutaneous	40, 8, and 32 D-Antigen units per 0.5 mL of types 1, 2, and 3, respectively	SC	Single-stranded RNA virus, g. Enterovirus	Inactivated	Whole viruses	Direct contact and fecal–oral route
Poliovirus trivalent (OPV) vaccine, oral	800,000, 100,000 and 500,000 viral particles per 0.5 mL of types 1, 2, and 3, respectively	Oral	Single-stranded RNA virus, g. Enterovirus	Live, attenuated	Whole viruses	Direct contact and fecal–oral route
Rabies vaccine	IM = NLT 2.5 IU/mL ID = 0.25 IU/0.1 mL	IM ID	Single-stranded RNA virus, g. Lyssavirus	Inactivated	Whole viruses	Saliva of rabid animal
Smallpox (vaccinia) vaccine	10^8 pock-forming units/mL	Percutaneous	Double-stranded DNA vaccinia virus, g. Orthopoxvirus	Live, attenuated	Whole virus	Respiratory discharges and skin lesions
Yellow fever vaccine	NLT 5.04 plaque-forming units per 0.5 mL	SC	Single-stranded RNA yellow-fever virus, family Plaviviridae	Live, attenuated	Whole virus	Aedes aegypti mosquito

APPENDIX 7: TEMPERATURE CONVERSION

Although the centigrade or Celsius scale has been used by scientists for many years and is now routinely employed in many countries as the standard means of describing temperature, the Fahrenheit scale is still in common use in the United States. Any temperature scale is arbitrary and the difference in temperature represented by "1 degree" varies from one system to another. Freezing and boiling temperatures for water are compared in Table A7.1.

TABLE A7.1 Major points on two temperature scales

Characteristic	Fahrenheit	Celsius
Bolling point of water	212°	100°
Freezing point of water	32°	0°
Difference	180°	100°

From the table we see that a difference of 180° on the Fahrenheit scale is the same as a difference of 100° on the Celsius scale. The formula for converting temperature in degrees Celsius (C) to Fahrenheit (F) is

$$°F = \frac{9}{5}°C + 32$$

By algebraic manipulation, this equation can be rearranged to

$$°C = \frac{5}{9}(°F - 32)$$

Normal body temperature is considered to be 98.6 °F (although many individuals have a body temperature that is above or below this value). To convert to °C:

$$°C = \frac{5}{9}(98.6 - 32) = 37\ °C$$

Ordinary room temperature is about 20°C. On the Fahrenheit scale, this would be

$$°F = \frac{9}{5}(20) + 32 = 68\ °F$$

Try these practice problems for yourself.

A. Convert 23 °F to Celsius.

B. Convert 177 °F to Celsius.

C. Convert –20 °C to Fahrenheit.

D. Convert 60 °C to Fahrenheit.

Solutions.

A. –5 °C

B. 80.6 °C

C. –4 °F

D. 140 °F

APPENDIX 8: PROOF STRENGTH

The standard measure of alcohol (C_2H_5OH) for purposes of taxation, is the *proof gallon*, which represents 1 gal of 50% alcohol. This strength is referred to as *proof spirit*, so that a solution containing less than 50% alcohol is "below proof," while one containing more than 50% alcohol is "above proof." The *proof strength* of alcoholic solutions is exactly double the percentage strength, v/v. Thus, proof spirit, which refers to 50% alcohol, is 100 proof, 95% alcohol is 190 proof, and 34% alcohol is 68 proof. The federal tax levied on alcohol is determined by the actual volume of C_2H_5OH contained. Thus the tax on 1 pt of 40% alcohol is equal to that on 2 pt of 20% alcohol. To compute the tax on an alcoholic solution (for payment or refund), it is necessary to dilute or concentrate the solution to 100 proof and then measure the number of gallons. Fortunately, this process takes place on paper and is not actually carried out. Our calculation involves determination of the volume of proof spirit containing the same quantity of C_2H_5OH as the solution in question.

To find the number of proof gallons, multiply the proof of the alcohol solution by the volume in gallons and divide by 100. For example, if a pharmacist has 5 gal of 30% alcohol, the number of proof gallons is

$$\frac{5\text{ gal} \times 60}{100} = 3\text{ proof gal}$$

The federal tax depends on the number of proof gallons. To find the federal tax (at $13.50 per proof gallon) on 2 qt of 85% alcohol, it is necessary first to determine the number of proof gallons.

$$\frac{1/2\text{ gal} \times 170}{100} = 0.85\text{ proof gal}$$

$$0.85\text{ proof gal} \times \$13.50/\text{proof gal} = \$11.48$$

Try these problems for practice:

A. A wine that is 29 proof contains what percentage of ethanol?

B. How many proof gallons are equivalent to 12 gal of 25% alcohol?

C. What is the federal tax (at $13.50 per proof gallon) on 6 gal of 70% alcohol?

D. What is the federal tax (at $13.50 per proof gallon) on 1 pt of 72% alcohol?

E. Calculate the federal tax at 13.50 per proof gallon on 1 gal of alcohol USP, which is 95% C_2H_5OH.

Solutions.

A. 14.5

B. 6 proof gal

C. $113.40

D. $2.43

E. $25.65

APPENDIX 9: HYDROPHILE LIPOPHILE BALANCE (HLB)

HLB, which stands for *hydrophile lipophile balance* is a designation applied to nonionic (uncharged) surface-active compounds. The literature accompanying commercial raw materials of this type (they are called *surfactants*) usually specifies the HLB value. It has been found over the years that certain properties of these compounds are related to their HLB value regardless of chemical makeup.

HLB is expressed as a dimensionless number. The scale ranges from zero to 20, depending on the balance between lipophilic and hydrophilic character. Real nonionic surfactants may have values from about 1 (very much toward the lipophilic side) to 19 (very much toward the hydrophilic side).

The basis of the HLB method for formulating emulsions is to match the HLB of the surfactant system with another number, called the *required HLB*, which is assigned to the oil phase of the emulsion. According to the theory, the best emulsion for a particular surfactant or mixture of surfactants is found when the surfactant HLB and the required HLB of the oil are the same. Just as the HLB numbers for surfactants are readily available, required HLB values for a number of commonly used oils have been published.

As emulsions are dispersions in which droplets of one liquid are surrounded by another, immiscible liquid, it is possible to combine these liquids so as to form two different dispersion types. One, in which oil droplets are surrounded by water, is called *oil-in-water* and abbreviated o/w. In the other type, droplets of water are surrounded by the oil. This is called *water-in-oil*, abbreviated w/o. If the surfactant HLB is below 5 or so, w/o emulsions tend to form. Higher HLB values, 10 or above, favor o/w emulsion formation.

For a mixture of surfactants, the combined HLB can be found from the individual values by the following equation:

$$\text{HLB} = \Sigma\ (f_i)\ (\text{HLB}_i)$$

f_i is the weight fraction of the *i*th surfactant. This is defined as the weight of that surfactant divided by the total weight of all surfactants. In a group of surfactants, the sum of all weight fractions must be 1. For two components, the equation would be

$$\text{HLB} = f_1 \times \text{HLB}_1 + f_2 \times \text{HLB}_2$$

A formula for 100 g of an emulsion contains 6 g of Span 65 (HLB = 2) and 2 g of Myrj 53 (HLB = 18). The combined HLB is

$$\text{HLB} = f_1 \times \text{HLB}_1 + f_2 \times \text{HLB}_2$$

$$\frac{6}{6+2} \times 2 + \frac{2}{6+2} \times 18 = 6$$

A similar equation applies to the required HLB of a mixture of oils:

$$\text{RHLB} = \Sigma\ (f_i)\ (\text{RHLB}_i)$$

in which RHLB is the required HLB of the mixture and RHLB_i is the required HLB of the *i*th oil. f_i is the weight fraction of the *i*th oil in the oil mixture.

An emulsion contains 30 g of light mineral oil (RHLB = 12) and 6 g of another oil (RHLB = 15). The required HLB of the mixture is

$$\text{RHLB} = \Sigma\ (f_i)\ (\text{RHLB}_i)$$

$$\frac{30}{30+6} \times 12 + \frac{6}{30+6} \times 15 = 12.5$$

Two surfactants have been selected for the emulsion described above. They are polysorbate 20, with an HLB of 16.7 and sorbitan mono-oleate, with an HLB of 4.3. What combination of these should be used for an optimized emulsion if the total amount of surfactant to be used is 10 g? (Remember that the surfactant HLB is supposed to match the required HLB of the oil phase.)

$$\text{HLB} = f_1 \times \text{HLB}_1 + f_2 \times \text{HLB}_2$$

Let j equal the amount of polysorbate 80.

$$12.5 = \frac{j}{10} \times 16.7 + \frac{10-j}{10} \times 4.3$$

$$j = 6.6\,\text{g of polysorbate 80}$$

3.4 g of sorbitan monooleate would also be needed.

Try these problems:

A. The HLB of sorbitan tristearate is 2.1 and that of polyoxyethylene (40) stearate is 16.9. Calculate the HLB of a 50 : 50 mixture of the two.

B. Referring to problem A, what is the HLB of a 5-gram mixture of the two surfactants containing 1.25 g of sorbitan tristearate?

C. The required HLB for light mineral oil in a w/o emulsion is 4; that of anhydrous lanolin is 8. What is the required HLB of the oil phase for an emulsion if it consists of 500 g of light mineral oil and 125 g of anhydrous lanolin?

D. The required HLB of the oil phase of an o/w emulsion is 13. What amounts of glyceryl monostearate (HLB = 3.8) and polysorbate 80 (HLB = 15.0) should be used if the total of the two is to be 150 g?

E. The required HLB of the oil phase of an emulsion is 10.8. What percentages of propylene glycol monostearate (HLB = 3.4) and polyethylene glycol 400 monostearate (HLB = 11.6) should be used if the total of the two surfactants is to be 8% of the total weight of the emulsion?

Solutions.

A. 9.5

B. 13.2

C. 4.8

D. Glyceryl monostearate: 26.8 g; polysorbate 80: 123.2 g

E. Propylene glycol monostearate: 0.78%; polyethylene glycol 400 monostearate: 7.22%

APPENDIX 10: VAPOR PRESSURE OF LIQUEFIED PROPELLANTS

Liquefied propellants are used in pressurized package systems ("aerosol" containers) to maintain constant pressure as the product is emptied. The vapor pressure is a function of temperature. In some cases, single propellants are utilized and the pressure within the product is simply the propellant's vapor pressure. When a combination of propellants is employed, each component contributes a portion of the total pressure. The partial pressure of each component may be approximated (assuming ideal behavior) as the product of the vapor pressure of the pure substance and its mole fraction. Thus

$$p_i = \mathrm{VP}_i \times X_i$$

where p_i is the partial pressure of ingredient i, VP_i is the vapor pressure of pure i, and X_i is the mole fraction of i in the propellant mixture. The vapor pressure of the mixture is the sum of the partial pressures

$$\mathrm{VP} = \Sigma\, p_i$$

The units commonly employed in the industry are pounds per square inch gauge (psig) and pounds per square inch absolute (psia). psig is the reading that a pressure gauge would indicate and represents the difference between the actual pressure and atmospheric pressure. The psia value is the actual pressure inside an aerosol container. Vapor pressures are commonly expressed in terms of psig, but all calculations should be carried out in psia. If atmospheric pressure is taken as 14.7 psi, the two units are related through the following equation:

$$\text{psia} = \text{psig} + 14.7$$

As an example, a propellant mixture is prepared by combining 150 g of propellant a, whose vapor pressure is 20 psig with 200 g of propellant b, whose vapor pressure is 45 psig. The molecular weight of propellant a is 178 while that of propellant b is 112. To calculate the mole fractions, we begin by determining the number of moles of each component:

$$\frac{150\,\text{g}}{178} = 0.843\,\text{mol} \tag{a}$$

$$\frac{200\,\text{g}}{112} = 1.7857\,\text{mol} \tag{b}$$

$$\mathrm{VP} = \left[(20 + 14.7) \times \frac{0.843}{0.843 + 1.7857}\right] + \left[(45 + 14.7) \times \frac{1.7857}{0.843 + 1.7857}\right]$$
$$= 51.7\,\text{psia} = 37.0\,\text{psig}$$

Try these practice problems:

A. What is the vapor pressure, in psig, of a mixture containing 50 g each of propellant 12 and propellant 114. Propellant 12 has a molecular weight of 121 and a vapor pressure of 84.9 psig while propellant 114 has a molecular weight of 171 and a vapor pressure of 27.6 psig.

B. What is the vapor pressure of a mixture of 20% w/w *n*-butane and 80% isobutane at 70°F? At this temperature, *n*-butane (MW 58.1) has a vapor pressure of 16.5 psig and isobutane (MW 58.1) has a vapor pressure of 30.4 psig.

C. What is the vapor pressure at 70°F of a mixture of 200 g of *n*-butane, 200 g of isobutane and 600 g of propane? At this temperature, propane (MW 44.1) has a vapor pressure of 110 psig. See problem B for information on the other two propellants.

Solutions.

A. 75.9 psia = 61.2 psig

B. 42.3 psia = 27.6 psig

C. 95.6 psia = 80.9 psig

APPENDIX 11: DRUG BINDING MEASURED BY DIALYSIS

Drug molecules may become attached to macromolecules, such as proteins or polymers. Dialysis is one of a series of techniques that separate molecules by size. In this technique, two chambers are separated by a semipermeable membrane. One chamber is loaded with the macromolecule, drug, water, and possibly other components, such as salts. The other chamber, which we will call the receiver, contains no drug or macromolecule. Individual drug molecules of relatively low molecular weight are able to diffuse across the membrane and reach the receiver. The macromolecule, and any drug molecules bound to the macromolecule, are not.

After remaining for some time, say 24 or 48 h, the dialysis system is essentially at equilibrium. The receiver side is sampled and assayed for drug. Based on the known volume and measured concentration in the receiver, the amount of drug remaining in the first chamber can be calculated. From this, and the fact that the concentration of free (unbound) drug is the same in both chambers at equilibrium, we can determine the extent of binding. Binding is commonly expressed in two ways. One is the percentage of the drug molecules bound; the second is the amount bound per macromolecule unit.

In most dialysis apparatus, the two chambers have the same volume, and we will make that assumption in our calculations. As an example, the aqueous solution introduced into the left chamber of a dialysis cell contains 0.0003 mol of a polymer and 0.05 mol of a drug. The right chamber contains only water. After standing for two days, the right chamber is analyzed and found to contain 1 mmol/mL of drug. Both chambers are of equal volume, 10 mL.

Before starting the calculation, note that if no binding had occurred, the drug placed in the left chamber would have been diluted to half its concentration if all of it had been available to diffuse across the membrane. The fact that the measured concentration is less than 0.025 mol/mL tells us that some binding has taken place.

The initial amount of drug in the left chamber was 50 mmol. The amount found by assay in the right chamber (receiver) was

$$10\,\text{mL} \times 1\,\text{mmol/mL} = 10\,\text{mmol}$$

The amount remaining in the left chamber is therefore

$$50\,\text{mmol} - 10\,\text{mmol} = 40\,\text{mmol}$$

Of this quantity, 10 mmol is in equilibrium with the drug in the receiver and is therefore unbound. The amount bound is thus

$$40\,\text{mmol} - 10\,\text{mmol} = 30\,\text{mmol}$$

The percent bound is given by the amount bound divided by the total amount of drug within the left chamber

$$\frac{30\,\text{mmol}}{40\,\text{mmol}} = 75\%$$

and the binding ratio, r, defined as $\frac{\text{amount drug bound}}{\text{amount polymer}}$ is given by

$$r = \frac{0.03\,\text{mol drug}}{0.0003\,\text{mol polymer}} = 100\,\text{mol drug/mol polymer}$$

Here are some problems for more practice.

A. The aqueous solution introduced into the left chamber of a diffusion cell contains 1.2% polyvinylpyrrolidone (PVP) and 0.30 mmol/mL drug. The right chamber contains only water. After standing for 2 days, the right chamber is analyzed and found to contain 0.08 mmol/mL of drug. Calculate the percentage of drug bound in the left chamber. Both chambers are of equal volume, 12 mL.

B. The 50-mL chamber of a dialysis apparatus contains 0.0010 mol of methylcellulose (a polymer) and a drug solution, 0.10 mol/L. A second chamber of equal volume containing only water is separated from the first by a semipermeable membrane. After equilibration, the drug concentration in the second chamber (no polymer present) was found to be 0.020 molar. Calculate the extent of a drug binding by polymer in moles drug/mole polymer and the percent of drug remaining in the first chamber that is bound.

C. The extent of binding of a drug to albumin, a blood protein, is tested by dialysis. The source solution contains a predetermined amount of albumin and 1 mmol/mL of a drug. After equilibration, the receiver is assayed and a drug concentration of 0.1 mmol/L is found. The volume of each chamber is 25 mL. What is the percentage bound?

Solutions.

A. 64%

B. 75% bound; $r = 3$ mol drug/mol polymer

C. 89%

APPENDIX 12: VISCOSITY MEASUREMENT BY THE CAPILLARY VISCOMETER

Viscosity, the resistance of a fluid to flow, is an important parameter of many pharmaceutical products. It is one of the factors controlling the rate of sedimentation of particles in dispersions. As part of stability monitoring of drug products, viscosity can be measured over time to see whether physical changes have occurred.

The unit of viscosity in the cgs system is the poise, which is equivalent to $g\,cm^{-1}\,s^{-1}$. The SI unit is Pa·s (pascal second). For convenience, derivatives of these units are commonly used. Thus 100 cP (centipoise) = 1 poise and 1000 mPa·s (millipascal seconds) = 1 Pa·s. It is possible to convert from one set of units to the other through the following relationship:

$$1\,\text{cP} = 1\,\text{mPa} \cdot \text{s}$$

A variety of instruments (called *viscometers*) are used to measure viscosity. For simple liquids, such as water, various oils, and glycerin, a capillary viscometer is frequently utilized. In this instrument, liquid is permitted to flow under gravity or pressure through a narrow capillary. The time required for liquid to fall between two marks is measured. For a particular volume of liquid in a particular instrument, the viscosity, η, is given by

$$\eta = K\,t\,\rho$$

where t is the measured time, ρ the liquid density and K an instrument constant that depends on a variety of factors including the geometry of the setup. The value of K can be determined by experimenting with a liquid of known density and viscosity. Once K has been determined, the same value applies for other liquids used in the same instrument under the same conditions.

The efflux time for water ($\rho = 1$ g/mL; $\eta = 1$ cP) in a capillary viscometer was 125 s. If the same volume of a second liquid whose density is 0.90 g/mL is placed in the same instrument and gives an efflux time of 180 s, we can calculate its viscosity by first determining K using the data for water.

$$K = \eta / t_\rho = \frac{1}{125 \times 1} = 0.008$$

$$\eta = Kt_\rho = 0.008 \times 180 \times 0.90 = 1.3\,\text{cP}$$

Here are some practice problems.

A. The efflux time for water in a capillary viscometer is 140 s. What is the viscosity of a second liquid whose efflux time in the same viscometer under the same conditions is 220 s? This liquid's specific gravity is 0.88.

B. The time required for efflux of ethyl alcohol in a capillary viscometer is 124 s. In the same viscometer, benzyl alcohol requires 438 s. Calculate the viscosity of benzyl alcohol. The viscosity of ethanol is 1.19 cP. The specific gravity of ethanol at the temperature of measurement is 0.789. That of benzyl alcohol is 1.043.

C. The equation for calculating viscosity from the falling ball technique is

$$\eta = K(\rho_B - \rho_L)t$$

where ρ_B is the density of the ball, ρ_L the density of the liquid and t the time required for the ball to fall a fixed distance. Using water (η = 1 cP) and a sphere of density 4.00 g/mL, the required time was 125 s. With a test liquid in the same apparatus, the same ball fell the same distance in 200 s. Calculate the viscosity of the test liquid if its density is 0.800 g/mL.

Solutions.

A. 1.38 cP

B. 5.56 cP = 5.56 mPa·s

C. 1.71 cP

APPENDIX 13: SEDIMENTATION KINETICS

Sedimentation rate is one aspect of the physical stability of suspensions, emulsions and other disperse systems. With certain assumptions, the rate of sedimentation, v, can be described by the *Stokes equation*:

$$v = \frac{2r^2(\rho - \rho_0)g}{9\eta}$$

In this equation, r is the particle radius (spherical particles are assumed), ρ the density of the particle, ρ_0 the density of the medium surrounding the particle, g the acceleration due to gravity (sometimes called the gravitational constant), and η the viscosity of the medium. A key assumption is that the particles do not interfere with each other as they settle, so that the equation applies to dilute dispersions.

Calculate the rate of sedimentation of drug particles (26 μm diameter) in water and the time required for the suspension to settle 1 cm. (Assume that particle density is 1.5 g/mL and that the viscosity of water is 1 cP.)

Using the cgs system of units yields

$$v = \frac{2(13\times10^{-4}\,\text{cm})^2(1.5-1\,\text{g/cm}^3)980\,\text{cm}^2\,\text{s}^{-1}}{9\times0.01\,\text{g}\,\text{cm}^{-1}\text{s}^{-1}}$$
$$= 0.018\,\text{cm/s}$$

Note that for the cgs system, viscosity is in terms of poises, with 100 cP = 1 P. Getting back to the calculation, we have

$$v = x/t$$

where x is the distance traveled and t the time.

$$t = \frac{x}{v} = \frac{1\,\text{cm}}{0.018\,\text{cm/s}} = 56\,\text{s}$$

Try these problems for practice.

A. The density of a solid suspended in water is 1.41 g/mL. (The viscosity of water is 1.0 cP.) If the particles are of spherical shape and the same diameter, 7 μm, calculate the time required for the suspension to settle 0.5 cm.

B. Cellulose particles that are suspended in a simple water vehicle settle at a rate of 1.33 cm/h. The formulation is changed by adding sorbitol to the suspension. Calculate the length of time for the particles in the modified suspension to settle 1 cm. (The density of cellulose is 1.35 g/mL; the density and viscosity of water are 1 g/mL and 1 cP, respectively; the density and viscosity of the modified suspension medium are 1.15 g/mL and 20 cP, respectively.)

Solutions.

A. 457 s

B. 26 h

APPENDIX 14: SUPPOSITORY MOLD CALCULATIONS

Suppositories are solid dosage forms used to administer drugs locally or systemically through the rectum, vagina, or urethra, where they melt or dissolve.

Suppositories are dosage forms that are relatively easy to prepare in a retail pharmacy because the preparation requires no sophisticated equipment. Suppositories are prepared in a variety of shapes and weights, which depend on the suppository base used and the suppository mold. Suppository molds may be made of aluminum, rubber, or disposable plastic strips or shells (also used as the dispensing wrapper). A pharmacist should always have each of the available suppository molds precalibrated for the suppository bases in use in the pharmacy. Due to the difference in the densities of suppository bases, a suppository mold calibrated for cocoa butter, for example, will not be calibrated for a mixture of polyethylene glycols.

General guidelines for *precalibration of suppository molds* include:

1. Melt a portion of suppository base alone and pour into suppository molds. Allow to congeal. Trim.
2. Remove suppositories from the molds and weigh.
3. Calculate the average weight of a blank suppository for this particular base by dividing the total weight by the number of suppositories.
4. Use this weight as the calibrated value for the specific mold using the specific base.

Prescriptions for extemporaneous medicated suppositories indicate only the amount of medicinal agent(s) in each suppository while the type and amount of base are frequently left to the discretion of the pharmacist. Once the suppository base material and compounding methods are selected, some calculations must be performed. Most calculations while preparing suppositories involve reduction and enlargement of formulas (Chapter 3), calibration of molds, and calculations related to the quantity of base that will be occupied by the various active ingredients.

When calculating the amount of base required in preparing a given medicated suppository, some consideration should be given to the volume occupied by the drug ingredients in each suppository. Often, suppository components are listed and measured by weight but compounded by volume. Therefore, not only mold calibration but density calculations are required to provide correct doses. It is generally accepted that, for a 2 g suppository, when the quantity of active ingredients is small ($\leq$100 mg/supp), the volume occupied is insignificant, and density calculations may be ignored. Usually, when an active drug is mixed with a suppository base with similar density, it will displace an equivalent weight of the base, while another drug may displace a proportionally smaller weight of base, if its density is greater than that of the base. Thus, the pharmacist needs to know the density factors for suppository bases and other ingredients/drugs to determine the correct weights of each to be used in a prescription.

Several methods are used to calculate the amount of base required in preparing a given medicated suppository. In practice, the density factor method is the most frequently used.

Density Factor Method

Density factors (amount of base displaced by the active drug) for usual ingredients and drugs in cocoa butter base (most versatile base) are available in several texts and a selec-

TABLE A14.1 Density factors (amount of base displaced by the active drug) for some ingredients/drugs in cocoa butter suppository base

Ingredient	Factor	Ingredient	Factor
Alum	1.7	Opium	1.4
Aminophylline	1.1	Paraffin	1.0
Aminopyrine	1.3	Pentobarbital sodium	1.2
Aspirin	1.1	Phenobarbital	1.2
Barbital sodium	1.2	Procaine	1.2
Benzoic acid	1.5	Progesterone	1.25
Bismuth salicylate	4.5	Quinine hydrochloride	1.2
Castor oil	1.0	Resorcinol	1.4
Codeine phosphate	1.1	Secobarbital sodium	1.2
Digitalis leaf	1.6	Spermaceti	1.0
Dimenhydrinate	1.3	Sulfathiazole	1.6
Diphenhydramine hydrochloride	1.3	Tannic acid	1.6
Glycerin	1.6	White wax	1.0
Ichthammol	1.1	Witch hazel fluid extract	1.1
Morphine hydrochloride	1.6	Zinc oxide	4.0

tion is presented in Table A14.1. When other suppository bases are used or when the density factor is unknown, other practical methods are employed.

When the density factor (amount of base displaced by the active drug) of a drug in a suppository base is known, the amount of base that will be displaced by the drug may be determined. For example, a prescription calls for suppositories, each containing 150 mg phenobarbital and the pharmacist decided to use cocoa butter (d = 0.9) as the base. If the suppository mold has been calibrated for 2 g cocoa butter, how many grams of phenobarbital and cocoa butter would be required to prepare 10 suppositories?

Total weight of 10 supps with cocoa butter only: $2\,g \times 10 = 20\,g$

Weight of phenobarbital required for the prescription:

$$150\,mg \times 10 = 1500\,mg = 1.5\,g$$

Density ratio of drug/base:

$$\frac{1.2}{0.9} = 1.33$$

Amount of cocoa butter displaced by the active drug:

$$\frac{1.5}{1.33} = 1.13\,g$$

Weight of cocoa butter required for the prescription:

$$20\,g - 1.13 = 18.87\,g$$

Answer: Use 18.87 g of cocoa butter and 1.5 g of phenobarbital

Another method that uses density factors when calculating how much of the base will be displaced by a drug is based on the following relationship:

$$\text{Density factor} = \frac{\text{weight of drug}}{\text{weight of base displaced}}$$

If a pharmacist needs to prepare twelve 200-mg aminophylline suppositories in cocoa butter, how many grams each of cocoa butter and aminophylline should be weighed?

If one suppository is to contain 200 mg (= 0.2 g) of aminophylline, it will replace:

$$\frac{0.2}{1.1} = 0.18\,\text{g of cocoa butter}$$

If a blank suppository weighs 2 g, then

$$2\,\text{g} - 0.18\,\text{g} = 1.82\,\text{g of cocoa butter will be needed for each suppository}$$

Each suppository will then weigh,

$$1.82\,\text{g} + 0.2\,\text{g} = 2.02\,\text{g}$$

Grams of aminophylline and cocoa butter required for 12 suppositories:

$$1.82\,\text{g} \times 12 = 21.84\,\text{g cocoa butter}$$

$$0.2\,\text{g} \times 12 = 2.4\,\text{g aminophylline}$$

If the density factor (amount of base displaced by the active drug) in cocoa butter is unknown or other base than cocoa butter is used, the pharmacist may experimentally determine the density factor using any of several practical methods available. The density factor of the drug in that base should then be recorded for future use. We will discuss two of such methods.

Double Casting Method

1. Determine the average weight of a blank suppository (base of interest).
2. Using the same mold mix the total amount of drug called for in a prescription with an amount of base not enough to fill the number of cavities.
3. Melt, mix and pour into molds, partially filling each cavity. Add pure base to complete the remaining portion of each cavity.
4. Cool, trim, and remove suppositories from the mold.
5. Remelt suppositories, mix and recast to distribute drug evenly in each suppository cavity.
6. Cool and remove suppositories from mold.
7. Calculate the *average weight of a medicated suppository* (divide total weight by number of suppositories).
8. Calculate the *weight of base in each medicated suppository* by subtracting the amount of drug in each suppository from the average weight of a medicated suppository.
9. Calculate the *weight of base displaced* by subtracting the weight of base in each medicated suppository from the average weight of a blank suppository.
10. Calculate the *density factor* dividing the weight of drug in each suppository by the weight of base displaced.

A prescription calls for 10 suppositories, each containing 280 mg of acetaminophen. The pharmacist decided to use Fattibase® and he has molds calibrated for 2g suppositories in Fattibase®. How many grams of Fattibase® should the pharmacist use?

The density factor of acetaminophen in Fattibase® is not known and once determined, it may be used to calculate the amount of base that will be displaced by the drug.

Calculation of density factor by double casting method:

Weight of base needed for 10 blank supp = $10 \times 2\,\text{g} = 20\,\text{g}$

Weight of drug needed: $10 \times 280\,\text{mg} = 2800\,\text{mg} = 2.8\,\text{g}$

Mix 2.8 g acetaminophen with Fattibase® in an insufficient amount to fill 10 cavities. Melt plain Fattibase® and overfill each cavity. Cool, trim, remove suppositories from mold, remelt and recast suppositories. Remove cooled suppositories, weigh and determine the average weight of a medicated suppository (e.g. 1.8 g).

Thus, weight of base in medicated suppository = 1.8 g − 0.28 g = 1.52 g

And, the weight of base displaced = 2 g − 1.52 g = 0.48 g

$$\text{Density factor of acetaminophen} = \frac{0.28}{0.48} = 0.58$$

To make 10 suppositories each containing 0.28 g of acetaminophen:

$$\text{Weight of base displaced by total amount of drug} = \frac{\text{amount of drug}}{\text{density factor}} = \frac{2.8\,\text{g}}{0.58} = 4.83\,\text{g}$$

Weight of base needed for medicated suppositories = 20 g − 4.83 g = 15.17 g

The pharmacist should weigh 2.8 g of acetaminophen and 15.17 g of Fattibase® to compound this prescription.

Blank Suppository Ratio

On some occasions, the density factor of a drug is calculated using the *ratio* of a blank suppository of an unknown base to a blank cocoa butter suppository.

1. Calibrate the mold with each base. Record average blank suppository weight of each.
2. Determine blank suppository ratio $= \dfrac{\text{weight of blank supp. of unknown base}}{\text{weight of blank cocoa butter supp.}}$ = density of unknown base relative to cocoa butter.
3. Calculate the amount of base displaced by the drug: divide the weight of drug by the density factor of the drug in cocoa butter, then, multiply by the ratio obtained above (density of unknown base relative to cocoa butter).
4. The amount of unknown base needed for the prescription will be the total amount of base needed to fill the mold (average blank supp.) minus the amount of base displaced by the drug.

A suppository mold was calibrated with Polybase® and the average blank suppository weighed 2.2 g. The same mold, when calibrated with cocoa butter base, provided blank suppositories weighing 1.88 g. How many grams of Polybase® will be needed to prepare 6 progesterone suppositories, each containing 120 mg?

Determine the density of Polybase® relative to cocoa butter:

$$\frac{\text{blank of Polybase}^{®}}{\text{blank cocoa butter}} = \frac{2.2\,\text{g}}{1.88\,\text{g}} = 1.17$$

Assume the density factor of progesterone in cocoa butter = 1.2

For each suppository,

$$0.12\,\text{g of progesterone will displace: } \frac{0.12\,\text{g}}{1.2\,\text{g}} \times 1.17 = 0.12\,\text{g of Polybase}^{®}.$$

Amount of Polybase® needed for each suppository = 2.2 g − 0.12 g = 2.08 g

Amount of Polybase® needed for prescription = 6 supp × 2.08 g/supp = 12.48 g

The pharmacist will need 0.72 g (6 × 0.12 g) of progesterone and 12.48 g of Polybase® to prepare the prescription.

Several variations of the calculations we presented here are described in the literature. The utmost concept to keep in mind when compounding suppositories is that when a drug is combined with a suppository base, it will displace an amount of base as a function of its density.

APPENDIX 15: POWDER VOLUME CALCULATIONS

Several solid dosage forms, when compounded, call for the determination of the volume occupied by the diverse ingredients. This is true for basically all preparations that need a mold and have therefore a fixed-volume. Because of the various ingredients' densities, different substances will occupy different volumes in the final product. Lozenges, suppositories and capsules are some examples of this principle.

When asked to compound one of these dosage forms, the pharmacist frequently faces the fact that the actual density of some ingredients is unknown. A ratio of an ingredient to another may then be used to calculate the volumes they occupy. The solution of the following problem illustrates the calculations required to encapsulate powders, which also may be applied to an array of dosage forms.

A pharmacist needs to prepare 30 capsules containing 50 mg phenobarbital. Using a #2 capsule, what quantities of phenobarbital and lactose are required for the prescription? Prepare sufficient powder for 2 extra capsules.

Average #2 capsule filled with phenobarbital = 250 mg

Average #2 capsule filled with lactose = 350 mg

Equivalent volume of lactose that is occupied by phenobarbital in one #2 capsule:

$$\frac{50\,\text{mg}}{250\,\text{mg}} = \frac{z}{350\,\text{mg}}$$

$$z = 70\,\text{mg}$$

Quantity of lactose required for the prescription (knowing that 50 mg of phenobarbital occupies the same volume as 70 mg of lactose):

Amount of lactose per capsule = 350 mg − 70 mg = 280 mg

For 32 capsules:

Lactose: 280 mg × 32 caps = 8960 mg = 8.96 g

Phenobarbital: 50 mg × 32 caps = 1600 mg = 1.6 g

Another method considers the percentage of the capsule occupied by each ingredient, and is illustrated below.

A prescription calls for 12–20 mg capsules of Drug A and the pharmacist needs to calculate the amount of drug and excipients required for each capsule.

The first step is to calculate the average weight of one capsule containing *only excipients* = 280 mg

Next, calculate the average weight of one capsule containing *only Drug A* = 375 mg

The percentage of the capsule occupied by Drug A is determined by dividing the amount of drug per capsule by the average weight of one capsule with drug only:

$$\frac{20\,\text{mg}}{375\,\text{mg}} \times 100\% = 5.3\%$$

The percentage of the capsule occupied by the excipients will be:

$$100\% - 5.3\% = 94.7\%$$

And the weight of the excipients per capsule:

$$94.7\% \times 280\,\text{mg} = 265.16\,\text{mg}$$

Total weight of one capsule:

$$20\,\text{mg (drug)} + 265.16\,\text{mg (excipients)} = 285.16\,\text{mg}$$

$$\text{Amount of Drug A for 12 capsules} = 20\,\text{mg} \times 12\text{ caps} = 240\,\text{mg}$$

$$\text{Amount of excipients for 12 caps} = 265.16\,\text{mg} \times 12\text{ caps} = 3181.9\,\text{mg}$$

Selection of Capsule Size

Capsules, one of the most versatile dosage forms, are frequently used by the compounding pharmacist to prepare specific products for individual patient needs. Frequently, ingredients incorporated into capsules are powders, weighed as a raw ingredient, from a crushed and triturated tablet, or taken from other capsules. On some occasions, the pharmacist may prepare a "capsule within a capsule", if one ingredient must be separated from the others in a formulation, or a "tablet within a capsule", if a small tablet of active ingredient is commercially available but enhancement of patient compliance is needed. In any of these situations, if the material encapsulated is not enough to fill completely the capsule shell, additional inert powder (e.g. lactose) should be added to make up the required volume and enhance the elegance of the dosage form. Occasionally, liquids that do not dissolve gelatin (e.g. oils, absolute alcohol) are placed into capsules using a micropipette or a calibrated dropper, with special attention placed on sealing to prevent leakage.

There are eight sizes of gelatin capsules for human use, ranging from number 5, the smallest, to number 000 (three zeroes), the largest. Each capsule size provides a relative volume capacity, which depends upon the density and characteristics of the encapsulated powders. Compounding capsules involves accurate weighing of powdered ingredients, mixing by geometric dilution and filling in appropriate size capsules. The selection of a capsule size depends on several factors including the amount of powder it will contain, pharmacist's judgment, trial and error, and individual patient needs. Although, as a general rule, the smallest capsule size that will hold the formulation should be selected, some patients have trouble swallowing larger capsules, and others, especially the elderly, find smaller capsules difficult to handle.

In practice, pharmacists use one of two general "rules of thumb" available in the literature to calculate the appropriate capsule size for a specific formulation.

1. The rule of sixes is useful, simple, though not very accurate, and is based on capsule average fill density (bulk density) of 0.6 g/mL. For formulations with different bulk density, this method will provide an approximate capsule size to be used. The box below summarizes the basic rule of sixes, which involves the following steps:
   (a) Organization of six "6s".
   (b) Recording of capsule sizes below each six: size #0 through #5
   (c) Determination of fill weight (in grains): subtract step *b* from step *a*
   (d) Conversion of grains to milligrams: 1 grain = 65 mg

6	6	6	6	6	6	Sixes
0	1	2	3	4	5	Capsule size
6	5	4	3	2	1	Fill weight (grains)
390	325	260	195	130	65	Fill weight (milligrams)

2. The rule of seven is based on the following steps:

(a) Convert the weight of the powder (formulation) for one capsule to grains.

(b) Subtract the number of grains from 7.

(c) Match the result with the following:

Resulting number	Capsule size
−3	000
−2	00
−1 or 0	0
1	1
+2	2
+3	3
+4	4
+5	5

For example, if the total weight (drug + excipients) of one capsule was determined to be 280 mg:

$$\frac{1\,\text{gr}}{65\,\text{mg}} \times 280\,\text{mg} = 4.3\,\text{gr}$$

$$7 - 4.3 = 2.7$$

thus, capsule sizes #2 or 3 may be appropriate.

The limitation of this method is that it does not work if the resulting number is lower than −3 or higher than +5.

One important concept to keep in mind when selecting a capsule size is that the calculated powder should fill just the body of the capsule, with the cap used only to retain the powder.

A pharmacist received the following prescription. Show all calculations and describe the procedure used to prepare the product.

℞ Amiloride 4 mg
Hydrochlorothiazide (HCTZ) 40 mg
M. Ft. caps DTD # 30

The formula is for one capsule and the prescription asks for 30 such capsules (calculated for two extra), thus,

$$4\,\text{mg} \times 32\text{ caps} = 128\,\text{mg of amiloride are needed}$$

$$40\,\text{mg} \times 32\text{ caps} = 1280\,\text{mg of HCTZ are needed}$$

Both drugs are unavailable at the pharmacy but there are amiloride 5 mg tabs and HCTZ 50 mg tablets, which may be crushed and triturated. The number of tablets necessary will be:

$$\frac{1\,\text{tab}}{5\,\text{mg amiloride}} \times 128\,\text{mg amiloride} = 25.6 = 26\,\text{tabs}$$

$$\frac{1\,\text{tab}}{50\,\text{mg HCTZ}} \times 1280\,\text{mg} = 25.6 = 26\,\text{tabs}$$

After weighing 26 tablets each, amiloride and HCTZ, the pharmacist obtained:

$$\text{Amiloride} = 520\,\text{mg}$$

$$\text{HCTZ} = 5200\,\text{mg}$$

Tablets should be crushed, mixed geometrically to yield a powder weighing 5720 mg

$$\text{Each capsule should contain: } \frac{5720\,\text{mg}}{32\,\text{caps}} = 178.75 = 179\,\text{mg of the tablet mixture}$$

Using the rule of sixes, the potential capsule size to use would be #3, which holds 195 mg. One or two capsules must be filled with 179 mg of powder to determine the appropriateness. Lactose can be calculated and added to make up the required volume if 179 mg does not fill one capsule.

Show the calculations involved and capsule size selected for the following prescription.

℞ Indomethacin 20 mg
Lactose qs ad 300 mg
M. Ft. caps dtd #12
Sig. i caps q AM

Available: 25 mg Indomethacin tablets weighing 200 mg each

Twelve capsules are needed so 14 caps should be prepared.

Indomethacin needed: 20 mg × 14 = 280 mg

$$\text{Tablets needed: } \frac{1\,\text{tab}}{25\,\text{mg}} \times 280\,\text{mg} = 11.2 = 12\,\text{tab must be crushed.}$$

$$11.2\,\text{tab} \times \frac{200\,\text{mg}}{\text{tab}} = 2240\,\text{mg of tablet powder (for 14 caps)}$$

$$\frac{2240\,\text{mg}}{14\,\text{tab}} = 160\,\text{mg drug/caps}$$

The prescription requires adding lactose for one tablet (the dose) to have a final weight of 300 mg.

Using the rule of sixes, capsule size #1 holds 325 mg, thus:

Amount of lactose needed per capsule: 325 mg − 160 mg = 165 mg

Total amount of lactose needed for 14 caps: 165 mg × 14 caps = 2310 mg

The pharmacist should weigh 2240 mg of indomethacin powder from 12 tablets and 2310 mg of lactose, mix well, and fill 12 capsules. Capsule size #1 will be used, each filled with 325 mg of powder mixture (indomethacin + lactose).

A compounding pharmacist was asked to prepare 10 capsules of cephalexin 120 mg for B.G., 85 years-old, who has experienced difficulty swallowing large pills. The physician requested the capsules to be of an adequate size to overcome this problem and enhance compliance. There are commercially available capsules of cephalexin 250 mg and 500 mg in the pharmacy. How should the pharmacist proceed with this order?

Need 10 capsules (+2 caps): 120 mg × 12 = 1440 mg cephalexin needed

$\frac{1\,\text{caps}}{500\,\text{mg}} \times 1440\,\text{mg} = 2.88 = 3$ caps of cephalexin 500-mg or 6 caps of cephalexin 250-mg should be opened and total powder weighed.

Weight of 3 caps of cephalexin 500 mg = 2180 mg

$\frac{2180\,\text{mg}}{3\,\text{caps}} \times 2.88\,\text{caps} = 2092.8\,\text{mg}$ of capsule powder will contain 1440 mg of cephalexin.

Capsule powder for *one* capsule to contain 120 mg cephalexin:

$$\frac{2092.8\,\text{mg}}{12\,\text{caps}} = 174.4\,\text{mg/caps}$$

By the rule of sixes, capsule #3 holds 195 mg of powder and is a good size for swallowing. However, because the prescription emphasized small capsules, the pharmacist decided to divide the dose into two capsules #2 (capacity = 130 mg) rather than filling the smallest existing size (too small for an 85 year-old patient to handle). Lactose will be added to make up the required volume.

Cephalexin capsule powder in each #2 capsule $= \frac{174.4\,\text{mg}}{2} = 87.2\,\text{mg}$

Lactose needed per capsule: 130 mg − 87.2 mg = 42.8 mg/caps

Total lactose needed: 42.8 mg × 24 caps = 1027.2 mg

The pharmacist should weigh and mix geometrically 2092.8 mg of cephalexin capsule powder with 1027.2 mg of lactose and fill 20 #2 capsules with 130 mg each. Label of product should state: 2 caps/dose.

LITERATURE CONSULTED

Allen, L.V. (2002). *The Art, Science, and Technology of Pharmaceutical Compounding, 2nd Edition.* American Pharmaceutical Association.

Ansel, H.C., Allen, L.V., Jr, and Popovich, N.G. (1999). *Pharmaceutical Dosage Forms and Drug Delivery Systems, 7th Edition.* Lippincott, Williams & Wilkins.

ASPEN Nutrition Support Practice Manual (1998). American Society for Parenteral and Enteral Nutrition.

Davis, N.M. (1997). *Medical Abbreviations: A Handbook of 12,000 Abbreviations, Symbols, and Acronyms, 8th Edition.* Neil M. Davis Associates.

Drug Facts and Comparisons® Pocket Version (2002). Facts and Comparisons, Inc.

Gennro, Alfonso R. (2000). *Remington: The Science and Practice of Pharmacy, 20th Edition.* Lippincott, Williams & Wilkins.

Grabenstein, J.D. (1995). *ImmunoFacts®: Vaccines and Immunologic Drugs.* Facts and Comparisons, Inc.

Shrewsbury, R. (2001). *Applied Pharmaceutics in Contemporary Compounding.* Morton Publishing Co.

Thompson, J.E. (1998). *A Practical Guide to Contemporary Pharmacy Practice.* Williams & Wilkins.

Trissel, L.A. (2003). *Handbook on Injectable Drugs, 12th Edition.* American Society of Health System Pharamacists, Inc.

United States Pharmacopeia and National Formulary. Official Monographs, Tables, and Supplements (1995 and 2003). United States Pharmacopeial Convention, Inc.

Valgus, J.M. et al. (2000). 2000 Guide to Cancer Chemotherapeutic Regimens. In *Pharmacy Practice News*, October 2000. Melvin E. Lister and S.J. Turco (editors), McMahon Publishing Group.

Pharmaceutical Calculations, Fourth Edition, By Joel L. Zatz and Maria Glaucia Teixeira
ISBN 0-471-67623-3

INDEX

T followed by the number denotes a table, App. denotes an Appendix section.

Pharmaceutical Calculations, Fourth Edition, By Joel L. Zatz and Maria Glaucia Teixeira
ISBN 0-471-67623-3